Why chemical reactions happen

Why chemical reactions happen

James Keeler

Lecturer in the Department of Chemistry and
Fellow of Selwyn College, Cambridge

and

Peter Wothers

Teaching Fellow in the Department of Chemistry and
Fellow of St Catharine's College, Cambridge

OXFORD
UNIVERSITY PRESS
Great Clarendon Street, Oxford OX2 6DP
Oxford University Press is a department of the University of Oxford.
It furthers the University's objective of excellence in research, scholarship,
and education by publishing worldwide in

Oxford New York

Auckland Cape Town Dar es Salaam Hong Kong Karachi
Kuala Lumpur Madrid Melbourne Mexico City Nairobi
New Delhi Shanghai Taipei Toronto

With offices in
Argentina Austria Brazil Chile Czech Republic France Greece
Guatemala Hungary Italy Japan Poland Portugal Singapore
South Korea Switzerland Thailand Turkey Ukraine Vietnam

Oxford is a registered trade mark of Oxford University Press
in the UK and in certain other countries

Published in the United States
by Oxford University Press Inc., New York

First published 2003
Reprinted 2004 (with corrections), 2005, 2006

British Library Cataloguing in Publication Data
Data available

Library of Congress Cataloging in Publication Data
Data available

ISBN 13: 978-0-19-924973-2
ISBN 10: 0-19-924973-3

5 7 9 10 8 6

Typeset by the authors
Printed in Great Britain on acid-free paper by
Antony Rowe Ltd, Chippenham, Wiltshire

Acknowledgements

We are grateful to Michael Rodgers, then commissioning editor for science texts, for being so enthusiastic about our original proposal for this book and for being responsible for its adoption by OUP. Michael has since moved on to higher things, leaving us in the capable hands of Jonathan Crowe, John Grandidge, Miranda Vernon and Mike Nugent, who have been very helpful throughout the writing and production process – we are most grateful to them for their assistance. James Brimlow and John Kirkpatrick were kind enough to read and comment on the final draft; we are very grateful to them for their suggestions and corrections.

It is something of a cliché to offer thanks to our students for what they have taught *us*. However, be it a cliché or not, it is true that having to think about, explain and re-explain this material over a number of years has improved our own understanding and, we hope, the clarity of our explanations. So in a very real sense we are indeed grateful to the many students we have had the privilege to teach.

JK wishes to thank Prof. A. J. Shaka and his colleagues in the Department of Chemistry, University of California Irvine, for being excellent hosts during a three-month sabbatical which made it possible for a great deal of progress to be made on the text. He also wishes to thank Prof. Jeremy Sanders for his invaluable advice, support and encouragement.

PW wishes to thank Del Jones for his support and encouragement during the production of this book.

This book has been typeset by the authors using LaTeX, in the particular implementation provided by MiKTeX (`www.miktex.org`); we are grateful to the many people who have been involved in the implementation of this typesetting system and who have made it free for others to use. The text is set in *Times* with mathematical extensions provided by Y&Y Inc., (Concord, MA). Most of the diagrams were prepared using *Adobe Illustrator* (Adobe Systems Inc., San Jose, CA) or *Mathematica* (Wolfram Research Inc., Champaign, IL), and chemical structures were drawn using *ChemDraw* (CambridgeSoft Corp., Cambridge, MA). Molecular orbital calculations were made using *HyperChem* (HyperCube Inc., Gainesville, FL).

Our intention is to maintain some pages on the web which will contain some additional material in the form of appendices, downloadable versions of many of the diagrams and corrections to the text. The address is

`www.oup.com/uk/best.textbooks/chemistry/keeler.`

Finally, we hope that you enjoy reading this book and that it helps in your understanding and exploration of the fascinating world of chemistry. How reactions happen, and why they happen, is the most fundamental question in chemistry – we hope that you find some answers here!

Cambridge, February 2003.

TO OUR PARENTS

AND OUR TEACHERS

Contents

1 What this book is about and who should read it

When we first come across chemical reactions they are presented to us as 'things that happen' – this book is about *why* and *how* chemical reactions happen. The ideas and concepts presented here will give you a set of tools with which you can begin to understand and make sense of the vast range of chemical reactions which have been studied.

Let's look at some examples of the kinds of things we are going to be talking about.

- In solution Ag^+ ions and Cl^- ions react together to give a precipitate of solid AgCl:

$$Ag^+(aq) + Cl^-(aq) \longrightarrow AgCl(s);$$

  on the other hand Ag^+ and NO_3^- do not form a precipitate but remain in solution. Why?

- Acyl chlorides react readily with water to give carboxylic acids:

$$H_3C{-}C({=}O){-}Cl \quad + \quad H_2O \longrightarrow H_3C{-}C({=}O){-}OH \quad + \quad HCl$$

  whereas amides are completely unreactive to water;

$$H_3C{-}C({=}O){-}NH_2 \quad + \quad H_2O \not\longrightarrow \text{no reaction}$$

  Why?

- In solution 2-bromo-2-methylpropane can form the positively charged ion $(CH_3)_3C^+$, but bromomethane cannot form CH_3^+:

$$(H_3C)_3C{-}Br \longrightarrow (H_3C)_3C^{\oplus} \quad + \quad Br^{\ominus}$$

$$H_3C{-}Br \not\longrightarrow CH_3^{\oplus} \quad + \quad Br^{\ominus}$$

  Why?

- Propene reacts with HBr to give the bromoalkane with the structure **A** and not **B**:

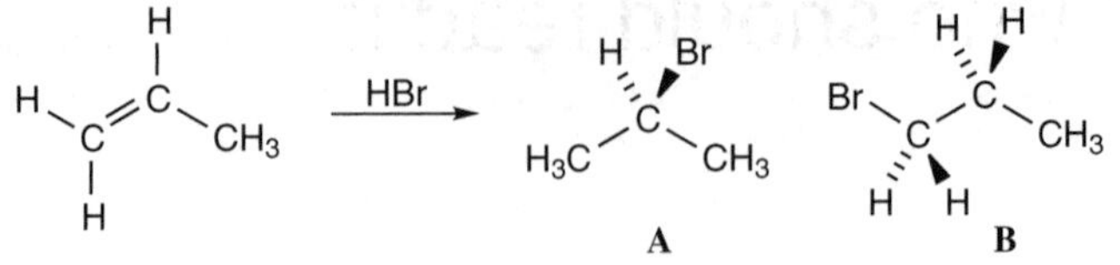

  Why?

You may already have come across explanations for some of these observations.

- A reason often given as to why a precipitate of AgCl is formed from Ag^+ and Cl^- ions in solution is that there is a large release of energy when these ions form a solid lattice.

  However, the precipitation of $CaCO_3$ from a solution of Ca^{2+} and CO_3^{2-} ions is endothermic, i.e. energy is taken in. This being the case, our explanation for what is going on in the case of Ag^+ and Cl^- cannot be correct! We will see in Chapter 12 why it is that both of these precipitation reactions are possible, even though one is exothermic and one endothermic.

- A reason often given for why acyl chlorides are more reactive than amides is that the chlorine substituent is electron withdrawing and so makes the carbonyl carbon 'more positive' and so more reactive towards nucleophiles such a water. The problem with this argument is that nitrogen has just about the same electronegativity as chlorine, so we would expect it to have the same electron withdrawing effect. Clearly something more complicated is going on here, and in Chapter 10 we will explain just what this is.

- You will find that the explanation often given for the easy formation of the $(CH_3)_3C^+$ ion is that the positive charge is stabilized by the electron-donating methyl groups. The question is, why, and how, do methyl groups donate electrons and why does this stabilize the cation? We will answer this in Chapter 11.

- The result of the reaction of propene with HBr may be predicted by Markovnikov's rule, which states that the bromine ends up on the carbon with the most alkyl groups attached to it. This just begs the question as to why Markovnikov's rule works – a question we will also answer in Chapter 11.

What we hope to do in this book is provide you with a logical and soundly based set of ideas which you can use to explain why these, and many other, reactions behave in the way they do.

How this book is laid out

It turns out that the most fundamental principle which determines whether or not a reaction will take place is the Second Law of Thermodynamics, so we will start out by discussing this very important and fundamental concept. Powerful though this law is, it does not really help us to understand what is going on in detail in a reaction – for example, which bonds are broken, which are formed and, most importantly, why. To understand these aspects of reactions we need to look in more detail at chemical bonding.

There are two basic interactions which lead to the formation of chemical bonds: electrostatic interactions between charged species, leading to ionic bonds, and covalent interactions in which electrons are shared between atoms. Of these, ionic bonds are perhaps the easiest to understand, so we consider them first. In Chapter 3 we will see that we can develop a simple picture of the bonding in ionic compounds which enables us to make some useful chemical predictions.

To understand how covalent bonds are formed, we need to go into quite a lot of detail about the behaviour of electrons, first in atoms and then in progressively more complex molecules. These topics are considered in Chapters 4, 5 and 6. Once we have mastered how to describe the bonding in reasonably sized molecules – which boils down to describing where the electrons are and what they are doing – we are ready to start considering reactions. It may seem like a long road to get to Chapter 7, but it is well worth it, as by then you will have all of the tools necessary to really understand what is going on.

In the subsequent chapters we will go on to look at different kinds of reactions, introducing along the way new concepts and ideas such as the idea of reaction mechanisms and the way in which chemical equilibrium is established and influenced. The final chapters look at the crucial role the solvent plays in a chemical reaction and then at what happens when there is more than one pathway open to a reaction.

How to read this book

This is not a textbook which aims to cover all the material of, say, a typical first-year university course. Rather, this book is aimed at giving you an overview of why and how chemical reactions happen. For this reason, the text cuts across the traditional divisions of the subject and we have freely mixed concepts from different areas when we have felt it necessary and appropriate.

Our intention is not to give a complete, detailed, description of what is going on, but rather to give an overview, bringing the key ideas into a clear perspective. If you continue your studies in chemistry you will no doubt learn about many of the ideas we introduce here in greater depth. We hope that having read this book you will have an overview of the subject so that you can see why you need to know more about particular ideas and theories and how these will help you to understand what is going on in chemical reactions.

This book will probably be most useful to those who are starting a university course in chemistry. It will also be of interest to those who are coming to the end of pre-university courses (such as A level) and who wish to deepen and extend their understanding beyond the confines of the syllabus. Teachers in schools and colleges may also find the text to be a useful source. We develop our argument from chapter to chapter, so the book is best read through from the beginning.

Some practical matters

Generally in the text we have used the systematic names for compounds, but have also given the common names, as you are likely to encounter these much more at university level. In any case, the molecules we have chosen as examples are all rather simple and so naming them is not a great problem. When it comes to drawing molecular structures, we have included all of the carbons and hydrogens rather than simply drawing a framework as is sometimes done. Although this makes the structures a little more cluttered, it avoids ambiguity and for simple structures does not seem to cause a problem. The only exception we have made to this rule is that benzene rings are drawn as a framework.

As you know, the four single bonds around a carbon atom are arranged so that they point towards the corners of a tetrahedron. The usual way of representing this is shown in Fig. 1.1. We imagine that two of the bonds lie in the plane of the paper – these bonds a represented by simple lines. The third bond is represented by a solid wedge with its point toward the carbon; this bond is the one coming out of the page. Finally, the fourth bond, which is going into the page, is represented by a thick, dashed line.

Fig. 1.1 The usual representation of the tetrahedral arrangement of groups around a carbon atom.

Where we show pictures of atomic and molecular orbitals we have been at particular pains to make sure that these are realistic. Often the pictures are based on actual calculations, and so can be relied upon. Even where we have drawn sketches of orbitals we have made sure that they are realistic. So, for example, the 2*p* orbital is not represented as an 'hourglass' simply because this is *not* its true shape, despite what you will see in many books!

The molecules we are going to talk about are the smallest we can find which contain the functional groups we are interested in. This makes is possible to focus on what is really important – the functional group and its electronic properties – without worrying about other details. We will also use just a few molecules to illustrate the ideas rather than introducing numerous different examples. Again, this will allow us to focus in detail on what is going on in these reactions.

Abbreviations

For quick reference, the abbreviations we are going to use are listed below. Some if these may be not be familiar to you as yet, but as you read on you will come across explanations of all of these terms.

2c-2e	two-centre, two-electron
AO	atomic orbital
HAO	hybrid atomic orbital
HOMO	highest occupied molecular orbital
LUMO	lowest unoccupied molecular orbital
MO	molecular orbital
DMF	dimethyl formamide
DMSO	dimethyl sulfoxide

2 What makes a reaction go?

From experience we know that, once started, some reactions simply 'go' with no further help from us. For example, gaseous ammonia reacts with gaseous hydrogen chloride to give white clouds of solid ammonium chloride:

$$NH_3(g) + HCl(g) \longrightarrow NH_4Cl(s).$$

Another example is magnesium burning in oxygen – once ignited the metal burns strongly, giving off heat and light:

$$Mg(s) + \tfrac{1}{2}O_2(g) \longrightarrow MgO(s).$$

In writing these equations we used the right arrow $\longrightarrow$ to symbolize that the reaction goes entirely to the products on the right-hand side.

Some reactions, rather than going entirely to products, come to equilibrium with reactants and products both present in significant amounts. For example, the dissociation of ethanoic acid in water is far from complete and at equilibrium only about 1% of the ethanoic acid in a 0.1 M solution has dissociated:

$$CH_3COOH(aq) + H_2O(l) \rightleftharpoons CH_3COO^-(aq) + H_3O^+(aq).$$

Another example is the dimerization of NO_2 to give N_2O_4:

$$2NO_2(g) \rightleftharpoons N_2O_4(g);$$

at 25 °C the equilibrium mixture consists of about 70% N_2O_4.

In these equations we use the equilibrium arrows $\rightleftharpoons$ to indicate that the reaction can go either way. Indeed, as you may know, at equilibrium the forward and back reactions have not stopped but are proceeding at equal rates.

There is another class of reactions which simply do not 'go' at all, at least not on their own. For example, ammonium chloride does not spontaneously dissociate into ammonia and hydrogen chloride, neither does magnesium oxide spontaneously break apart to give magnesium and oxygen.

Taking all of these reactions together we see that the real difference between them lies in the *position* of equilibrium, illustrated in Fig. 2.1. For some reactions the equilibrium lies so completely to the side of the products that to all intents and purposes the reaction is complete. For other reactions, there are significant amounts of products and reactants present at equilibrium. Finally, for some reactions the equilibrium lies entirely on the side of the reactants, so that they appear not to take place at all.

So, the question to answer is not 'what makes a chemical reaction go?' but 'what determines the position of equilibrium?'. We will begin to answer this question in this chapter.

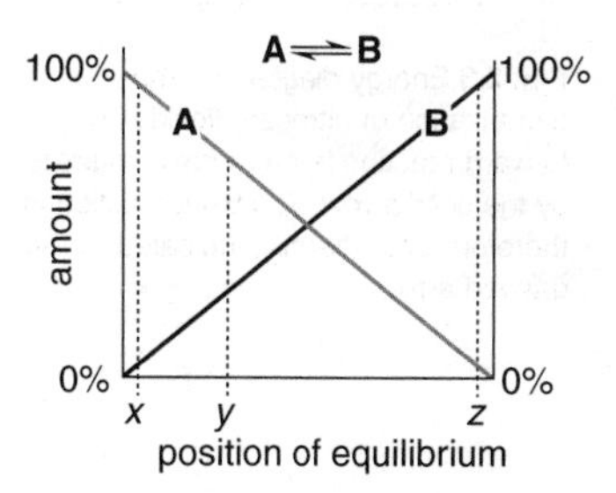

Fig. 2.1 Illustration of the concept of position of equilibrium for the simple equilibrium between **A** and **B**; the percentage of **A** and **B** in the equilibrium mixture is given by the grey and black lines, respectively. If the position of equilibrium is at *x* the equilibrium mixture contains almost entirely **A** – the reaction has not gone to a significant extent. At *y* the equilibrium mixture contains significant amounts of **A** and **B**. Finally, at *z* the reaction has gone almost entirely to products.

2.1 Energy changes

The reaction between magnesium and oxygen is strongly *exothermic*, meaning that heat is given out; we can represent this on the energy diagram shown in Fig. 2.2. The products are lower in energy than the reactants, and the energy difference between them, shown by the arrow, is the amount of energy that appears as heat.

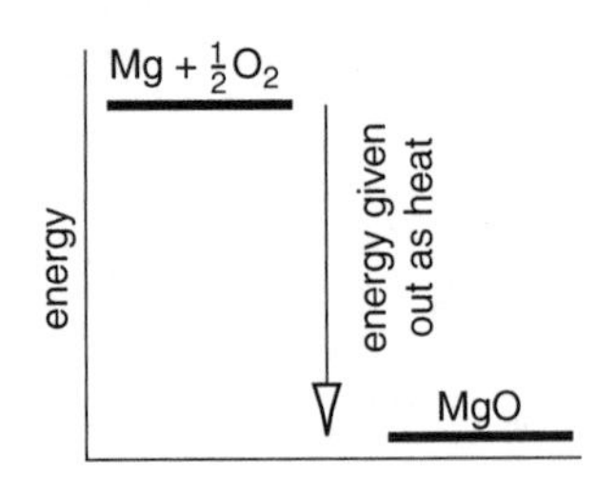

Fig. 2.2 Energy diagram for the reaction between magnesium and oxygen to produce MgO. The reaction is exothermic and the heat given out is represented by the arrow.

As the products are lower in energy than the reactants we often say that 'the products are more *stable* than the reactants'. In this context, 'more stable' means 'lower in energy'. It would be tempting to conclude that magnesium oxide is formed from magnesium and oxygen *because* magnesium oxide is more stable (i.e. lower in energy) than magnesium + oxygen. Generalizing this, we might say that a reaction goes if the products are more stable, that is lower in energy, than the reactants. This train of thought is very tempting but, as we shall see, it is at best incomplete and at worst completely wrong!

The best way to see that this tempting argument is wrong is to find a clear example in which it fails. We need look no further than the dimerization of nitrogen dioxide:

$$2NO_2(g) \rightleftharpoons N_2O_4(g).$$

As we commented on above, this reaction comes to a position of equilibrium which involves a mixture of reactants and products. The reaction going from left to right involves making a bond between the two nitrogen atoms – it is the formation of this bond which leads to a reduction in energy and hence the reaction being exothermic. The reverse, in which the bond is broken, therefore has to be *endothermic*, that is heat is absorbed; Fig. 2.3 illustrates these points.

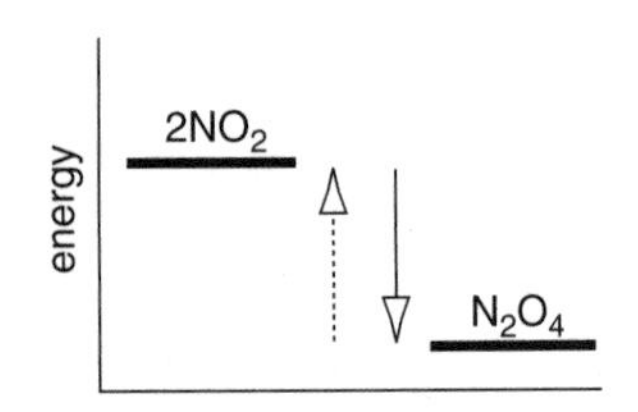

Fig. 2.3 Energy diagram for the dimerization of nitrogen dioxide. The forward reaction is exothermic, indicated by the solid arrow. The back reaction is therefore *endothermic*, indicated by the dashed arrow.

Suppose we start with pure NO_2 and then allow the system to come to equilibrium. We can actually do this experiment by first lowering the pressure, as this favours dissociation (dissociation involves increasing the number of moles of gas); at sufficiently low pressures it is found that very little N_2O_4 is present. If we then increase the pressure up to, say, one atmosphere, the position of equilibrium changes and some N_2O_4 is formed; this is illustrated in Fig. 2.4. The reaction involved is the exothermic process:

$$2NO_2(g) \longrightarrow N_2O_4(g).$$

On the other hand, we can approach equilibrium by starting with the pure product, N_2O_4. Experimentally, this can be achieved by increasing the pressure on the system; this favours the dimer and at high enough pressures dimerization is essentially complete. If we then decrease the pressure back to one atmosphere some of the dimer dissociates to NO_2 as the equilibrium position is approached. The reaction involved is the endothermic process:

$$N_2O_4(g) \longrightarrow 2NO_2(g).$$

We see that the equilibrium position can be approached *either* starting from solely reactants or solely products. Approaching equilibrium from one side involves an exothermic process and approaching from the other side involves an endothermic process.

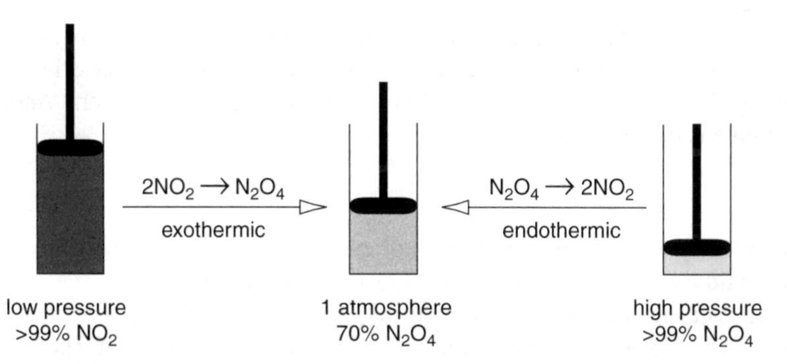

Fig. 2.4 Illustration of how the equilibrium between N_2O_4 and NO_2 can be altered by altering the pressure. Since NO_2 is dark brown and N_2O_4 is colourless the shift in the position of equilibrium can easily be detected by observing the brown colour.

Clearly, it cannot be the case that the only reactions which 'go' are exothermic ones in which the products are lower in energy than the reactants. Endothermic reactions, in which the products are higher in energy than the reactants, also take place and in general the approach to chemical equilibrium will involve both types of reactions. The tempting idea that reactions which 'go' are ones in which the products are 'more stable' has to be abandoned.

The reaction between ammonia and hydrogen chloride provides us with another example. If we apply sufficient heat, ammonium chloride will dissociate to ammonia and hydrogen chloride:

$$NH_4Cl(s) \xrightarrow{\text{heat}} NH_3(g) + HCl(g).$$

This is another example of an endothermic reaction which can be made to 'go'.

The true driving force for a reaction is not moving to lower-energy products but is concerned with a quantity known as *entropy* and is governed by the Second Law of Thermodynamics; we will discuss both of these in the next section.

2.2 Entropy and the Second Law of Thermodynamics

The *Second Law of Thermodynamics* allows us to predict which processes will 'go' and which will not. It is therefore the key to understanding what drives chemical reactions and what determines the position of equilibrium. The Second Law is expressed in terms of *entropy* changes. Entropy is a property of matter, just like density or energy, and we need to understand what it is if we are going to apply the Second Law.

The simplest description of entropy is to say that it is a quantity associated with randomness or disorder. A gas has greater entropy than a solid, as in the gas the molecules are free to move around. A liquid has higher entropy than a solid, but lower entropy than a gas. We can rationalize this by arguing that the molecules in a liquid are more free to move than those in a solid, but not as free as those in a gas; Fig. 2.5 illustrates these points.

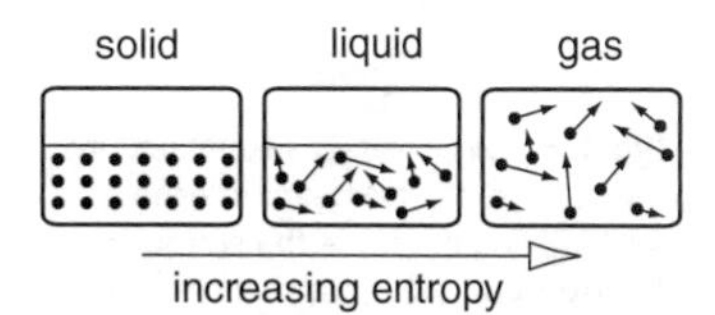

Fig. 2.5 The entropy of a gas is greater than that of a liquid which in turn is greater than that of a solid. This increase in entropy can be associated with the increasing freedom with which the molecules move as we go from a solid to a gas.

You may have come across the statement that 'entropy always increases'. The idea is that any process which takes place is always associated with an increase in entropy, that is an increase in chaos or disorder. However, a moment's thought reveals that there must be something wrong with this statement. After all, if it were true, how could ice form from water? This is a process in which the entropy decreases, as ice has lower entropy than liquid water. Another example is the reaction

$$NH_3(g) + HCl(g) \longrightarrow NH_4Cl(s)$$

in which there is clearly a reduction of entropy as gaseous reactants give rise to a solid product. Yet we know that this reaction takes place.

In fact the statement 'entropy always increases' is almost correct, it just needs to be expressed more carefully to give the Second Law of Thermodynamics:

In a spontaneous process, the entropy of the Universe increases.

This statement introduces two important ideas. Firstly, the idea of a spontaneous process. Secondly, that it is the entropy of the Universe that we need to consider, not just the entropy of the chemical system that we are thinking about. We will look at each of these ideas in turn.

A *spontaneous process* is one which takes place without continuous intervention from us. A good example of this is water freezing to ice – a process about which we will have a lot more to say in later sections. If we put water in a freezer set to −20 °C we know that ice will form; this is an example of a spontaneous process. The reverse will never happen: at −20 °C ice will *never* melt to give water. Another example of a spontaneous process is the mixing of gases: once two unreactive gases are released into a container they will mix completely. The reverse, which would involve the separation of the two gases, is never observed to take place.

We can bring about the reverse of spontaneous processes by intervening. For example, we can melt ice by applying heat and we can separate gases by using chromatography or liquefying them followed by distillation. The key point is, however, that the reverse of a spontaneous process never happens on its own.

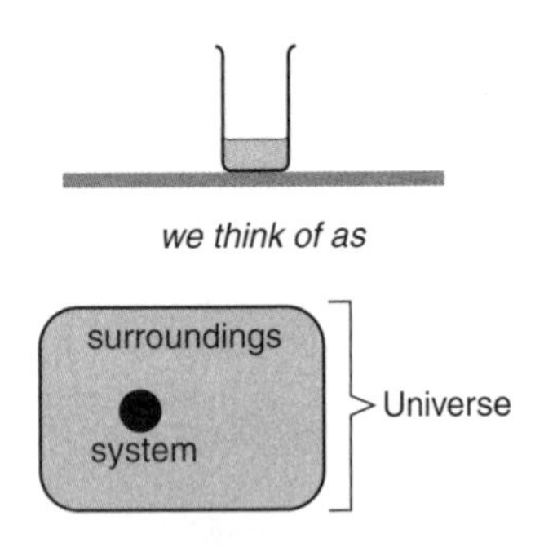

Fig. 2.6 The Universe is separated into the part we are interested in – the system, and the rest – the surroundings. For example, the system would be our chemical reactants in a beaker and the surroundings would be the bench, the laboratory and everything else in the Universe.

The Second Law tells us that we have to consider the entropy change of the Universe, not just the entropy change of the process we are interested in. This sounds like a tall order, but it turns out not to be as difficult as it at first seems. We simplify things by separating the Universe into two parts: the *system*, which is the part we are concentrating on, such as the species involved in a chemical reaction, and the *surroundings*, which is everything else (Fig. 2.6). The entropy change of the Universe can then be found by adding together the entropy changes of the system and the surroundings:

entropy change of Universe =
entropy change of surroundings + entropy change of system

or, in symbols,

$$\Delta S_{\text{univ}} = \Delta S_{\text{surr}} + \Delta S_{\text{sys}}$$

where ΔS_{univ}, ΔS_{surr} and ΔS_{sys} are the entropy changes of the Universe, the surroundings and the system, respectively.

We shall see in the next section that in an exothermic process the heat given out increases the entropy of the surroundings. This increase can compensate for a decrease in the entropy of the system so that *overall* the entropy of the Universe increases. This is why water can freeze to ice and NH_3 can react with HCl. These processes involve a reduction in entropy of the system but they are sufficiently exothermic that the surroundings increase in entropy by enough to make the entropy of the Universe increase.

These are rather subtle points, which we will tease out in the next few sections. The first thing to discuss is how the exchange of heat can lead to an entropy change and how the temperature affects the size of this entropy change.

2.3 Entropy and heat

Imagine an exothermic reaction which gives out heat to a nearby object. What effect does this have on the entropy of the object? We know that absorbing heat increases the energy of the molecules in the object and so causes them to move around more.

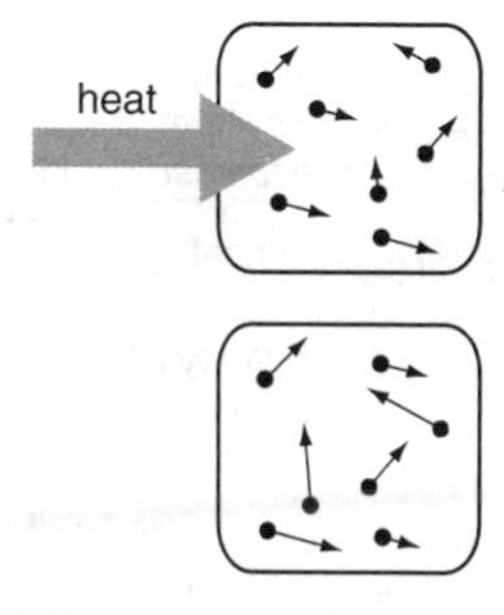

Fig. 2.7 When an object absorbs heat the molecules move more rapidly and so its entropy increases.

We have seen before that more motion – increasing disorder – is associated with an increase in entropy. It is therefore not surprising to find that when an object absorbs heat its entropy increases (Fig. 2.7). A further consequence of the heat being absorbed is that the temperature of the object increases, so we can also see that raising the temperature of an object increases its entropy. Conversely, if heat flows out of an object, its entropy decreases.

What happens, then, if the object we are supplying heat to is very large – for example the surroundings, which comprise all of the Universe apart from the system? The temperature of such a large object as the rest of the Universe is not going to be affected by the amounts of heat which typical chemical processes produce – put another way, we are not going to heat up the rest of the Universe significantly by leaving the oven door open! Nevertheless, although the temperature of the surroundings has not changed, they have absorbed energy (in the form of heat) and, as described above, this leads to an increase in the entropy. In summary:

*In an **exothermic** process the system gives out heat to the surroundings and so the entropy of the surroundings is **increased**.*

*In an **endothermic** process the system takes in heat from the surroundings and so the entropy of the surroundings is **decreased**.*

We can apply these ideas right away to understand how it is that water can freeze to ice. To melt ice to water we need to apply heat, so ice $\rightarrow$ water is an endothermic process. This makes sense, as in melting ice we are beginning to break the hydrogen bonds between the molecules. The reverse process, water $\rightarrow$ ice, in which bonds are being made, is therefore exothermic.

Ice has lower entropy than liquid water, so on freezing the entropy of the system decreases. However, the process is exothermic, so the entropy of the surroundings increases. It must be that the increase in entropy of the surroundings is sufficient to compensate for the decrease in entropy of the system so that

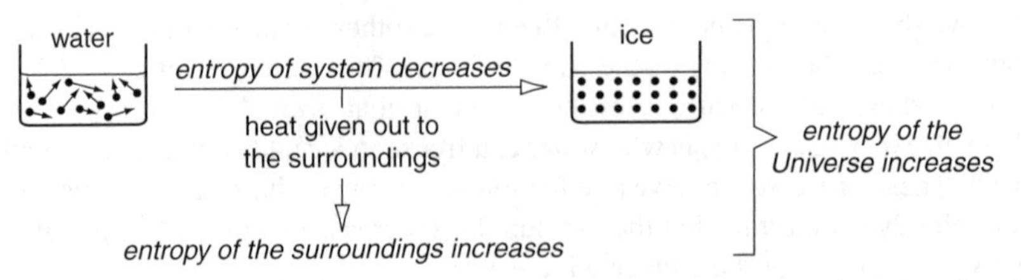

Fig. 2.8 When water freezes to ice its entropy decreases; however, the process is exothermic and the heat given out results in an increase in the entropy of the surroundings. For this process to be spontaneous, the increase in entropy of the surroundings must outweigh the decrease in entropy of the system so that overall the entropy of the Universe increases.

overall the entropy of the Universe increases. The overall process is illustrated in Fig. 2.8.

Coming back to the reaction we discussed above

$$NH_3(g) + HCl(g) \longrightarrow NH_4Cl(s)$$

we see that there is a decrease in the entropy of the system as a solid is formed from gases. In solid NH_4Cl there are strong interactions between NH_4^+ and Cl^- ions and so it is not surprising to find that the reaction is exothermic; it follows that the entropy of the surroundings increases. These entropy changes are illustrated in Fig. 2.9. As the reaction is spontaneous it follows from the Second Law that the increase in entropy of the surroundings must outweigh the decrease in entropy of the system, giving an overall increase in the entropy of the Universe.

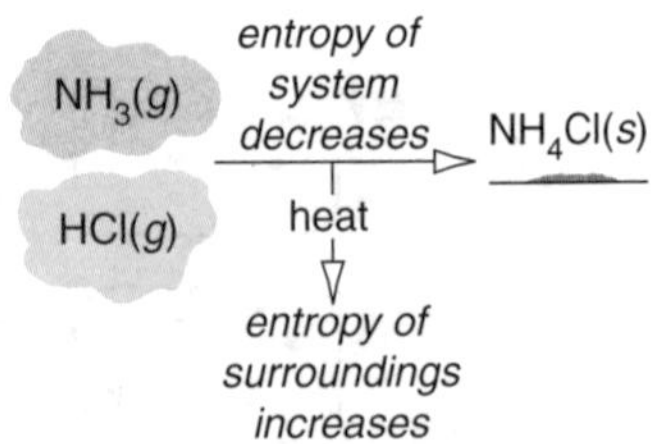

Fig. 2.9 When gaseous HCl reacts with gaseous NH_3, solid NH_4Cl is formed; there is consequently a reduction in entropy of the system. However, the reaction is exothermic and the heat given out increases the entropy of the surroundings. Overall, the entropy increase of the surroundings outweighs the decrease in entropy of the system, so the entropy of the Universe increases and the process is spontaneous.

There is one more subtlety to sort out here. Water freezes to ice only if the temperature is below 0 °C; at higher temperatures the water remains liquid. Water solidifying to ice is *always* exothermic, and as a result the entropy of the surroundings would always increase. However, it seems that this increase in entropy is only sufficient to compensate for the decrease in entropy of the system when the temperature is sufficiently low. We will see in the next section just exactly how and why temperature affects the entropy change of the surroundings.

2.4 Role of temperature

The entropy increase resulting from supplying a certain amount of heat to an object depends on its temperature. We can understand this in the following way.

Suppose that the object is very cold so that the molecules are not moving around very much. Supplying some heat will make the molecules move around more, and so the entropy increases. Now suppose that we apply the same amount of heat to a much hotter object in which the molecules are already moving around a great deal. The heat will cause the molecules to move even more vigorously, but the change is much less significant in the hotter object than in the cooler one, simply because in the hotter object the molecules are already moving more vigorously in the first place. We conclude that supplying heat to a hotter object causes a smaller increase in entropy than supplying the same amount of heat to a cooler object; this is illustrated in Fig. 2.10.

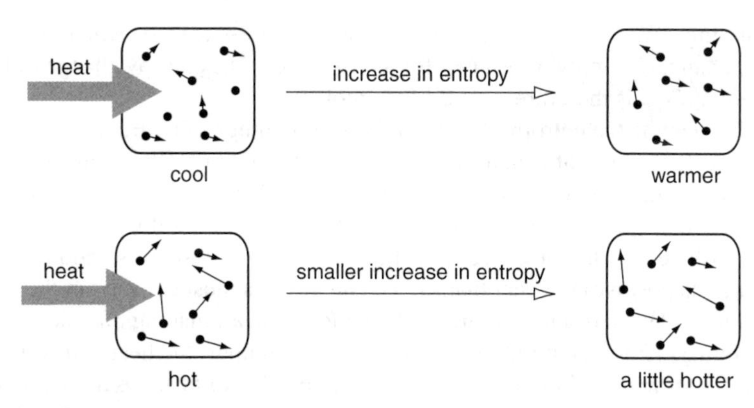

Fig. 2.10 Supplying heat to a cool object, in which the molecules are not moving very much, causes a larger increase in entropy than does supplying the same amount of heat to a hot object in which the molecules are already moving vigorously.

For the surroundings, which are very large, the entropy change can be computed very simply from:

$$\text{entropy change of the surroundings} = \frac{\text{heat absorbed by the surroundings}}{\text{temperature of the surroundings}}.$$

Using symbols we can write this as

$$\Delta S_{\text{surr}} = \frac{q_{\text{surr}}}{T_{\text{surr}}} \quad (2.1)$$

where q_{surr} and T_{surr} are the heat absorbed by and absolute temperature of the surroundings, respectively.

Heat is normally measured in joules and temperature in kelvin, so it follows from Eq. 2.1 that the units of the entropy change (and of entropy) are J K^{-1}. In chemistry we often quote molar quantities, that is the entropy change per mole, in which case the units are J K^{-1} mol^{-1}.

Equation 2.1 can only be used to compute the entropy change of the surroundings. Finding the entropy change of the system requires a different approach which we will not discuss here but simply note that the *absolute entropies* (as they are called) of many substances have been determined and tables of these can be found in any standard text. Such tables also list the entropy change associated with phase changes, such as ice melting to water. With the aid of such data we shall, in the next section, use the Second Law to explain why it is that the temperature has to be below 0 °C for water to freeze.

Some typical values of absolute entropies at 298 K:
He(*g*): 126 J K^{-1} mol^{-1}
H_2(*g*): 131 J K^{-1} mol^{-1}
Hg(*l*): 76 J K^{-1} mol^{-1}
Cu(*s*): 33 J K^{-1} mol^{-1}

How to make ice

The approach we will adopt is to compute the entropy change of the system and of the surroundings; adding these together gives us the entropy change of the Universe, ΔS_{univ}. If this is positive, the entropy of the Universe increases and from the Second Law we know that the process can take place. If ΔS_{univ} is negative, that is the entropy decreases, we know that the process cannot take place, as it is forbidden by the Second Law.

From tables we can find that the entropy change for ice $\rightarrow$ water is 22.0 J K^{-1} mol^{-1}. As expected, this is a positive number since disorder is increasing. We are interested in the change water → ice, for which the entropy change is simply minus that for ice → water. The entropy change of the system, ΔS_{sys}, is therefore $-22.0\ \text{J K}^{-1}\ \text{mol}^{-1}$.

To compute the entropy change of the surroundings using Eq. 2.1 we need to know the amount of heat absorbed by the surroundings. This heat can only come from the system. If the system gives out heat in an exothermic process (i.e. q_{sys} is negative), the surroundings must be absorbing this heat, making q_{surr} positive. On the other hand, if the process in the system is endothermic (i.e. q_{sys} is positive) the heat that the system absorbs must come from the surroundings. The surroundings are therefore losing heat, making q_{surr} negative. In summary, from the point of view of the surroundings, the heat is travelling in the *opposite* direction to that of the system; Fig. 2.11 illustrates these points. We can therefore write:

Fig. 2.11 Illustration of the heat flow between the system (the black circle) and the surroundings. An exothermic process results in the flow of heat into the surroundings; an endothermic process results in the flow of heat out of the surroundings.

$$q_{\text{surr}} = -q_{\text{sys}}$$

where q_{surr} is the heat *absorbed* by the surroundings and q_{sys} is the heat *absorbed* by the system.

From tables we can easily find that the heat needed to melt ice to water (called the heat of fusion) is 6010 J mol^{-1}; as expected, this is positive. We are interested in the process water → ice, which is the reverse of melting, so the heat for this is simply minus that for melting ice to water: $q_{\text{sys}} = -6010\ \text{J mol}^{-1}$. Therefore, when water freezes to ice, this heat is given to the surroundings so $q_{\text{surr}} = 6010\ \text{J mol}^{-1}$.

We can now compute the entropy change of the surroundings. First, we will do this at 5 °C, which is 278 K:

$$\begin{aligned}\Delta S_{\text{surr}} &= \frac{q_{\text{surr}}}{T_{\text{surr}}} \\ &= \frac{6010}{278} \\ &= 21.6\ \text{J K}^{-1}\ \text{mol}^{-1}.\end{aligned}$$

We already know that $\Delta S_{\text{sys}} = -22.0\ \text{J K}^{-1}\ \text{mol}^{-1}$ so we can find the entropy change of the Universe as follows:

$$\begin{aligned}\Delta S_{\text{univ}} &= \Delta S_{\text{surr}} + \Delta S_{\text{sys}} \\ &= 21.6 - 22.0 \\ &= -0.4\ \text{J K}^{-1}\ \text{mol}^{-1}.\end{aligned}$$

The entropy change of the Universe is negative, meaning that the entropy has decreased. Such a process is forbidden by the Second Law and so we conclude that water cannot freeze at 5 °C.

Now, let us repeat the calculation at −5 °C, which is 268 K:

$$\begin{aligned}\Delta S_{\text{surr}} &= \frac{q_{\text{surr}}}{T_{\text{surr}}} \\ &= \frac{6010}{268} \\ &= 22.4\ \text{J K}^{-1}\ \text{mol}^{-1},\end{aligned}$$

$$\begin{aligned}\Delta S_{\text{univ}} &= \Delta S_{\text{surr}} + \Delta S_{\text{sys}} \\ &= 22.4 - 22.0 \\ &= 0.4 \text{ J K}^{-1} \text{ mol}^{-1}.\end{aligned}$$

At this lower temperature the entropy change of the Universe is positive, meaning that the entropy has increased. So, the process is allowed by the Second Law and at this lower temperature water can freeze to ice.

It is important to notice that what has changed between 5 °C and −5 °C is the entropy change of the *surroundings*. Furthermore, the reason that ΔS_{surr} has changed in size is simply due to the temperature change; the heat given to the surroundings has remained the same. Due to the way in which we compute ΔS_{surr} (Eq. 2.1), for a given amount of heat the size of the entropy change can be altered simply by changing the temperature of the surroundings.

It is now clear why we put water in a freezer in order to make ice (Fig. 2.12). The process of water going to ice is associated with a reduction in the entropy of the system. For this to be allowed by the Second Law, the reduction in entropy must be compensated for by an increase in the entropy of the surroundings. Such an increase comes about when the surroundings absorb heat, as will be the case for an exothermic process such as water freezing. However, the entropy change of the surroundings depends on their temperature. The lower the temperature, the greater the entropy change. So, water will only freeze once the temperature of the surroundings is low enough that the increase in their entropy is large enough to compensate for the decrease in entropy of the water as it freezes.

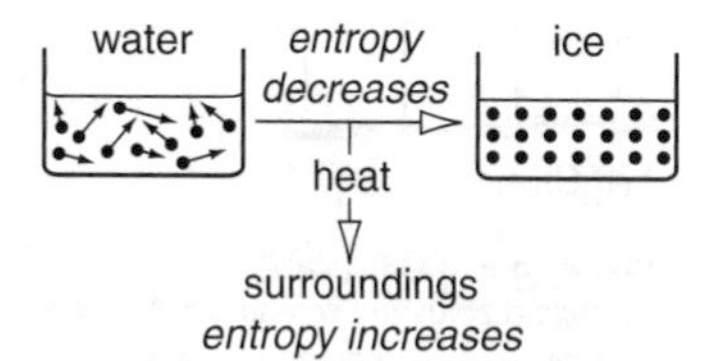

Fig. 2.12 When water solidifies to give ice, the entropy of the system decreases. However, the process is exothermic and so the entropy of the surroundings will be raised as they absorb the heat. The process will only be spontaneous if the surroundings are cold enough that their entropy increase outweighs the entropy decrease of the system, thus making the entropy of the Universe increase.

So, the reason why we put water in the freezer when we want to make ice is so that the entropy change of the surroundings (inside the freezer) is large enough to compensate for the entropy decrease of water freezing. It can be hard to explain this to people who think that the freezer 'sucks the heat out of the water'!

Why all this fuss about making ice?

You may well be asking yourself why it is that we are discussing making ice in a book devoted to understanding chemical reactions! The reason for this is that it is easier to describe how the Second Law applies to a physical process such as water freezing to ice than it is to apply it to chemical equilibrium. However, we can use the tools we have developed so far to understand some reactions which either go entirely to products or do not go at all; these are the subject of the next section. Later, in Chapter 8, we will return to the discussion of chemical equilibrium.

2.5 Exothermic and endothermic reactions

Using the Second Law and the idea of separating the entropy change of the Universe into the entropy change of the system and the surroundings, we can begin to understand why some chemical reactions go. We have already discussed the reaction

$$NH_3(g) + HCl(g) \longrightarrow NH_4Cl(s)$$

in which there is a reduction in entropy of the system. However, as the reaction is exothermic the entropy of the surroundings increases, and this compensates for the decrease in entropy of the system.

Another example is the precipitation of solid AgCl from solution (illustrated in Fig. 2.13):

$$Ag^+(aq) + Cl^-(aq) \longrightarrow AgCl(s).$$

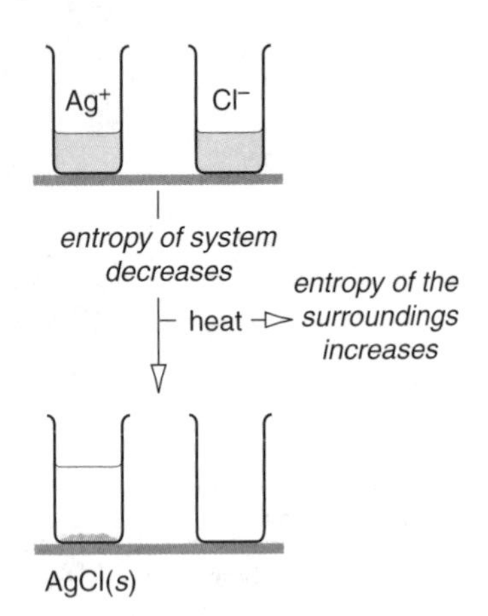

Fig. 2.13 The precipitation of solid AgCl by mixing solutions containing Ag^+ and Cl^- ions is a spontaneous process. The entropy of the system decreases as a solid is formed from two solutions; however, the process is exothermic, resulting in an increase in the entropy of the surroundings which must outweigh the decrease in entropy of the system so that overall the entropy of the Universe increases.

Again, there is a reduction in entropy of the system, as a solid is being formed. In solid AgCl there are strong interactions between the Ag^+ and Cl^- ions; however in solution these ions have favourable interactions with the polar water molecules. Which of these two sets of interactions are stronger depends on the particular ions in question but it turns out for AgCl that the interactions in the solid are the stronger, so the above reaction is exothermic. As a result there is an increase in the entropy of the surroundings. Since we know that the precipitation takes place it must be allowed by the Second Law and so the increase in entropy of the surroundings must outweigh the decrease in entropy of the system. We can simply turn this discussion round and use it to explain why solid AgCl does not dissolve in water as such a process would involve a decrease in the entropy of the Universe and so cannot occur.

Reactions which involve an increase in the entropy of the system do not necessarily need to be exothermic to be allowed by the Second Law. For example, dissolving solid ammonium nitrate in water, illustrated in Fig. 2.14, involves an increase in the entropy of the system because the ions which were held in the crystal lattice are free to move in solution – there is an increase in disorder. For ammonium nitrate it turns out that the interactions between the ions in the solid are stronger than those between the ions and solvent water, so forming the solution is an endothermic process. This means that the entropy of the surroundings decreases. We know that ammonium nitrate dissolves, so it must be that the entropy increase of the system overcomes the entropy decrease of the surroundings so that the Second Law is obeyed.

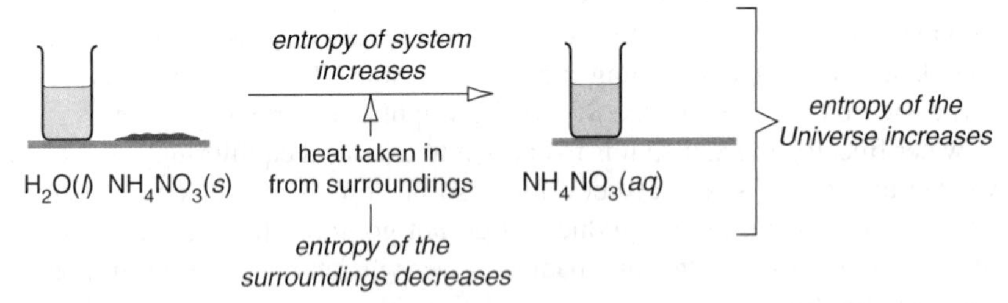

Fig. 2.14 When ammonium nitrate dissolves in water there is an increase in the entropy of the system as the solution is more 'random' than the solid plus liquid which we start with. Dissolving this salt turns out to be endothermic, which means that heat is absorbed from the surroundings resulting in a decrease in their entropy. The process is spontaneous at room temperature, so we conclude that the entropy increase of the system must outweigh the entropy decrease of the surroundings so that the entropy of the Universe increases.

A contrast to ammonium nitrate dissolving in water is when solid sodium hydroxide dissolves. As before, the entropy change of the system is positive but this time the process is exothermic so the surroundings also increase in entropy. The entropy of the Universe increases unconditionally and so the Second Law is obeyed.

As was discussed in Section 2.4 on p. 10, temperature plays an important role in determining the size of the entropy change of the surroundings. A good example of the effect of this is in dissociation reactions such as

$$N_2O_4(g) \longrightarrow 2NO_2(g).$$

Since two moles of gas are produced from one there is an increase in disorder and the entropy of the system increases. This reaction is endothermic, resulting in a decrease in the entropy of the surroundings. The key thing is that the size of this decrease depends on the temperature of the surroundings, as can be seen directly from the way in which the entropy change of the surroundings is calculated: $\Delta S_{surr} = q_{surr}/T_{surr}$. Increasing the temperature decreases the size of ΔS_{surr} simply because of the division by T_{surr}; Fig. 2.15 illustrates this point.

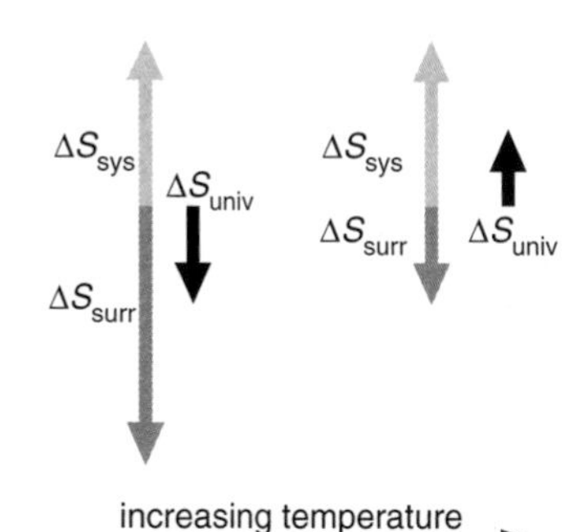

Fig. 2.15 Illustration of the way in which the entropy change of the Universe is affected by an increase in temperature. On the left, the entropy change of the system is positive (upward pointing arrow), but the entropy change of the surroundings is sufficiently negative that it outweighs the entropy increase of the system and so the entropy of the Universe decreases. Increasing the temperature reduces the magnitude of the entropy change of the surroundings and, as shown on the right, eventually the entropy change of the Universe becomes positive.

For the dissociation reaction, when the temperature is low the entropy decrease of the surroundings outweighs the entropy increase of the system and so the reaction does not take place as it would result in a decrease in the entropy of the Universe, in violation of the Second Law. At higher temperatures, the entropy change of the surroundings is smaller in size and eventually it ceases to outweigh the entropy increase of the system. Then the dissociation can take place as it is allowed by the Second Law.

Another example of a dissociation reaction is solid MgO going to gaseous atoms:

$$MgO(s) \longrightarrow Mg(g) + O(g).$$

This is a rather strange looking reaction, but at high enough temperatures it will take place. The argument goes as follows: firstly, we note that the entropy of the system increases as gases are formed from a solid. Secondly, the reaction is endothermic as in forming the gaseous atoms strong interactions in the crystal lattice are broken up. So, the reaction is analogous to the dissociation of N_2O_4 and will only take place when the temperature is high enough. In the case of MgO this turns out to be a *very* high temperature as the reaction is much more endothermic than is the dissociation of N_2O_4.

In summary, to be allowed by the Second Law a process which leads to a decrease in the entropy of the system must be exothermic. A process which leads to an increase in the entropy of the system can be either exothermic or endothermic. The temperature of the surroundings is also important as this affects the magnitude of the entropy change of the surroundings.

2.6 How do we know all this is true?

In this chapter we have simply asserted the truth of the Second Law of Thermodynamics and have also introduced a definition of entropy without much justification. How do we know that these ideas and definitions are correct, and where do they come from? To answer these questions is by no means a simple task – it would involve a detailed study of thermodynamics, and we simply do not have room to go into this now. So, for now we ask you simply to accept these assertions.

Like all physical laws, the truth of the Second Law of Thermodynamics rests on experimental observations and these show that the Second Law is a very strong law indeed. In fact no process which is in violation of the Second Law has ever been observed. We can happily apply the Second Law to understanding chemical reactions in the knowledge that it will not let us down.

2.7 The Second Law and Gibbs energy

It is not really convenient to have to think about the entropy change of the system and the surroundings in order to apply the Second Law. A more convenient approach is to define a quantity called the *Gibbs energy* which can be computed from properties of the system alone. We will show how the change in Gibbs energy is related simply to the change in the entropy of the Universe so that it is possible to apply the Second Law once the change in Gibbs energy is known.

We saw before that we can separate the entropy of the Universe into a part referring to the surroundings and a part referring to the system:

$$\Delta S_{\text{univ}} = \Delta S_{\text{surr}} + \Delta S_{\text{sys}}. \tag{2.2}$$

We also saw that the entropy change of the surroundings can be computed from the amount of heat absorbed by the surroundings

$$\Delta S_{\text{surr}} = \frac{q_{\text{surr}}}{T_{\text{surr}}}, \tag{2.3}$$

and that this heat absorbed is opposite to that absorbed by the system

$$q_{\text{surr}} = -q_{\text{sys}}. \tag{2.4}$$

If we used q_{surr} from Eq. 2.4 in Eq. 2.3 and then put the resulting expression for ΔS_{surr} into Eq. 2.2 we have a new expression for ΔS_{univ}

$$\Delta S_{\text{univ}} = \frac{-q_{\text{sys}}}{T_{\text{sys}}} + \Delta S_{\text{sys}} \tag{2.5}$$

in which we have assumed that the system and surroundings are at the same temperature, so $T_{\text{sys}} = T_{\text{surr}}$ (we will do this from now on). The nice thing about Eq. 2.5 is that all of the quantities on the right-hand side refer to the system – we do not have to worry explicitly about the surroundings.

Under conditions of constant pressure, q_{sys} is identical to the *enthalpy* change of the system, given the symbol ΔH. Chemists virtually always work under conditions of constant pressure, so the identification of q_{sys} with the enthalpy change is certainly appropriate. Using this, Eq. 2.5 becomes

$$\Delta S_{\text{univ}} = \frac{-\Delta H_{\text{sys}}}{T_{\text{sys}}} + \Delta S_{\text{sys}}. \tag{2.6}$$

We now introduce the Gibbs energy, given the symbol G. For a process the change in the Gibbs energy of the system, ΔG_{sys}, is defined as

$$\Delta G_{\text{sys}} = \Delta H_{\text{sys}} - T_{\text{sys}}\Delta S_{\text{sys}} \tag{2.7}$$

where T_{sys} is the absolute temperature, in kelvin. For reasons which will become clear in a moment, we divide both sides of Eq. 2.7 by $-T_{sys}$ to give

$$\frac{-\Delta G_{sys}}{T_{sys}} = \frac{-\Delta H_{sys}}{T_{sys}} + \Delta S_{sys}. \tag{2.8}$$

Comparing Eq. 2.8 with Eq. 2.6 we have

$$\Delta S_{univ} = \frac{-\Delta G_{sys}}{T_{sys}}.$$

In words, $-\Delta G_{sys}/T_{sys}$ is exactly the same thing as the change in entropy of the Universe. A process which is allowed by the Second Law has a positive value for ΔS_{univ} and therefore has a positive value for $-\Delta G_{sys}/T_{sys}$; hence ΔG_{sys} must be *negative*. A positive value for ΔG_{sys} implies a negative value for ΔS_{univ}, and so such a process would not be allowed by the Second Law.

Remember that T_{sys} is in kelvin and so is always positive.

We see that once ΔG_{sys} is known, we can simply inspect its sign to determine whether or not the process is allowed by the Second Law. The convenient thing about this approach is that ΔG_{sys} is computed only from properties of the system – we do not need to worry directly about the surroundings. Indeed, from now on we will drop the 'sys' subscript and simply assume that we are dealing with the properties of the system.

Processes which are allowed by the Second Law of Thermodynamics have $\Delta S_{univ} > 0$ which is the same thing as having $\Delta G_{sys} < 0$.

Making ice – again!

To illustrate the use of the Gibbs energy we will go through the same calculation as in Section 2.4 on p. 10 about making ice. All we need to do is to compute ΔG for the process water $\rightarrow$ ice using

$$\Delta G = \Delta H - T\Delta S.$$

We already know that the heat absorbed by the system in going from water to ice is -6010 J mol^{-1}; in fact this value is for the change at constant pressure and so is equal to ΔH. We also know that the entropy change is -22.0 J K^{-1} mol^{-1} and so it is easy to compute ΔG. At 5 °C (278 K) this is

$$\begin{aligned}\Delta G &= \Delta H - T\Delta S \\ &= -6010 - 278 \times (-22.0) \\ &= 106 \text{ J mol}^{-1}.\end{aligned}$$

Note that the units of the Gibbs energy are joules, or for molar quantities, joules per mole. We see that at 278 K, ΔG is positive; this implies that the process *cannot* take place as it is in violation of the Second Law.

At -5 °C (268 K) the calculation is

$$\Delta G = -6010 - 268 \times (-22.0) = -114 \text{ J mol}^{-1}.$$

Now ΔG is negative and so the process can take place as it is in accord with the Second Law. The conclusions are as before, but the calculations are somewhat easier as we only have to think about the system, not the surroundings.

2.8 Gibbs energy in reactions and the position of equilibrium

In Section 2.1 on p. 6 we spent some time pointing out that energy changes are not the criterion for deciding whether or not a reaction will 'go'. We have now seen that the criterion for whether or not a process can take place is the sign of ΔS_{univ} which is reflected in the sign of ΔG. If it is negative, the process is allowed by the Second Law and so can take place; if it is positive, the process is not allowed by the Second Law and so cannot take place.

For a chemical reaction ΔG means 'Gibbs energy of the products – Gibbs energy of the reactants', just in the same way that ΔH is defined. So, if ΔG is negative, it means that the products have lower Gibbs energy than the reactants. The driving force for a chemical reaction can therefore be described as going to products which have lower *Gibbs energy* than the reactants. Note that it is the *Gibbs energy* which has to be lowered, not the energy.

There are two contributions to the Gibbs energy: an enthalpy term, ΔH, and an entropy term, ΔS:

$$\Delta G = \Delta H - T\Delta S.$$

A negative value of ΔH (i.e. an exothermic reaction) and a positive value of ΔS both contribute to making ΔG negative, and hence a reaction which 'goes'. If ΔH is large and negative, the sign of the entropy term is unimportant as the $-T\Delta S$ term is swamped by the ΔH term. This is why it seems that having a large negative ΔH (i.e. a strongly exothermic reaction) results in a reaction which goes to products.

It is for this reason that in much of the discussion that follows we will be concentrating on energy (enthalpy) changes and trying to identify reaction pathways which lead to the greatest reduction in energy. The chances are that such reactions will be favoured because of their negative ΔH values. The entropy term plays a much more important role when it comes to reactions involving ions in solution, a topic which we will consider in Chapter 12.

The position of equilibrium

At the beginning of this chapter we pointed out that the important thing to understand about a reaction is the position of equilibrium. So far we have introduced the idea that by inspecting the sign of ΔG we can determine whether or not a process is allowed by the Second Law. We have not, however, extended this to the more subtle point of determining the position of equilibrium. We will do this in Chapter 8, but for the moment we will simply concentrate on looking at reactions from the point of view of energy changes.

2.9 Where does the energy come from?

The energy change which accompanies a reaction is due to the making and breaking of chemical bonds: energy is given out when a bond is formed, and taken in when one is broken. As energy changes are very important in determining whether or not a reaction will go, we need to understand the factors which influence the energies involved in making and breaking bonds.

There are two extremes of bonding: one is when electrons are shared equally between two atoms in a *covalent* bond, the other is when two charged groups are held together by electrostatic interactions (i.e. the attraction between opposite charges). The vast majority of bonds fall somewhere between these two extremes, with the electrons in the bond being shared unequally between the two atoms, thus giving a polar bond.

Later on we will discuss the nature of covalent bonding, but first we will look at the bonding due to purely electrostatic interactions. As we shall see, such bonds are particularly strong and so their formation is a powerful driving force for a reaction.

3 Ionic interactions

In the previous chapter we saw how energy changes are crucial in determining whether or not a reaction will take place. There are two major sources of energy in chemical reactions: interactions between charged species (ions) and the making and breaking of covalent bonds. Ionic interactions are probably the easiest to understand so we will discuss these first before moving on to the more complex topic of covalent bonding. This chapter will only consider the interactions between ions in *solid* materials; ions in solution are of course very important in chemistry but are more complex to describe and so we will delay this until Chapter 12.

3.1 Ionic solids

That solid sodium chloride is made up of positively charged sodium ions and negatively charged chloride ions is such a familiar idea that we are perhaps in danger of taking it for granted, so we will spend a few moments reviewing the evidence for the existence of ions in such solid materials.

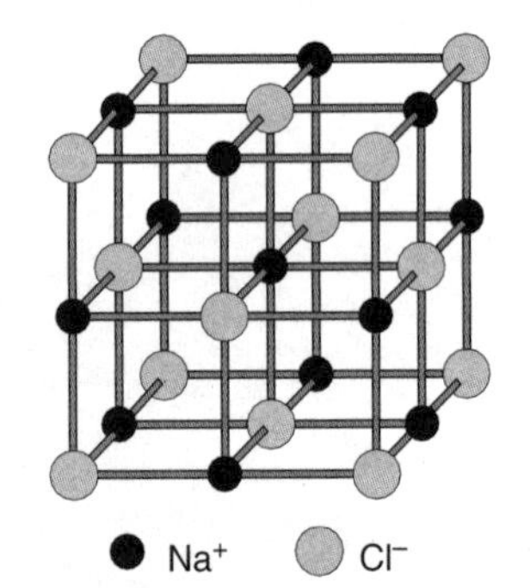

Fig. 3.1 Representation of part of the lattice of solid NaCl as determined by X-ray diffraction.

The best evidence comes from X-ray diffraction studies which can both locate the positions of the nuclei and also map the electron density between them. Such studies show that the sodium and the chlorine atoms are arranged in a regular lattice, as depicted in Fig. 3.1, and that the charge on the chlorine and sodium atoms is close to one unit. Furthermore, it is found that between adjacent sodium and chlorine nuclei the electron density falls to a very low value. These observations are consistent with the almost complete transfer of an electron from sodium to chlorine, leading to the formation of ions with little electron density between them.

X-ray diffraction studies also allow us to measure the distances, $r(\text{MX})$, between the ions in the lattice. If we look at this data for the series of salts MBr and MCl, where M is any of the Group I metals (Li, Na, K, Rb, Cs), we find that the difference $r(\text{MBr}) - r(\text{MCl})$ varies by only a few percent as we change M. Similarly, if we look at the difference $r(\text{RbX}) - r(\text{KX})$, where X is one of the halogens, there is little variation from halogen to halogen.

These observations lead to the idea that each ion can be assigned an *ionic radius* which is more or less independent of the salt in which the ion is present. So, for example, the Na–Cl distance can be found by adding together the ionic radii of Na^+ and Cl^-; the same value of the ionic radius of Cl^- can be used to find the K–Cl distance in solid KCl. That tables of such ionic radii can be drawn up suggest that, to a large extent, the ions can be thought of as behaving independently in the solid.

There are some physical properties which are often taken as being indicative of the presence of ions in a solid. For example, such solids are generally insulators but, on melting, the electrical conductivity rises sharply; this is attributed to the ions being immobile in the solid but free to move in the liquid.

The energy of interaction between charged species, such as ions, has a simple form and we will be able to exploit this in the next section to develop a straightforward expression for the total energy of the ions in a lattice. We will then go on to see how this expression can be used to understand the energetics of the formation of a lattice and rationalize the outcome of some reactions involving solids.

3.2 Electrostatic interactions

In this section we are going to calculate the energy of interaction between the ions in a lattice just by considering the ions as charged objects. As you know, oppositely charged objects attract one another – that is there is a *force* between them; we call this the *electrostatic* or *Coulomb* interaction. This force is found to be proportional to the charges involved and inversely proportional to the square of the distance, r, between the two charges. The force is given by

$$\text{force} = \frac{q_1 q_2}{4\pi\varepsilon_0 r^2}$$

where q_1 and q_2 are the two charges (in units of coulombs), and ε_0 is a fundamental constant called the vacuum permittivity (it has the value $8.854 \times 10^{-12}\,\text{F m}^{-1}$).

The *energy* of interaction of two charges, which is what we are interested in, is given by

$$\text{energy} = \frac{q_1 q_2}{4\pi\varepsilon_0 r}, \qquad (3.1)$$

which is properly described as the electrostatic (or Coulomb) potential energy. Note that it varies inversely with the distance in contrast to the force which varies as the inverse square of the distance.

The charge on an ion can be thought of as the result of an atom gaining or losing a whole number of electrons, which means that the charge can be expressed as a multiple of the charge on the electron, e (which is 1.602×10^{-19} C). So for a singly charged negative ion the charge is $-e$, for a doubly charged negative ion the charge is $-2e$ and so on. Similarly, singly and doubly charged positive ions have charges of $+e$ and $+2e$, respectively.

We need to be careful to distinguish the *numerical* charge on the ion, which is an integer, and the actual charge on the ion, which is some multiple of e. The numerical charge of a positive ion will be written as z_+ and that of a negative ion as z_-; z_+ and z_- are *always* positive numbers, even for negatively charged ions. The charge on a positive ion is z_+e and on a negative ion is $-z_-e$; note the introduction of the minus sign for the negatively charged species.

Putting these values for the charges q_1 and q_2 into Eq. 3.1, the energy of interaction of a positive and negative ion with numerical charges z_+ and z_-, respectively, is

$$\text{energy} = -\frac{z_+ z_- e^2}{4\pi\varepsilon_0 r}. \qquad (3.2)$$

The expression is *negative*, indicating that there is a *lowering* of the energy due to this favourable interaction between opposite charges.

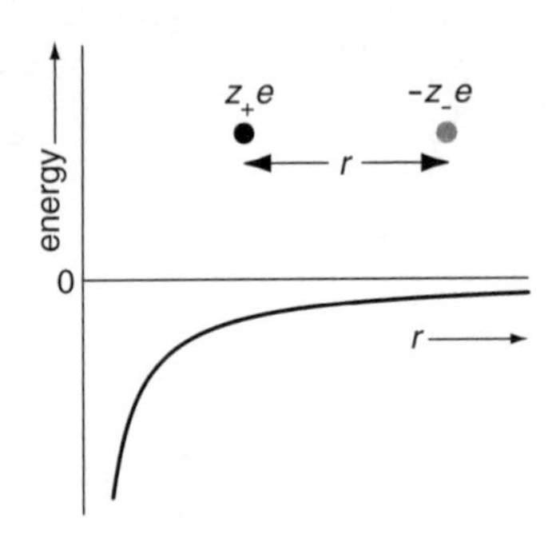

Fig. 3.2 Plot showing how the energy of interaction of two opposite charges varies with the distance between them. The energy is negative as the interaction is favourable, and the closer the ions are to one another the more negative the energy becomes.

Figure 3.2 shows a plot of this energy as a function of the distance between the charges; the closer the two charges come, the more negative the energy. At large values of r the energy goes to zero, but on account of the $1/r$ dependence, this approach to zero is rather slow. The graph shows us that two oppositely charged ions will move towards each other, if they are able to, since this will lead to a lowering in the energy. We will now use this expression to discuss the energetics of the formation of an ion pair.

Forming an ion pair

Imagine we have sodium and chlorine *atoms* in the gas phase. Given what we know about these two elements, we might reasonably expect the sodium to lose an electron to form Na^+ and the chlorine to pick this electron up to form Cl^-. The two oppositely charged ions would then be attracted to one another, and so might form an NaCl molecule which we imagine to be an ion pair, Na^+Cl^-. The whole process can be written:

$$\mathrm{Na}(g) + \mathrm{Cl}(g) \xrightarrow{E_1+E_2} \mathrm{Na}^+(g) + \mathrm{Cl}^-(g) \xrightarrow{E_3} \mathrm{Na}^+\mathrm{Cl}^-(g).$$

E_1 is the ionization energy of sodium, and has the value 502 kJ mol^{-1}; E_2 is the electron affinity of chlorine, which is the energy of the process

$$\mathrm{Cl}(g) + e^- \longrightarrow \mathrm{Cl}^-(g).$$

The value of the electron affinity is -355 kJ mol^{-1}, so E_1+E_2 is 147 kJ mol^{-1}. The fact that $E_1 + E_2$ is positive tells us that when a sodium and chlorine atom are far apart it is not favourable simply to remove an electron from the sodium and add it to the chlorine.

However, when the two oppositely charged ions come together there is a favourable energy of interaction, as shown by Fig. 3.2 – this means that E_3 will be negative. The closer the ions come together the more negative E_3 will become, eventually cancelling out the 147 kJ mol^{-1} (which is $E_1 + E_2$) and so making the overall process exothermic.

The usual expectation for molecules is that there will be a particular distance, the equilibrium distance, at which the energy of interaction is a minimum. However, this will not be the case for these two ions experiencing a Coulomb interaction; all that happens to them is that the energy goes on and on decreasing as the ions approach – there will be no minimum in the energy, which is simply not a realistic situation.

What is missing from our model is that at short enough separations even ions of opposite charge will begin to repel one another due to interactions between their electrons. This repulsive interaction is much shorter range than the Coulomb interaction and typically varies inversely as some high power of r:

$$\text{repulsion energy} \propto \frac{1}{r^n}$$

where n is between 5 and 12; it is around 8 for Na and Cl.

So what we have is a balance between the electrostatic interaction which makes the energy more negative as the ions get closer, and this repulsive interaction which makes the energy increase. The result is that there will be some

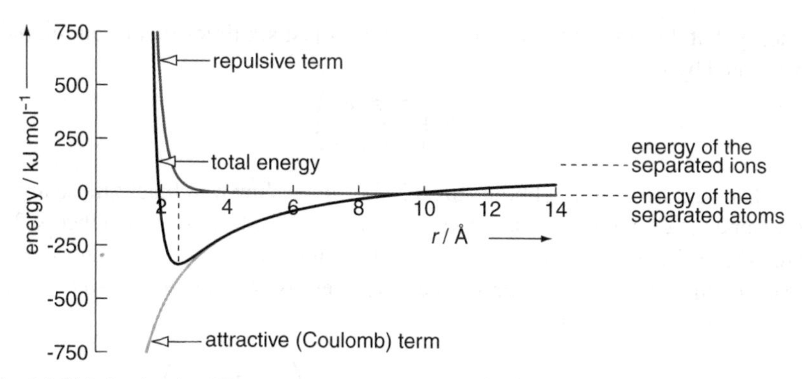

Fig. 3.3 Plot showing the different contributions to the energy of an Na^+Cl^- ion pair as a function of the distance between the two ions. The attractive Coulomb term is always negative and becomes increasingly so as the distance decreases. In contrast, the repulsive term is only significant at much shorter internuclear distances. Adding together these two contributions results in a total energy curve which has a minimum at around 2.5 Å, shown by the dashed line. Zero on the energy scale is defined as the energy of the two separated *atoms*. Since we are considering the interaction of two *ions* the energy at very large separations tends to 147 kJ mol^{-1} (shown by the dashed line) which, as was calculated on p. 22, is the energy needed to form the ions from the atoms.

distance at which the total energy is a minimum; this will be the equilibrium separation of the two ions. These points are illustrated in Fig. 3.3 which shows how the attractive Coulomb term, the repulsive term and the total energy vary with internuclear distance for the Na^+Cl^- pair.

As noted before, the Coulomb term is always negative and falls as the internuclear separation decreases. The repulsive term is much shorter range and, for this ion pair, only makes a significant contribution when the distance is less than about 3 Å; however, once it does get going it rises very steeply. Looking at the curve for the total energy we see that at large distances it follows the attractive Coulomb curve and falls as the distance decreases. Then, once the repulsive term starts to become significant the fall is slowed and eventually reversed leading to the formation of a minimum at around 2.5 Å; this corresponds to the equilibrium separation of the ion pair.

1 Å (Ångström) = 10^{-10} m or 100 pm.

From the graph we can see that forming the ion pair from the atoms is significantly exothermic and so, although the entropy change is likely to be unfavourable, we can expect that the reaction will go. Our simple electrostatic model therefore accounts very nicely for the formation of NaCl molecules in the gas phase as ion pairs. However, this was not really what we set out to do – we are trying to work out the energy of a solid lattice – but as we will see in the next section, this can be done by quite a simple modification of the ion pair case.

Forming a lattice

In a lattice there are many ion–ion interactions which we need to take into account if we are going to estimate the energy using this model. However, as the ions are in a regular array it turns out not to be too difficult to work out what the total electrostatic energy is.

Look at the NaCl lattice depicted in Fig. 3.4 and concentrate on the Na^+ ion at the very centre, which is shown cross-hatched. Its closest neighbours, at a distance r_0, are six Cl^- ions arranged at the corners of an octahedron. The

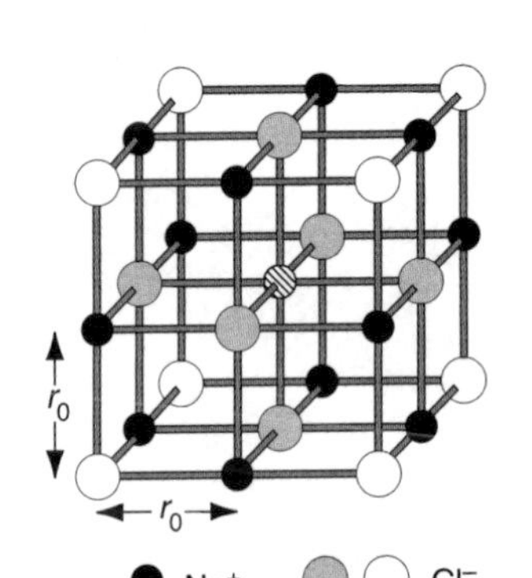

Fig. 3.4 Part of the lattice of solid NaCl. The closest neighbours of the central Na^+ (shown cross-hatched) are the six Cl^- ions shown in grey, next are the 12 Na^+ ions shown in black and next after that are the eight Cl^- ions shown in white. The distances between all of these ions and the central Na^+ can be expressed in terms of the separation, r_0, of adjacent ions.

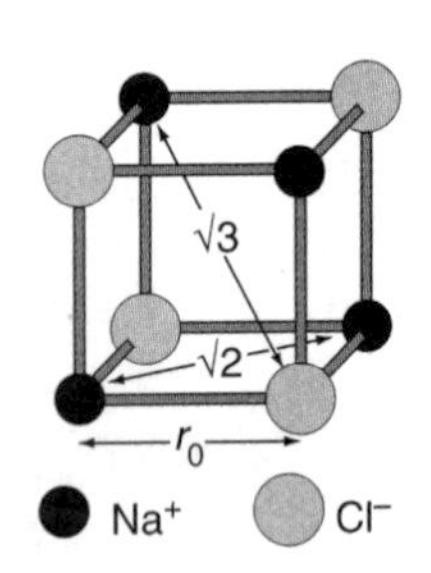

Fig. 3.5 Part of the lattice of NaCl showing how the distances can all be expressed in terms of the distance, r_0, between adjacent ions. The closest distance between Na^+ ions is across the diagonal of a square of side r_0, giving a distance of $\sqrt{2}r_0$. The distance between Na^+ ions and Cl^- ions which are at opposite corners of a cube is the length of the body diagonal, which is $\sqrt{3}r_0$.

energy that these six interactions contribute is just six times that of an ion pair separated by r_0:

$$-6 \times \left(\frac{z_+ z_- e^2}{4\pi\varepsilon_0 r_0}\right).$$

The next nearest neighbours are 12 Na^+ ions, shown in black; these are at the opposite corner of a square to the central Na^+ and so are at distance $\sqrt{2}r_0$ (see Fig. 3.5). As these ions are positive their interaction with the central Na^+ leads to an increase in the energy (i.e. it is a repulsive interaction), so the term we have to add is positive:

$$\underbrace{-6 \times \left(\frac{z_+ z_- e^2}{4\pi\varepsilon_0 r_0}\right)}_{\text{interaction with six Cl}^- \text{ at } r_0} \quad \underbrace{+12 \times \left(\frac{z_+ z_- e^2}{4\pi\varepsilon_0 \sqrt{2} r_0}\right)}_{\text{interaction with 12 Na}^+ \text{ at } \sqrt{2}r_0}.$$

The next set of ions are the eight Cl^-, and these are across the body diagonal from the central Na^+ and so the separation is $\sqrt{3}r_0$. These ions are opposite in charge to the central ion and so contribute a negative term to the energy:

$$\underbrace{-6 \times \left(\frac{z_+ z_- e^2}{4\pi\varepsilon_0 r_0}\right)}_{\text{interaction with six Cl}^- \text{ at } r_0} \quad \underbrace{+12 \times \left(\frac{z_+ z_- e^2}{4\pi\varepsilon_0 \sqrt{2} r_0}\right)}_{\text{interaction with 12 Na}^+ \text{ at } \sqrt{2}r_0} \quad \underbrace{-8 \times \left(\frac{z_+ z_- e^2}{4\pi\varepsilon_0 \sqrt{3} r_0}\right)}_{\text{interaction with eight Cl}^- \text{ at } \sqrt{3}r_0}.$$

We can tidy this expression up by taking out the common factor:

$$\text{electrostatic energy} = -\frac{z_+ z_- e^2}{4\pi\varepsilon_0 r_0}\left(6 - \frac{12}{\sqrt{2}} + \frac{8}{\sqrt{3}} + \ldots\right) \qquad (3.3)$$

where the ... indicate that we can go on adding more terms as we move further away from the central ion. The series in the bracket in Eq. 3.3 turns out, after many terms, to converge on a value of 1.748.

This number is called the *Madelung constant*, M, and is characteristic of the particular arrangement of ions in the lattice; typical values of M are between 1.6 and 2.5. Using the value of M we can write the total electrostatic (Coulomb) energy as

$$\text{electrostatic energy} = -\frac{M N_A z_+ z_- e^2}{4\pi\varepsilon_0 r}$$

where we have multiplied by Avogadro's constant, N_A, so as to obtain the molar energy. Also, we have written the lattice spacing more generally as r, rather than as the equilibrium separation, r_0.

As with the ion pair, we need to include a term to account for the repulsion between ions; writing this as $+B/r^n$ we have our final expression for the total energy as

$$\text{total energy} = -\frac{M N_A z_+ z_- e^2}{4\pi\varepsilon_0 r} + \frac{B}{r^n}.$$

We do not know the value of the constant B, nor of the exponent, n. It turns out that the total energy is not particularly sensitive to the value of n so we can safely set it at some reasonable value. A value for B can be found by recognizing that the energy must be a minimum at the equilibrium separation, r_0 (we will not go into the details here). The final expression for the energy of the lattice at the equilibrium separation is

$$\text{energy at equilibrium} = -\frac{MN_A z_+ z_- e^2}{4\pi\varepsilon_0 r_0}\left(1 - \frac{1}{n}\right) \quad (3.4)$$

which is known as the *Born–Landé equation*.

The importance of the lattice

If instead of forming a lattice we form ion pairs, the energy would be given by Eq. 3.4 with M set to 1. This is because the only interaction is between ion pairs and so in Eq. 3.3 the quantity in the bracket will just be 1, representing the single interaction between oppositely charged ions.

For a lattice, M is greater than 1 which means that its formation is *more* energetically favourable than the formation of ion pairs. The reason for this is that in the lattice it is not just two ions which are interacting but many. True, some of the interactions in the lattice are between ions of like charge and so are repulsive, but on balance the favourable interactions between opposite charges win.

The example of forming Na^+Cl^- ion pairs which we discussed on p. 22 illustrates these points nicely:

$$Na(g) + Cl(g) \xrightarrow{E_1+E_2} Na^+(g) + Cl^-(g) \xrightarrow{E_3} Na^+Cl^-(g).$$

As explained before, $E_1 + E_2$ is 147 kJ mol^{-1}. If we assume that the bond length in Na^+Cl^- is the same as the equilibrium separation in the lattice (2.8 Å) then we can use Eq. 3.4 with $M = 1$ to find $E_3 = -417$ kJ mol^{-1}. These values make the overall process exothermic. However, if we allow a lattice to form:

$$Na(g) + Cl(g) \xrightarrow{E_1+E_2} Na^+(g) + Cl^-(g) \xrightarrow{E_4} NaCl(s).$$

E_4 is greater in magnitude than E_3 by a factor of M (1.75) which is rather significant. So, forming the lattice is a much more exothermic process than forming the ion pairs.

The case of forming MgO shows up this difference even more dramatically:

$$Mg(g) + O(g) \xrightarrow{E_1+E_2} Mg^{2+}(g) + O^{2-}(g) \xrightarrow{E_3} Mg^{2+}O^{2-}(g).$$

Now $E_1 + E_2$ is 2807 kJ mol^{-1} and using an internuclear separation of 2.1 Å we find E_3 is -2266 kJ mol^{-1}. Thus, the process of forming the ion pairs is endothermic and so unlikely to be favourable. However, if we form a lattice rather than the ion pairs

$$Mg(g) + O(g) \xrightarrow{E_1+E_2} Mg^{2+}(g) + O^{2-}(g) \xrightarrow{E_4} MgO(s),$$

we find that E_4 is now -3961 kJ mol^{-1}, making the overall process exothermic. So, forming ion pairs is endothermic, but forming the lattice is exothermic

– a nice illustration of the large extra lowering in energy to be had from forming the lattice.

We should end on a note of caution here. Species such as NaCl and MgO are known in the gas phase, but their internuclear separations are significantly less than the corresponding separations in the crystal. This is taken to imply that the bonding in these simple diatomics is not entirely ionic but has a significant covalent component.

3.3 Estimating lattice energies

The Born–Landé equation, Eq. 3.4, is an expression for the energy of interaction between the ions in a crystal lattice. Usually, when we are thinking about the energetics of solid lattices we talk in terms of the *lattice energy* which is the energy needed to take the ions from the lattice into the gas phase where they are so far apart that they are not interacting. For NaCl the process is

$$\mathrm{NaCl}(s) \longrightarrow \mathrm{Na}^+(g) + \mathrm{Cl}^-(g).$$

The energy of the ions in the lattice is given by Eq. 3.4 and as the gaseous ions are not interacting at all their electrostatic energy is zero. Hence the lattice energy is just minus the energy given by Eq. 3.4:

$$\text{lattice energy} = \frac{M N_A z_+ z_- e^2}{4\pi \varepsilon_0 r_0}\left(1 - \frac{1}{n}\right). \tag{3.5}$$

Strictly speaking this expression gives the change in what is called the *internal energy*, rather than the enthalpy; what is more the energy is for a process taking place at 0 K. It is possible to correct the value to give the usual enthalpy change at 298 K, but the correction is small and not really of any significance for the calculations we are going to do here. We will simply use Eq. 3.5 as an expression for the lattice enthalpy, which is the enthalpy change of going from the solid to the gaseous ions.

Let us use Eq. 3.5 to estimate the lattice energy of NaCl. For this crystal the value of r_0 determined by X-ray diffraction is 2.81 Å, the appropriate value of n is 8 and, as already mentioned, the Madelung constant is 1.748. These data are all we need to compute the lattice energy as 754 kJ mol^{-1}, a value which compares very favourably with the accepted value of 773 kJ mol^{-1}. It is remarkable that this very simple model produces such a good estimate of the lattice energy.

A similar calculation for MgO ($r_0 = 2.00$ Å) gives the lattice energy as 4043 kJ mol^{-1}. This value is much larger than that for NaCl on account of the much stronger interactions between the doubly charged Mg^{2+} and O^{2-} ions in MgO. It is striking how large these lattice energies are, indicating that the interaction between ions is a very significant contributor to the energy changes in chemical processes.

Ionic radii

We mentioned right at the start of this chapter (p. 20) that in a crystal it is possible to assign a radius to each ion, the value of which is broadly independent of the lattice in which the ion is found. This is very useful to us as it means that we can compute the equilibrium separation, r_0, needed in Eq. 3.5 as the sum of the radii of the positive and negative ions, r_+ and r_-, respectively:

$$r_0 = r_+ + r_-.$$

So, if we want to work out the lattice energy of a compound, rather than needing to know the value of r_0 (which would require X-ray diffraction experiments on the crystal) we can simply look up the ionic radii from tables. Some typical values of ionic radii (in units of Å) are shown in the table below:

cations				anions			
$z_+ = 1$		$z_+ = 2$		$z_- = 1$		$z_- = 2$	
Li^+	0.68	Mg^{2+}	0.68	F^-	1.33	O^{2-}	1.42
Na^+	1.00	Ca^{2+}	0.99	Cl^-	1.82	S^{2-}	1.84
K^+	1.33	Sr^{2+}	1.16	Br^-	1.98	Se^{2-}	1.97
Rb^+	1.47	Ba^{2+}	1.34	I^-	2.20	Te^{2-}	2.17

There are some general trends evident from this table:

- as we go down a group the radii increase;
- for elements in the same row of the Periodic Table, 2+ ions are smaller than 1+ ions;
- anions are generally larger than cations;
- for elements in the same row of the Periodic Table, 2− ions are larger than 1− ions.

It makes sense that 2+ ions are smaller than 1+ ions, as when an electron is removed the remaining electrons are held more tightly and so are pulled in by their attraction to the nucleus. Similarly, the increased size of anions is due to the electrons experiencing less attraction to the nucleus. In the following chapter we will look in more detail at the interactions which are responsible for these changes.

We should be aware of the limitations of such tables of ionic radii. Usually they have been 'adjusted' to give, on average, the best estimates of the lattice energy. Different people have different ideas of the best adjustments to make, so you will find that the values of ionic radii quoted in one book are often different to those quoted in another! For a particular lattice the value of r_0 determined by using these tabulated ionic radii is best regarded as a reasonable estimate, and so the value of the lattice energy we determine must be treated with caution. It will not be more precise than a few percent, at best.

The Kapustinskii equations

The Russian scientist A. F. Kapustinskii noticed that the ratio M/ν, where M is the Madelung constant and ν is the number of ions in the molecular formula, varied rather little from crystal structure to crystal structure. He proposed that a good compromise was to take the value for this ratio as 0.874, thus enabling the value of M to be written as 0.874ν. Kapustinskii also proposed that a typical value of $n = 9$ should be used in Eq. 3.5. With these two simplifications the expression for the lattice energy becomes:

$$\text{lattice energy} = \frac{0.874\nu N_A z_+ z_- e^2}{4\pi\varepsilon_0(r_+ + r_-)}\left(1 - \frac{1}{9}\right)$$

where we have replaced r_0 by $(r_+ + r_-)$. Putting in the values of all the constants we find the following rather simple expression for the lattice energy

$$\text{lattice energy/ kJ mol}^{-1} = \frac{1070\ \nu\, z_+ z_-}{(r_+ + r_-)} \qquad (3.6)$$

where the ionic radii are in Å.

For NaCl there are two ions in the molecular formula, so $\nu = 2$; from tables we find the radius for Na^+ to be 1.00 Å and for Cl^- to be 1.82 Å. Putting these numbers into Eq. 3.6 gives us a value of the lattice energy of 759 kJ mol^{-1}, which is quite close to the accepted value.

Equation 3.6 makes it clear that the lattice energy is dominated by two things: the charges on the ions and the ionic radii. A large lattice energy will result if the charges on the ions are large or the ions small.

The Kapustinskii expression for the lattice energy is rather simple and we will find it very useful in the following examples. However, we should recognize that the values it gives will be quite approximate and so we must be cautious about interpreting the results of our calculations in too detailed a way.

3.4 Applications

In this section we are going to look at two reactions in which trends can be understood using calculated lattice energies.

Stabilization of high oxidation states by fluorine

For many metals it has been observed that whereas the fluorides of high oxidation states are relatively easy to prepare, it is often the case that the other halides are more difficult or impossible to form. For example, CuF_2 is known, whereas attempts to prepare the corresponding copper (II) iodide have not been successful. Similarly, for cobalt in oxidation states III and IV, only the fluorides have been prepared. The story is similar for manganese where for oxidation state II all of the halides are known but for oxidation state III only the fluoride has been prepared. There are numerous other examples of this tendency of high oxidation states only to be found as the fluoride.

It is thought that the difficulty in preparing these high oxidation state halides comes about due to the tendency for the halide to decompose into the lower oxidation state with the release of the halogen. For the case of an oxidation state III metal the reaction we are talking about is

$$MX_3(s) \longrightarrow MX_2(s) + \tfrac{1}{2}X_2(g) \tag{3.7}$$

where M is the metal and X a halogen.

What we are going to do is to estimate the energy change for this reaction and see what effect changing the halogen has. The analysis will be done using the following cycle

$$\begin{array}{ccc} MX_3(s) & \xrightarrow{\Delta H_{\text{decomp}}} & MX_2(s) + \tfrac{1}{2}X_2(g) \\ \downarrow \Delta H_{\text{lattice}}(MX_3) & & -\Delta H_{\text{lattice}}(MX_2) \uparrow -\tfrac{1}{2}D(X_2) \\ M^{3+}(g) + 3X^-(g) & \xrightarrow[-EA(X)]{-\Delta H(M^{2+}\rightarrow M^{3+})} & M^{2+}(g) + 2X^-(g) + X(g) \end{array}$$

Starting on the left the solid MX_3 is dissociated into its constituent ions in the gas phase: the energy needed is the lattice enthalpy of MX_3. The M^{3+} ion then gains an electron from one of the X^- ions to give M^{2+} and an X atom: the energy needed is minus the third ionization energy of M, minus the electron affinity of X, EA(X). Finally, the M^{2+} and X^- ions are recombined to form solid MX_2, for which the energy change is minus the lattice enthalpy of MX_2, and the X atom recombines to form an X_2 molecule, for which the energy change is minus half the dissociation energy of X_2, $D(X_2)$.

Overall, the enthalpy change for decomposition, ΔH_{decomp}, is

$$\begin{aligned} \Delta H_{\text{decomp}} = & \Delta H_{\text{lattice}}(MX_3) - \Delta H(M^{2+} \rightarrow M^{3+}) \\ & - EA(X) - \Delta H_{\text{lattice}}(MX_2) - \tfrac{1}{2}D(X_2). \end{aligned} \tag{3.8}$$

We can find the dissociation energies, ionization energies and electron affinities from tables. The lattice energies can be estimated using the Kapustinskii formula, Eq. 3.6. For MX_3, the number of ions, ν, is 4, $z_+ = 3$ and $z_- = 1$, so the lattice energy is given by

$$\Delta H_{\text{lattice}}(MX_3)/\ \text{kJ mol}^{-1} = \frac{1070 \times 4 \times 3 \times 1}{(r_{M^{3+}} + r_{X^-})} = \frac{12840}{(r_{M^{3+}} + r_{X^-})}.$$

For MX_2, $\nu = 3$, $z_+ = 2$ and $z_- = 1$, so the lattice energy is given by

$$\Delta H_{\text{lattice}}(MX_2)/\ \text{kJ mol}^{-1} = \frac{1070 \times 3 \times 2 \times 1}{(r_{M^{2+}} + r_{X^-})} = \frac{6420}{(r_{M^{2+}} + r_{X^-})}.$$

We will complete the calculation for the specific case of cobalt, for which the 3+ ion has a radius of 0.58 Å, and the 2+ ion a radius of 0.71 Å. These data enable us to compute the required lattice energies and thus we can complete the calculation for all the halogens as shown in the table below.

		F	Cl	Br	I
	r_{X^-} / Å	1.33	1.82	1.98	2.20
1	$\Delta H_{lattice}(CoX_3)$ / kJ mol^{-1}	6723	5350	5016	4619
2	$\Delta H(Co^{2+} \rightarrow Co^{3+})$ / kJ mol^{-1}	3238	3238	3238	3238
3	$EA(X^-)$ / kJ mol^{-1}	−334	−355	−331	−301
4	$\Delta H_{lattice}(CoX_2)$ / kJ mol^{-1}	3147	2538	2387	2206
5	$\frac{1}{2}D(X_2)$ / kJ mol^{-1}	79	121	112	107
6	ΔH_{decomp} / kJ mol^{-1}	592	−192	−390	−631

In accordance with Eq. 3.8, the value on line 6 is computed by subtracting the values on lines 2 – 5 from that on line 1; these different contributions to ΔH_{decomp} are visualized in Fig. 3.6. We see clearly from the table that the decomposition reaction

$$CoX_3(s) \longrightarrow CoX_2(s) + \tfrac{1}{2}X_2(g) \tag{3.9}$$

is exothermic for all of the halogens except fluorine. This provides a neat explanation as to why it is possible to prepare CoF_3 but not any of the other trihalides as these can decompose to the lower oxidation state via an exothermic reaction.

In this reaction there is certainly an increase in the entropy as a gas is formed from solid reactants. However, this favourable entropy change is unlikely to outweigh the large positive enthalpy change in the case of the fluoride. For the other halogens, both the increase in entropy and the exothermicity imply that the reaction will go to products.

Just why it is that the fluoride is so different to the other halides can be appreciated by looking at Fig. 3.6. What the diagram shows is that in compar-

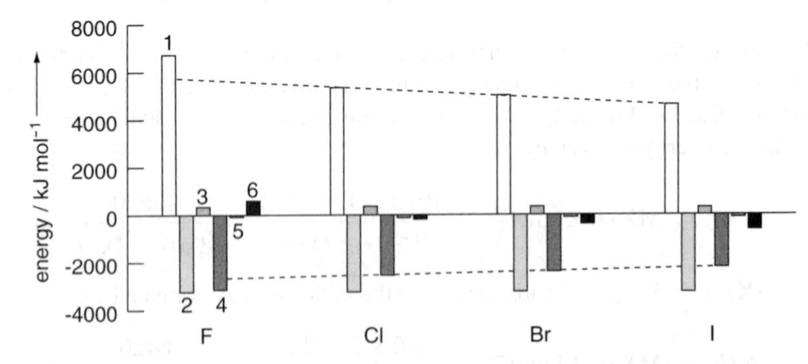

Fig. 3.6 Visualization of the different contributions to ΔH_{decomp} for the decomposition of CoX_3 according to Eq. 3.9. The bars indicate the size of the different energy terms given in the table above, with the numbers on the bars corresponding to the lines in the table. The quantities from lines 2 – 5 are plotted as minus the values shown in the table, as this is how they appear in the calculation of ΔH_{decomp} according to Eq. 3.8. Note how the lattice energies (numbered 1 and 4) and the ionization energy (numbered 2) are the dominant terms. The dashed lines show how the lattice energies of the chloride, bromide and iodide follow a steady trend, whereas the values for the fluoride are out of line with the others, being rather larger than the trend would suggest. It is this which makes the decomposition reaction for the fluoride endothermic.

ing one halide with another the lattice energies (lines 1 and 4 in the table) are the dominant terms (the ionization energy of the cobalt is significant, but is a constant for this series of compounds). We see that for the iodide, bromide and chloride there is a steady rise in the lattice energies which is simply due to the ions getting smaller. However, the increase in the lattice energy for the fluoride is much greater than the trend would suggest – something we can attribute to the unusually small size of the fluoride ion compared to the other halogens.

Figure 3.6 also shows us that as the halide ion gets smaller the increase in the lattice energy of CoX_3 is greater than for CoX_2; the different slopes of the dashed lines highlight this observation. It is easy to see why there is this difference by looking at the expressions for the lattice energies:

$$\Delta H_{\text{lattice}}(CoX_3) = \frac{12840}{(r_{Co^{3+}} + r_{X^-})},$$
$$\Delta H_{\text{lattice}}(CoX_2) = \frac{6420}{(r_{Co^{2+}} + r_{X^-})}.$$

On account of the larger number in the numerator for CoX_3, changing the radius of the negative ion will lead to a greater change in the lattice energy than for CoX_2. So, when it comes to the fluoride the increase in the lattice energy is significantly greater for CoF_3 than for CoF_2; ultimately it is this large increase which tips the balance making the decomposition reaction endothermic for the fluoride.

Peroxides and superoxides of the Group I metals

Our final example is concerned with the range of different compounds that are produced when Group I metals combine with oxygen. Lithium gives the simple oxide Li_2O, sodium gives mainly the peroxide, Na_2O_2, and metals further down the group give mainly the superoxide MO_2 (the anion present in the peroxide is O_2^{2-} and in the superoxide the anion is O_2^-). The trend is for the metals with the larger ions to form the superoxide – an observation which we will be able to rationalize using lattice energies.

Let us consider the reaction in which the superoxide decomposes to the peroxide plus oxygen gas; this is presumably the reaction by which the superoxides of sodium and lithium would decompose, were they to be formed:

$$2MO_2(s) \longrightarrow M_2O_2(s) + O_2(g).$$

We can analyse this reaction using the following cycle:

$$\begin{array}{ccc} 2MO_2(s) & \xrightarrow{\Delta H_{\text{decomp}}} & M_2O_2(s) + O_2(g) \\ \downarrow 2\Delta H_{\text{lattice}}(MO_2) & & -\Delta H_{\text{lattice}}(M_2O_2) \uparrow \\ 2M^+(g) + 2O_2^-(g) & \xrightarrow{x} & 2M^+(g) + O_2^{2-}(g) + O_2(g) \end{array}$$

The quantity simply denoted x is the enthalpy change on going from the superoxide anion to the peroxide anion plus gaseous O_2; for the present discussion we do not need to know its value.

The enthalpy change of the decomposition reaction, ΔH_{decomp}, can be computed from:

$$\Delta H_{\text{decomp}} = 2\Delta H_{\text{lattice}}(\text{MO}_2) + x - \Delta H_{\text{lattice}}(\text{M}_2\text{O}_2). \tag{3.10}$$

As before, we can use the Kapustinskii expression (Eq. 3.6 on p. 28) to estimate the lattice energies:

$$\Delta H_{\text{lattice}}(\text{MO}_2)/\text{ kJ mol}^{-1} = \frac{1070 \times 2 \times 1 \times 1}{(r_{\text{M}^+} + r_{\text{O}_2^-})} = \frac{2140}{(r_{\text{M}^+} + r_{\text{O}_2^-})}$$

$$\Delta H_{\text{lattice}}(\text{M}_2\text{O}_2)/\text{ kJ mol}^{-1} = \frac{1070 \times 3 \times 1 \times 2}{(r_{\text{M}^+} + r_{\text{O}_2^{2-}})} = \frac{6420}{(r_{\text{M}^+} + r_{\text{O}_2^{2-}})}.$$

Substituting these expressions into Eq. 3.10 we have

$$\Delta H_{\text{decomp}} = \left[\underbrace{\frac{4280}{(r_{\text{M}^+} + r_{\text{O}_2^-})}}_{2\Delta H_{\text{lattice}}(\text{MO}_2)} - \underbrace{\frac{6420}{(r_{\text{M}^+} + r_{\text{O}_2^{2-}})}}_{\Delta H_{\text{lattice}}(\text{M}_2\text{O}_2)} \right] + x.$$

The radii of the O_2^{2-} and O_2^- ions are not that much different, so the quantity in the square bracket is negative on account of the larger numerator of the second fraction. Now imagine what happens as the size of the cation M^+ is increased; both lattice energy terms decrease but the one for M_2O_2 decreases more than the one for MO_2 on account of the larger numerator in the second fraction. The term in the bracket therefore becomes *less negative* as the radius of the metal ion increases.

We therefore conclude that the trend is for the decomposition reaction of the superoxide to become less favourable in energetic terms as the size of the metal ion increases. This rationalizes the observation that the superoxide is not found for the early members of Group I, but is found for the elements toward the bottom of the group which have the larger ions.

The argument here is not as detailed as in the previous example. We have not computed explicit values of the ΔH_{decomp} for the different metals and then compared them with one another. Rather, recognizing from the first example that the variation in the lattice energies is usually the crucial factor, we have simply looked at the way in which these terms will vary as the ionic radii are changed.

3.5 Ionic or covalent?

We mentioned at the start of this chapter that X-ray diffraction studies of crystals such as NaCl reveal that the electron density falls to a very low value between adjacent ions, and we cited this as evidence for the existence of ions in such crystals. However, if we look more closely at the electron density in 'ionic' solids such as LiF, NaCl and KCl we find two things. Firstly, although the density between ions of opposite charge does become very low it does not

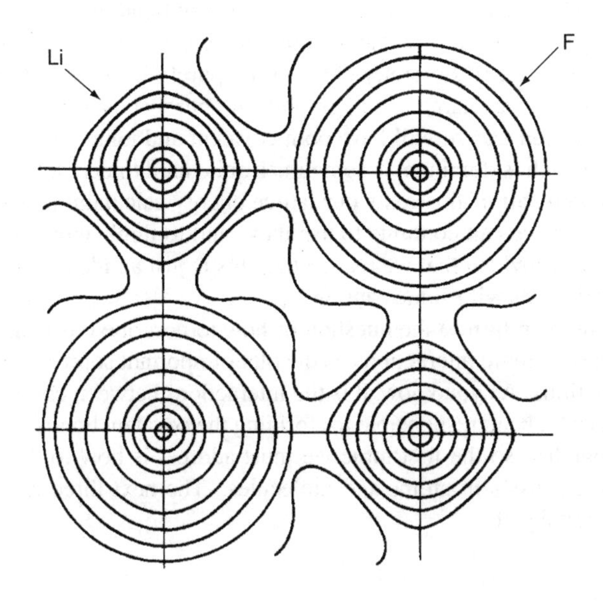

Fig. 3.7 Experimental electron density map for LiF determined using X-ray diffraction. The electron density is indicated by contours and the map shows the density in what would be a horizontal plane through the schematic structure shown in Fig. 3.1. Note that the electron distribution around the lithium is not spherical, which is taken as evidence of some covalent character in the bonding. (Reproduced, with permission, from H. Witte and W. Woelfel, *Reviews of Modern Physics*, **30**, 53, (1958). Copyright (1958) by the American Physical Society.)

actually fall to zero. Secondly, the distance between the nucleus and the minimum in the electron density does not correspond to the tabulated value of the ionic radius. These two observations cast doubt on whether these solids really do contain ions in the way we have assumed.

Close inspection of the electron density map, such as the one shown in Fig. 3.7 for LiF, reveals further problems. The contours of electron density around the lithium are not circular; if the lithium really was an ion its electron density would be spherical and so the contours would be circular. This distortion is taken as evidence that the lithium is not truly an ion but that there is some sharing of electrons with the fluorine – in other words there is some *covalent* contribution to the bonding.

Further evidence for there being some covalent contribution to the bonding in these apparently 'ionic' solids comes from comparison of the calculated and experimentally determined values of the lattice energies (remember that such energies can be determined experimentally using a Born–Haber cycle). Whereas compounds such as NaCl and KI show good agreement between the experimental and calculated values, for compounds of Ag(I) (e.g. AgI) and Tl(I) (e.g. TlBr) the experimental lattice energy is significantly in excess of the calculated value. As the calculation only considers the electrostatic contribution, this discrepancy is taken as indicating the presence of additional covalent contributions.

In reality what we have is a continuum of types of bonding. At one extreme there is purely ionic bonding and at the other there is purely covalent bonding. Compounds such as LiF lie pretty close to the purely ionic end of the scale, whereas solids such as diamond and graphite are purely covalent . Everything else lies in between and so will have both covalent and ionic contributions to the bonding; which is dominant is sometimes quite hard to decide!

Unfortunately, all of this casts our lattice energy calculations into further doubt. However, we can continue to use these and to think about such solids as being 'ionic' provided that we realize that this is just an idealization and so treat our conclusions with some caution.

We now need to turn to the question of how to describe covalent bonding which, along with ionic interactions, is the other important source of energy in chemical reactions. As you know, it is the interactions between charged nuclei and electrons which are responsible for forming the covalent bonds which hold molecules together, so the first topic we must address is how to describe the behaviour of electrons in atoms and molecules. The next three chapters are devoted to this subject.

4 Electrons in atoms

A chemical bond involves the sharing of electrons, so to understand how bonds are formed we need to understand where the electrons are and what they are doing; we will start this process by first talking about electrons in atoms. You probably know quite a lot about this already and are used to describing the arrangement of electrons in an atom in terms of which *orbitals* are occupied. Each orbital has a distinct energy and a particular shape, and we will see that when it comes to describing how bonds are formed these two properties are crucial. The material covered in this chapter is therefore vital to our understanding of chemical bonding.

The theory we need to describe the behaviour of electrons in atoms and molecules is *quantum mechanics*. We certainly do not have the space in this book to go into the background and mathematical formulation of this theory. Rather, we will simply introduce the key ideas and then present the results of quantum mechanical calculations in a pictorial way. This approach will bring us quickly and without too much effort to the really important points – the shapes and energies of the orbitals.

Many of the ideas of quantum mechanics seem rather unusual, even counter-intuitive. This is because the world which we experience directly is governed by Newton's Laws (classical mechanics) which are quite different from the principles of quantum mechanics which apply to tiny particles such as electrons. For the moment, we will simply ask you to accept the sometimes rather strange world of quantum mechanics. If you go on to study this in more detail you will see where these ideas come from and how the predictions of the theory have been tested against experiments and found to be correct.

4.1 Atomic energy levels

One of the most important predictions of quantum mechanics is that the energy of an electron in an atom or molecule cannot take any value but is restricted to a specific set of values; the energy is said to be *quantized*. This prediction is entirely at odds with our direct experience of the world. For example, when we ride a bicycle we expect to be able to go at any speed (that is, have any kinetic energy). We do not expect to find that our speed, and hence energy, is restricted to particular values – but this is exactly what happens to the electron. Each allowed value of the energy is called an *energy level*, and we imagine that a particular electron 'occupies' one of these levels.

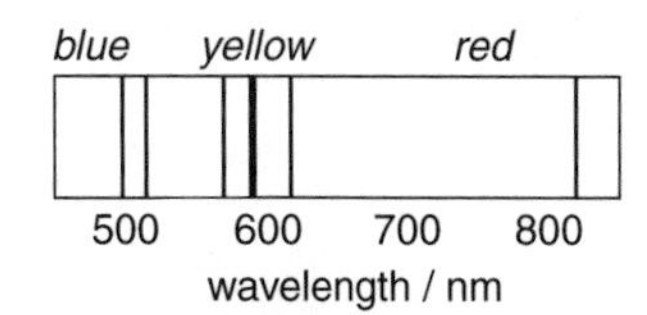

Fig. 4.1 Schematic emission spectrum from atomic sodium, such as would be seen by passing the light from a sodium discharge lamp through a prism or reflecting it off a diffraction grating. The most striking thing is that the emissions occur at a series of sharply defined wavelengths. In the case of sodium, the strongest by far is the yellow line at 589 nm.

The most direct evidence for the existence of atomic energy levels comes from the study of atomic spectra, which involves measuring the wavelength of the light emitted by excited atoms. Such spectra show that light is only emitted at certain wavelengths, and that these wavelengths are highly characteristic of the atom being studied. Figure 4.1 shows part of such an emission spectrum from sodium; such spectra are often called *line spectra* to indicate that the emissions are confined to a small number of well-defined wavelengths.

To understand how line spectra provide evidence of quantization of energy we first need to recall the relationship between the wavelength of light, λ, and its frequency, ν:

$$\lambda\nu = c$$

where c is the speed of light. Light can be thought of as being composed of photons whose energy, E, depends on their frequency in the following way:

$$E = h\nu$$

where h is Planck's constant. It therefore follows that the energy and wavelength of a photon are related by

$$E = h\frac{c}{\lambda}.$$

So, light of a particular wavelength corresponds to photons of a particular energy.

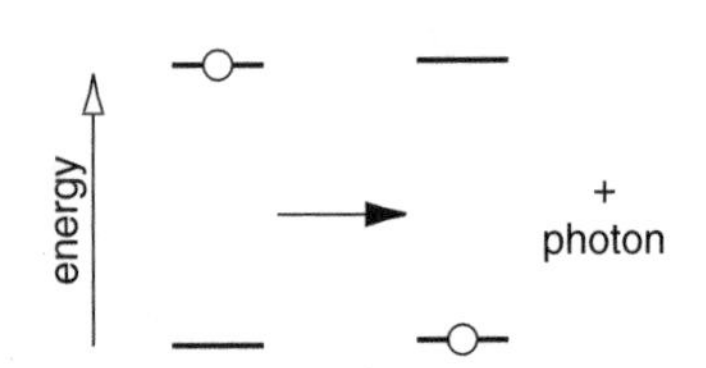

Fig. 4.2 Atomic emission spectra are produced when an electron (represented by the circle) drops down from a higher energy level to a lower one. This releases a photon whose energy matches the energy difference between the two levels. The energy of the photon therefore depends on the energy levels in the atom and the observation that only photons of specific energies are produced is taken to imply the existence of energy levels.

Our interpretation of atomic line spectra is that an electron in an excited atom drops down from one energy level to another, releasing a photon whose energy matches the *difference* in energy between the two levels (Fig. 4.2). As the photons emitted from a particular atom are at sharply defined energies, the energies of the levels themselves must also have fixed values. If the electrons in an atom were allowed to have *any* energy, any change in the energy of the electron would be possible and so light would be emitted at all wavelengths. The observation of atomic line spectra is thus direct evidence for the existence of atomic energy levels.

Atomic orbitals

The energy levels which the electrons can occupy in an atom are called *atomic orbitals*. The *electronic configuration* for an atom is generated by filling up the orbitals, starting from the lowest energy one and moving up in energy until all the electrons are used up; each orbital can accommodate two electrons, one with spin 'up' and one with spin 'down'.

The lowest energy orbital is the $1s$, and this single orbital forms the first shell, or K shell. This shell can accommodate up to two electrons giving the possible configurations $1s^1$ and $1s^2$ which correspond to hydrogen and helium, respectively.

The next in energy is the L shell, which contains the $2s$ orbital and the three $2p$ orbitals. Lithium (atomic number, $Z = 3$) has the configuration $1s^2 2s^1$ and neon ($Z = 10$), in which the L shell is full, has the configuration $1s^2 2s^2 2p^6$.

The next shell is labelled M and contains the $3s$ orbital, three $3p$ orbitals and five $3d$ orbitals. The elements of the third row of the Periodic Table have electronic configurations which involve filling this shell.

We will see in the next section that we can use a special kind of spectroscopy to find direct evidence for the existence of these orbitals and to measure their energies. Then, in the following section, we will look more closely at just exactly what an orbital is and what is meant by the labels $2s$, $3p$, etc.

4.2 Photoelectron spectroscopy

Photoelectron spectroscopy (PES) provides a particularly direct way of probing the energy levels of electrons in atoms and molecules. Such spectra are recorded by irradiating the sample with light of a fixed frequency high enough for the photons to ionize electrons from the sample; what we actually measure are the energies of these ionized electrons. As energy is conserved, it follows that

$$\text{energy of photon} = \text{energy of ionized electron} + \text{ionization energy of that electron.}$$

So, the *higher* the energy of the ejected electron, the *lower* the ionization energy of the electron which has been ejected; the whole process is illustrated in Fig. 4.3.

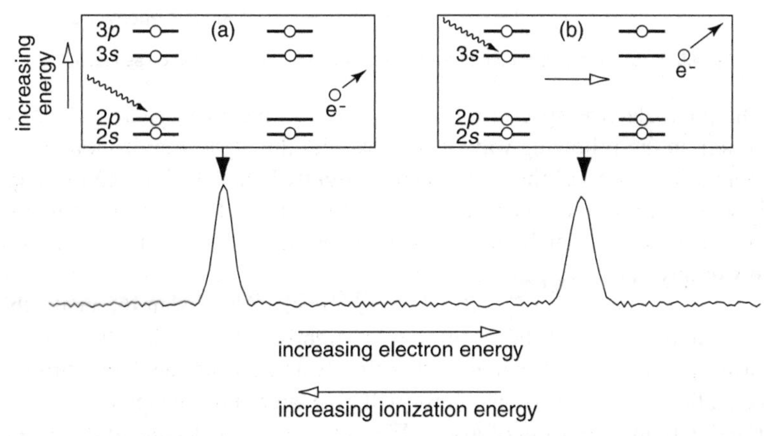

Fig. 4.3 Schematic representation of how a photoelectron spectrum is generated. In box (a) we see a photon (represented by the curly line) coming in and causing an electron (e^-) to be ionized from a $2p$ orbital. In box (b) a photon of the *same energy* ionizes a $3s$ electron which is not so tightly held; as a result the electron has more energy than is the case in (a) and the corresponding peak therefore appears at higher electron energy. Note that the scale on the spectrum can be marked in terms of electron energy or ionization energy, and that these two scales run in opposite directions.

What we see in a typical photoelectron spectrum is a number of peaks at different electron energies which we interpret as corresponding to the ionization of electrons from different energy levels of the atom. It is important to understand that these peaks are not due to the successive ionization of electrons to give M^+, M^{2+}, M^{3+} ... ions but are due to the formation of M^+ by the ionization of electrons from *different* levels.

It is usual to quote the energies of the electrons and their ionization energies in units of *electron-volts*, given the symbol eV. This is a unit of energy, with 1 eV being equal to 1.60×10^{-19} J or 96.5 kJ mol^{-1}.

Photoelectron spectra of atoms

The photoelectron spectrum of helium, shown in Fig. 4.4, shows a single peak at around 25 eV; this corresponds to ionizing an electron from the $1s$ orbital. There are no further peaks in the spectrum, which makes sense as there are no other occupied orbitals.

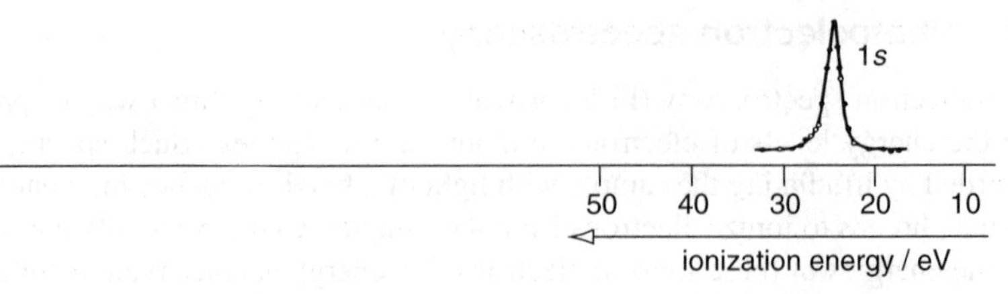

Fig. 4.4 Photoelectron spectrum of helium recorded using X-ray photons as the ionizing radiation. (Adapted from Fig. 4.1 of *ESCA Applied to Free Molecules* by K. Siegbahn *et al.*, p. 24, Copyright (1969), with permission from Elsevier Science.)

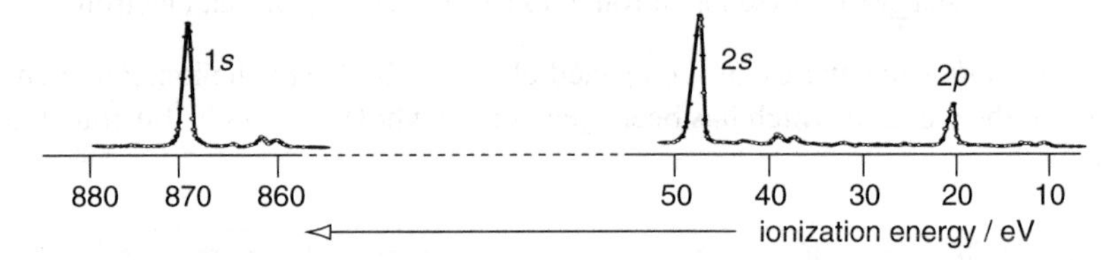

Fig. 4.5 Photoelectron spectrum of neon. (Adapted from the same source as Fig. 4.4.)

The heights of the peaks in photoelectron spectra cannot be interpreted in a simple way.

The photoelectron spectrum of neon, Fig. 4.5, has more peaks, which we can assign in the following way. The peak with the highest ionization energy is from the 1*s* electrons; these are the most tightly held. The ionization energy of these 1*s* electrons is much higher than that of the 1*s* electrons in helium because neon has a much higher nuclear charge and so attracts the 1*s* electrons more strongly. The two peaks at around 49 eV and 21 eV are due to ionizing electrons from the 2*s* and 2*p* orbitals, respectively. We assign the peaks this way round as we expect the 2*s* to be more tightly held than 2*p*. Indeed, this spectrum provides direct evidence that the 2*s* and 2*p* orbitals have different energies, the explanation of which we will give later in this chapter.

Finally, in the spectrum of argon, Fig. 4.6, we see peaks due to the 2*s* and 2*p* electrons at around 325 eV and 250 eV. These are at much higher ionization energies than the corresponding electrons in neon due to the increased nuclear charge of argon; at much lower energies there are peaks due to 3*s* and 3*p*. The peak at around 250 eV is split into two; it turns out that this is due to the influence of the electron spin, but this is a rather complex matter which we will not go into here. The peak due to the 1*s* electrons is at a very much higher ionization energy (about 3000 eV) and is not shown.

These photoelectron spectra show us in a direct way how the electrons fill

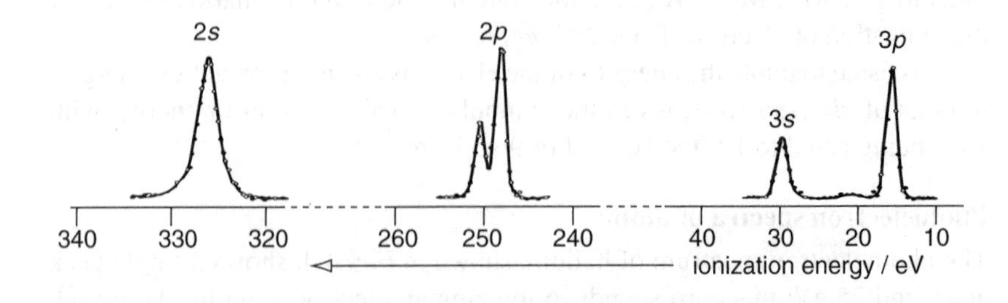

Fig. 4.6 Photoelectron spectrum of argon. (Adapted from the same source as Fig. 4.4.)

the various orbitals and the way in which the energies of these orbitals vary from element to element.

4.3 Quantum mechanics

Quantum mechanics is the theory which enables us to predict the energy levels, and much more besides, of electrons in atoms and molecules. Each energy level has associated with it a *wavefunction* which is also predicted by the theory. These wavefunctions are very important, as using them we can calculate *any* property of the system. We will focus on how wavefunctions predict the spatial distribution of the electrons – something which is crucial to our understanding of chemical bonding.

What is a wavefunction and what does it tell us?

Before we start thinking about wavefunctions we should just remind ourselves what a mathematical function is. Let us take as an example the quadratic function $f(x)$:

$$f(x) = ax^2 + bx + c. \tag{4.1}$$

The notation $f(x)$ means that f is a function of the variable x; a, b and c are just numbers, sometimes called *coefficients*. If we make a plot of $f(x)$ against x the appearance of the graph will depend on the values of these coefficients. Figure 4.7 shows such a graph for the case $a = 1$, $b = -2$ and $c = -8$; with these values, the function is zero for $x = 4$ and $x = -2$.

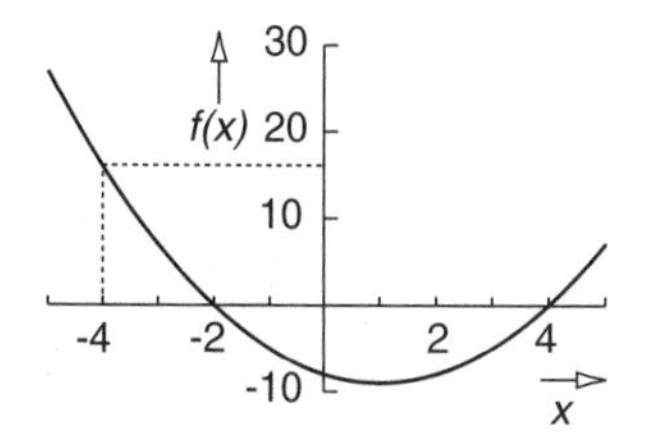

Fig. 4.7 A plot of the quadratic function given in Eq. 4.1 for the particular values of the coefficients $a = 1$, $b = -2$, $c = -8$. For any value of x the function can be evaluated to give a number; for example when $x = -4$, the function evaluates to 16.0.

Provided a, b and c are known, any value of x can be substituted into the right-hand side of Eq. 4.1 and, with a calculator, we can evaluate the expression to give a number; the function $f(x)$ therefore just returns a number for any value of the variable x. The same idea can be extended to a function of more variables, for example $g(x, y, z)$ which for any values of the three variables x, y and z can be evaluated to give us a number.

Quantum mechanics predicts that for the hydrogen atom there is a set of energy levels each with an associated wavefunction; in fact, these wavefunctions *are* the familiar orbitals. Not surprisingly, the wavefunctions are three-dimensional and depend on the distances x, y and z from the nucleus.

What, then, does the wavefunction tell us? One interpretation, called the *Born interpretation*, is as follows. Imagine a very small box hovering in space and centred at a point with coordinates (x, y, z) (Fig. 4.8). The probability of finding the electron in this small box is proportional to the *square of the wavefunction* at this point. The greater the value of the wavefunction the higher the probability of finding the electron in the box.

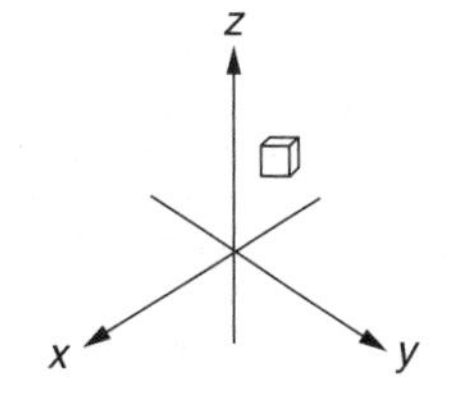

Fig. 4.8 The probability of finding the electron in a small box centred at point (x, y, z) is proportional to the square of the wavefunction at that point. The nucleus is at the centre of the coordinate system and the size of the box has been greatly exaggerated for clarity.

We have to think about a 'very small box' so that we do not need to worry about the wavefunction changing value from one side of the box to another. If you are familiar with calculus you will know that the very small box can be described as a *volume element*.

The strange thing about the quantum mechanical description of the electron is that the theory predicts a *probability* of the electron being in a small box at a particular position, rather than saying that the electron *is* at a particular position. You may have come across the Bohr model of the atom in which the electron orbits around the nucleus at a specific distance determined by the energy. The quantum mechanical picture is quite different to this; there are no specific orbits, but only a probability that the electron will be found at a particular distance from the nucleus.

The square of the wavefunction is properly described as the *probability density*. This is because the probability of finding the electron in our small box is equal to the square of the wavefunction multiplied by the volume of the box:

$$\text{probability of being in the box} = (\text{wavefunction})^2 \times \text{volume of the box}. \quad (4.2)$$

This makes sense, as if we increase the volume of the box we expect the probability of finding the electron in it to increase. Equation 4.2 is analogous to the one we would use to calculate the mass of an object of known volume and density:

$$\text{mass} = \text{density} \times \text{volume}. \quad (4.3)$$

Comparing Eqs. 4.2 and 4.3 we can see that the square of the wavefunction takes the same role as the density of the material – hence the description of the square of the wavefunction as a probability density.

4.4 Representing orbitals

The hydrogen atom wavefunctions (the orbitals) are three-dimensional functions and so we have something of a problem to represent them on paper. It is very important to understand the shapes of the orbitals as when it comes to forming bonds these shapes are crucial. So, in this section we will spend some time discussing different ways of representing the orbitals on paper and the problems that each of these representations has.

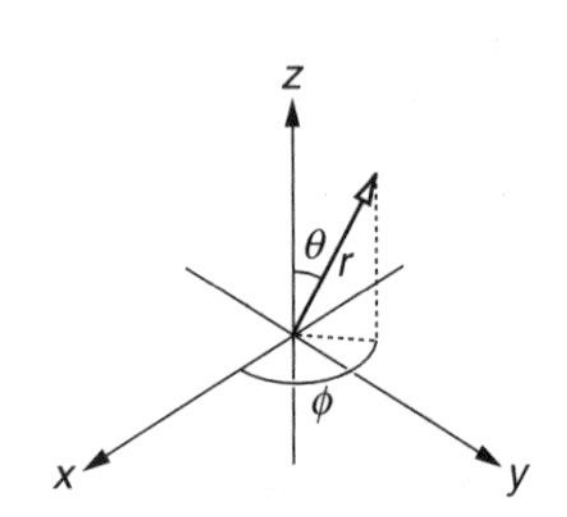

Fig. 4.9 In atoms it is convenient to specify the position in space using r, the distance from the nucleus, and the two angles θ (which is like the latitude) and ϕ (which is like the longitude). These are called *spherical polar coordinates*.

We will start with the wavefunction of the 1s orbital which, like all of the hydrogen atomic orbitals, is a function of the position in three-dimensional space. It turns out that for atoms it is more convenient to think of the position not in terms of x, y and z but in terms of r, the distance from the nucleus (r is always positive), and the two angles θ and ϕ as shown in Fig. 4.9. These two angles are like the latitude and the longitude, respectively. The coordinates r, θ and ϕ are called *spherical polar coordinates*.

Expressed in these coordinates, the mathematical form of the wavefunction of the 1s orbital, $\Psi_{1s}(r)$, is

$$\Psi_{1s}(r) = A\,\mathrm{e}^{-Br};$$

this particular wavefunction does not depend on θ or ϕ. The values of A and B depend on fundamental physical constants in a way predicted by the theory; their precise values do not concern us here.

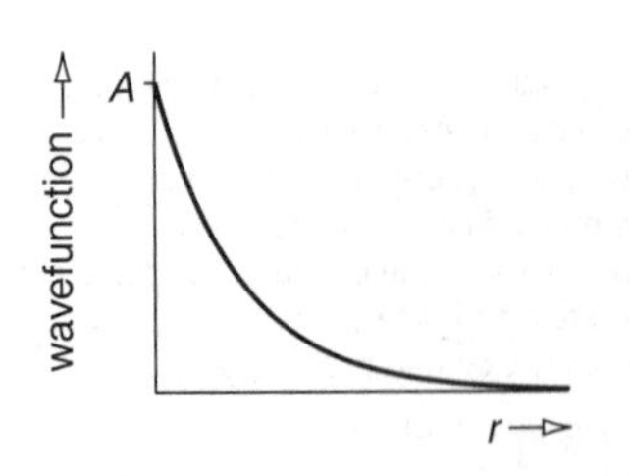

Fig. 4.10 A plot of the wavefunction for the 1s orbital against the distance, r, from the nucleus. Recall that r is always positive.

When $r = 0$ (i.e. at the nucleus) e^{-Br} is equal to 1 (as $\mathrm{e}^0 = 1$) and so the wavefunction has the value A. As r increases, e^{-Br} decreases from 1 and tails off towards zero as r gets large enough. These features can be seen in the plot of $\Psi_{1s}(r)$ as a function of r shown in Fig. 4.10.

The wavefunction falls off as we go away from the nucleus; the same is true of the square of the wavefunction, so the implication is that as we move away from the nucleus the probability of finding the electron falls off, eventually reaching a negligible level. The electron is not localized to a particular region nor is it going round a particular orbit; rather it is smeared out over space.

Contour plots and shaded plots

Another way of representing the wavefunction is to draw a *contour plot* of a cross-section through the wavefunction. Imagine taking the three-dimensional wavefunction and cutting a slice through it, say in the *xy*-plane. Then, on a flat surface, we draw lines connecting points which have the same value of the wavefunction. This is the same idea as representing the height of the surface of the Earth using contour lines on a map.

Figure 4.11 (a) shows a contour plot of the 1*s* wavefunction. We can see that points with the same value of the wavefunction come out as circles on the plot. This is what we expect since the value of the wavefunction depends only on the distance from the nucleus.

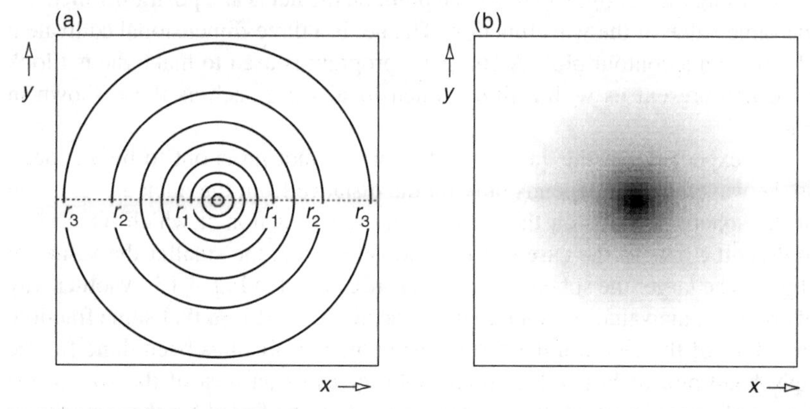

Fig. 4.11 Two different representations of the 1*s* orbital. Each plot is taken through the *xy*-plane and is centred on the nucleus. In (a) contours are drawn connecting points at which the wavefunction has the same value; the distances r_1, r_2 and r_3 marked along the dotted line refer to Fig. 4.12. In (b) the value of the wavefunction is represented using a grey scale; the darker the shade of grey, the larger the value of the wavefunction.

The way in which this contour plot is related to the plot of the wavefunction as a function of r is indicated in Fig. 4.12. Here, three distances (r_1, r_2 and r_3) are indicated, and these same distances are marked on the contour plot. The wavefunction clearly decreases in value as we go from r_1 to r_2 to r_3, so the circular contours with the *larger* radii correspond to *lower* values of the wavefunction.

Another way of representing this two-dimensional slice is to shade in the plot in proportion to the value of the wavefunction; Fig. 4.11 (b) shows such a representation of the 1*s* orbital. The darker the shade of grey, the larger the value of the wavefunction. Such shaded plots often give a more easily appreciated picture of the wavefunction than do contour plots.

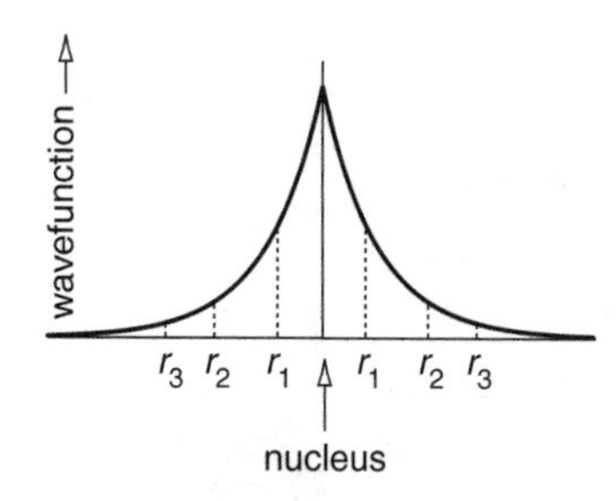

Fig. 4.12 A plot of the 1*s* wavefunction showing its value at three distances r_1, r_2 and r_3 from the nucleus. The same distances are marked on Fig. 4.11. This plot also shows that the wavefunction only depends on the distance, *r*, from the nucleus; the direction in which we measure *r* is unimportant.

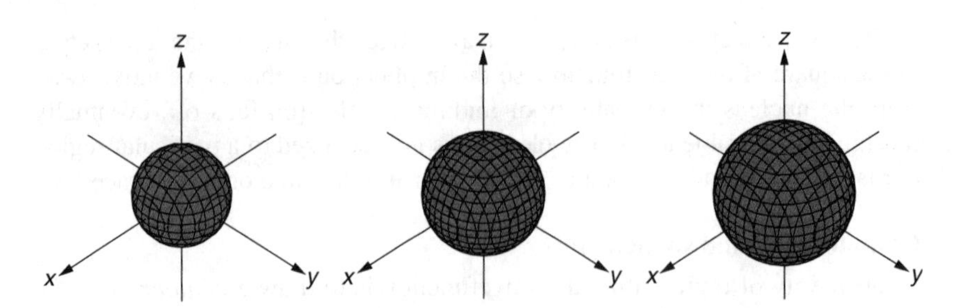

Fig. 4.13 Surface plots of the 1*s* orbital. The value of the wavefunction at which the surface is drawn decreases going from left to right, and in the right-hand plot the surface has been chosen so that 90% of the electron density is within it. The maximum distance along any of the axes is 265 pm from the nucleus.

Surface plots

The final representation we will consider is a three-dimensional version of the contour plot, sometimes called a *surface plot*; examples of these for the 1*s* orbital are shown in Fig. 4.13. Imagine throwing a net over the function and then pulling the net tight so that each point on the net is at a position which has the same value of the wavefunction. The net is a three-dimensional equivalent of a line on a contour plot. A computer program is used to make the net look solid and present us with a three-dimensional view, such as those shown in Fig. 4.13.

As expected, for the 1*s* orbital the surface plot turns out to be a sphere, as the wavefunction depends only on the distance from the nucleus. The size of the sphere depends on the value of the wavefunction at which we choose to draw the net. In the case of the 1*s* wavefunction, the smaller the value we choose, the larger the sphere will be, as is illustrated in Fig. 4.13. Another way of choosing the value at which to draw the net is to set it so that some fraction, say 90%, of the electron density is inside the net; this has been done for the right-hand plot of Fig. 4.13. Such a plot gives us an idea of the size of the region of space in which the electron is likely to be found for the majority of the time.

4.5 Radial distribution functions

It is often useful to know the *total* probability of finding the electron at a set distance from the nucleus, rather than the probability of finding the electron in a small box at a certain position. The way to do this is to imagine a thin spherical shell of radius r and centred on the nucleus (Fig. 4.14). The total probability of finding the electron at distance r is equal to the square of the value of the wavefunction at this distance multiplied by the volume of the shell:

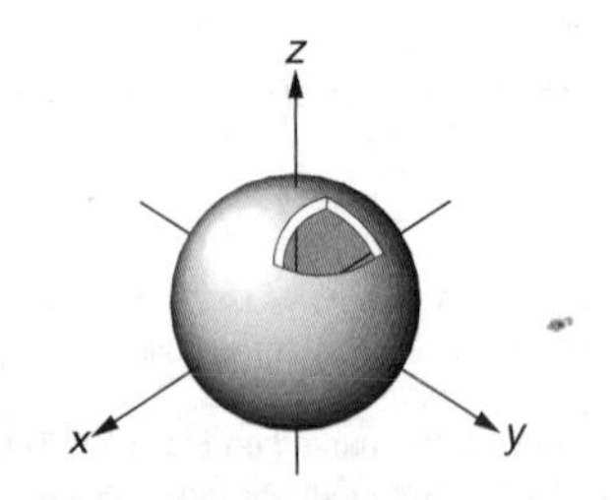

Fig. 4.14 Rather than thinking about the probability of finding the electron in a small box it is sometimes more useful to think about the probability of finding the electron in a thin shell of radius *r*. The diagram shows this shell, with a section cut out so that we can see inside.

$$\text{probability of being in the shell} = (\text{wavefunction})^2 \times \text{volume of the shell.}$$

It is important to understand that the volume we are talking about is *not* the volume inside the shell but of the shell itself.

When we were thinking of the probability of finding the electron in a small box we argued that the probability was proportional to the volume of the box.

For the same reasons, the probability of finding the electron in the shell is proportional to the volume of the shell. This brings us to an important point which is that if we keep the thickness of the shell constant its volume increases as the radius r increases. We need to take this into account when working out the probability of finding the electron in the shell.

As we have assumed that the shell is very thin, its volume is simply given by the surface area of the sphere multiplied by the thickness of the shell. Given that the surface area of a sphere is $4\pi r^2$ we can write

$$\text{probability of being in the shell} = (\text{wavefunction})^2 \times \left[4\pi r^2 \times \text{thickness of the shell}\right]; \quad (4.4)$$

where the quantity in the square brackets is the volume of the shell.

We now define the *radial distribution function* (RDF) as

$$\text{RDF} = (\text{wavefunction})^2 \times 4\pi r^2. \quad (4.5)$$

Comparing Eqs. 4.4 and 4.5 we see that

$$\text{probability of being in the shell} = \text{RDF} \times \text{thickness of the shell}.$$

It follows that the RDF tells us the probability of the electron being found in a thin shell of radius r. It is the most useful function for discussing how the electron density varies with distance from the nucleus.

Figure 4.15 compares the $1s$ wavefunction with the corresponding RDF; the two plots are quite different. The first thing we note is that the RDF goes to zero at the nucleus ($r = 0$) even though the wavefunction is not zero at this point. It is the r^2 factor in the definition of the RDF (Eq. 4.5) which forces it to zero at $r = 0$. Put another way, the r^2 factor is there to account for the volume of the shell; as r decreases the volume of the shell decreases, falling to zero when $r = 0$, making the RDF zero at this point.

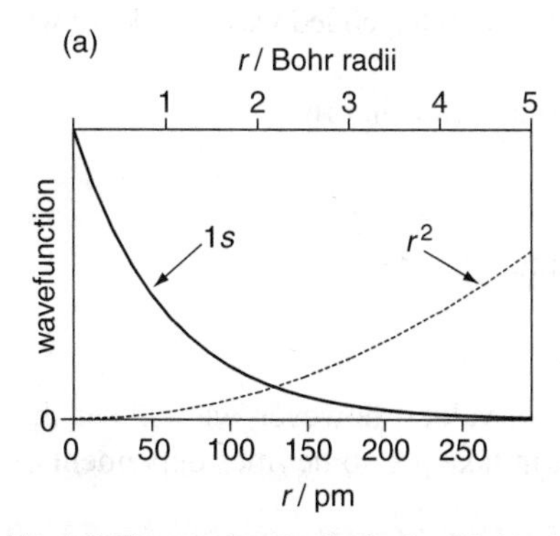

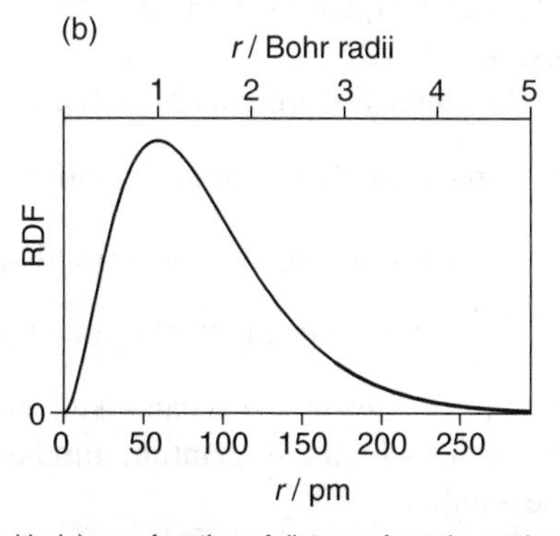

Fig. 4.15 The wavefunction for the $1s$ orbital is plotted in (a) as a function of distance from the nucleus, r, in units of pm (1 pm = 10^{-12} m); the dashed line is a plot (on a different vertical scale) of the function r^2. Plot (b) shows the radial distribution function (RDF) for the same orbital; as described in the text the RDF gives the probability of finding the electron in a thin shell at distance r from the nucleus. The most striking feature of the RDF for $1s$ is the presence of the maximum at r = 53 pm; this distance is called the *Bohr radius*. At the top of both plots the scale for r is shown in units of the Bohr radius.

The second thing to note is that whereas the wavefunction drops off from a maximum at $r = 0$, the RDF has a maximum at 53 pm from the nucleus. The maximum comes about because although the square of the wavefunction falls off as r increases, the r^2 factor (from the volume of the shell) increases (this is shown as the dashed line in Fig. 4.15 (a)). At small values of r the r^2 term is dominant and so the RDF increases with increasing r; at large values of r the exponential drop-off of the wavefunction dominates, forcing the RDF to zero. The overall result is the formation of a maximum.

The maximum occurs at $r = 53$ pm, a radius which is the same as the lowest energy orbit in the Bohr model of the atom; for this reason this distance is often called the *Bohr radius*. From now on we will plot the wavefunctions and RDFs against the distance expressed in units of the Bohr radius.

It is interesting to contrast this picture which the wavefunction gives us with the Bohr model. The latter says that the $1s$ electron is in an orbit of radius 53 pm, whereas quantum mechanics simply says that the electron is most likely to be found at a radius of 53 pm.

4.6 Hydrogen orbitals

In this section we will describe the atomic orbitals which are going to be important in bonding. To start with we will just consider the orbitals of hydrogen, and then in subsequent sections we will see how these can be adapted to describe the orbitals of other atoms.

Using quantum mechanics we can calculate the mathematical form of the wavefunctions (the orbitals) of hydrogen along with the associated energies. It turns out that these mathematical forms are specified by a set of three *quantum numbers*. These numbers are just integers (1, 2, 3, ...) and – perhaps surprisingly – it turns out that we can say a lot about the shape and energy of an orbital *just* from a knowledge of these numbers.

You have already been using quantum numbers, but without perhaps realizing it, as the familiar names of the orbitals, such as '$1s$' and '$2p$', give the values of the quantum numbers, albeit in a somewhat coded way which we will describe shortly.

The three quantum numbers needed to describe an orbital are:

the principal quantum number, n

the orbital angular momentum quantum number, l

the magnetic quantum number, m_l

Each is an integer which can take a range of values; however, we will see that the values which the quantum numbers can take are sometimes dependent on one another.

We can draw an analogy between these quantum numbers and the coefficients a, b and c in the quadratic function of Eq. 4.1 on p. 39. The values of these coefficients determine the exact form of the graph of $f(x)$ against x, i.e. where the curve crosses zero and the position of the minimum. In the same way, knowledge of the values of the quantum numbers enables us to say a lot about the form of the corresponding wavefunction. We will discuss the meaning of each of these quantum numbers in turn.

The principal quantum number

The principal quantum number takes integer values starting from one: $n = 1, 2, 3, \ldots$ and its value determines the energy of the orbital, E_n:

$$E_n = \frac{-R_H}{n^2}.$$

R_H is the *Rydberg constant* whose value is predicted by the theory and has been confirmed by experiment.

There are two things to note about this expression for the energy. Firstly, the energies are negative; this is because they are being measured *downwards* from the energy of the electron when it is infinitely far away from the nucleus, i.e. when it has been ionized.

The second point is that as n increases the energies increase (they become less negative). So the lowest energy orbital has $n = 1$, the next has $n = 2$ and so on. Figure 4.16 illustrates how the energies of these orbitals vary with n. In describing an orbital as being 1*s*, 2*p* or 3*d* the number is the value of n, the principal quantum number.

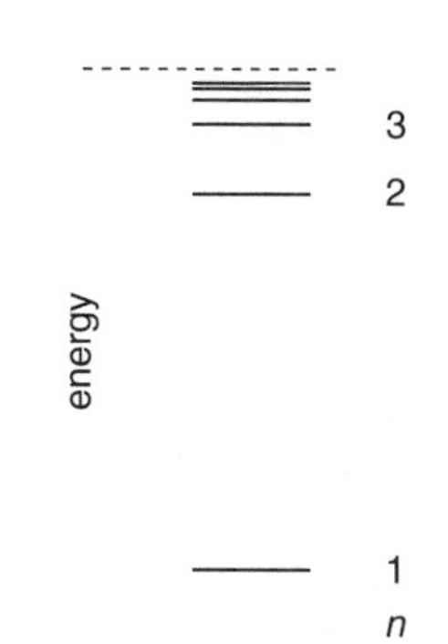

Fig. 4.16 The energies of the hydrogen orbitals depend only on the value of the principal quantum number, n, and are measured downwards from the energy of the ionized electron, represented by the dashed line. The diagram shows the energies of the orbitals with $n = 1$ to 6, and is drawn to scale.

The orbital angular momentum quantum number

The orbital angular momentum quantum number, l, takes values from $(n - 1)$ down to zero in integer steps. The table summarizes the possible values of l for some commonly encountered values of the principal quantum number, n.

principal q. n. (n)	orbital angular momentum q. n. (l)
1	0
2	0, 1
3	0, 1, 2

The value of l is usually represented by a letter: $l = 0$ is denoted by '*s*', $l = 1$ by '*p*' and $l = 2$ by '*d*'. So, when we talk of a 3*d* orbital we are referring to an orbital with $n = 3$ and $l = 2$; likewise a 2*s* orbital has $n = 2$ and $l = 0$. Taken together, the values of n and l tell us about the three-dimensional shape of the orbital, which we will discuss later in this section.

l	0	1	2	3
letter	*s*	*p*	*d*	*f*

The table shows the letters used to denote different values of the orbital angular momentum quantum number, l.

The magnetic quantum number

Finally, there is the magnetic quantum number, m_l, which takes values from l to $-l$ in integer steps. This means that there are $(2l + 1)$ separate values of m_l; of these, there are l positive and l negative values plus $l = 0$.

For example, if $l = 1$ the values of m_l are 1, 0 and -1; if $l = 0$ the only value of m_l is zero. The value of m_l tells us about the orientation of the orbital in space.

As the value of the principal quantum number, n, determines the energy, it is usual to think of orbitals with the same value of n as forming a *shell*. The shells with $n = 1, 2, 3$ are sometimes called the *K*, *L* and *M* shells, respectively. We will now discuss the orbitals in each of these shells in turn.

The K shell ($n = 1$)

The K shell has $n = 1$; for this value of the principal quantum number the only value of l is zero and hence the only value of m_l is zero. So, we have just one orbital with $n = 1$, $l = 0$ and $m_l = 0$ which is the $1s$ orbital. This orbital has been pictured in detail in Section 4.4 (Figs. 4.11 and 4.13 on p. 41 and p. 42, respectively).

The $1s$ wavefunction is spherical (it only depends on r), and we have already seen in Fig. 4.15 on p. 43 that the RDF has a single maximum at the Bohr radius.

The L shell ($n = 2$)

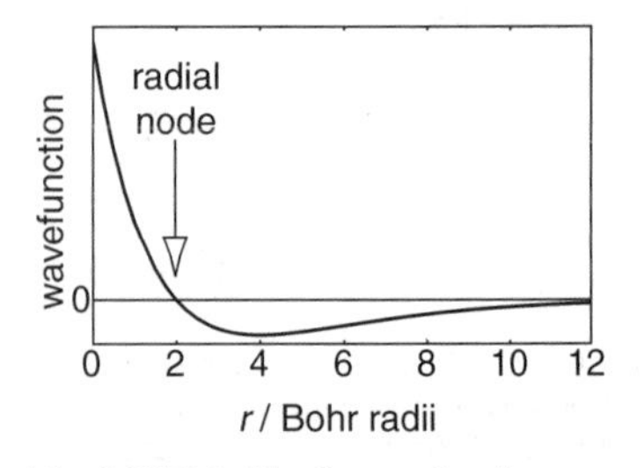

Fig. 4.17 Plot of the $2s$ wavefunction against the distance from the nucleus, r. Note the radial node, which is the point at which the wavefunction goes to zero, at $r = 2$ Bohr radii.

The L shell has $n = 2$; this means that l can be 0 or 1.

The orbital with $l = 0$ is the $2s$ orbital, and there is just one of these corresponding to $m_l = 0$. Like the $1s$ orbital, the $2s$ orbital does not depend on the angles θ and ϕ; it is therefore possible to plot the wavefunction against the distance r as is shown in Fig. 4.17.

The new feature we see here is that the wavefunction has a *radial node*; this is a value of r (in this case 2 Bohr radii) at which the wavefunction is zero. This radial node can also be seen in the contour and shaded plots shown in Fig. 4.18.

For $n = 2$ we can also have $l = 1$, and for this value of l there are three possible values of m_l: 1, 0 and -1. These are the three $2p$ orbitals, depicted as surface plots in Fig. 4.19. When compared to $1s$ and $2s$, the new feature of the $2p$ orbitals is that the wavefunctions are no longer spherical but depend on the angles θ and ϕ.

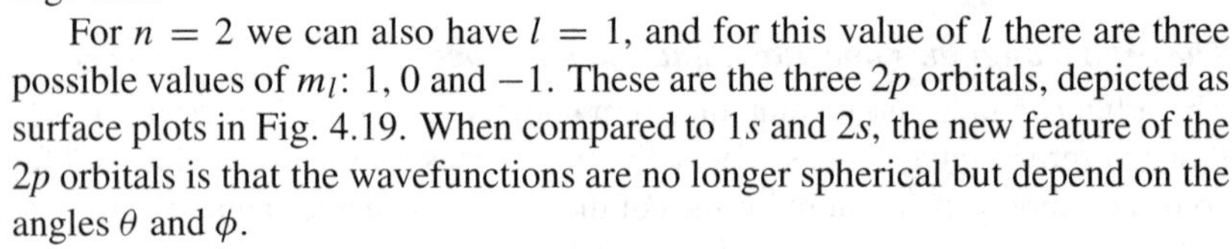

The $2p_z$ orbital is positive for positive values of the coordinate z, and negative for negative values of z. When z is zero, the wavefunction is zero for all values of x and y; the xy-plane is therefore called a *nodal plane*. Another way of describing this nodal plane is to look at Fig. 4.9 on p. 40 and note that

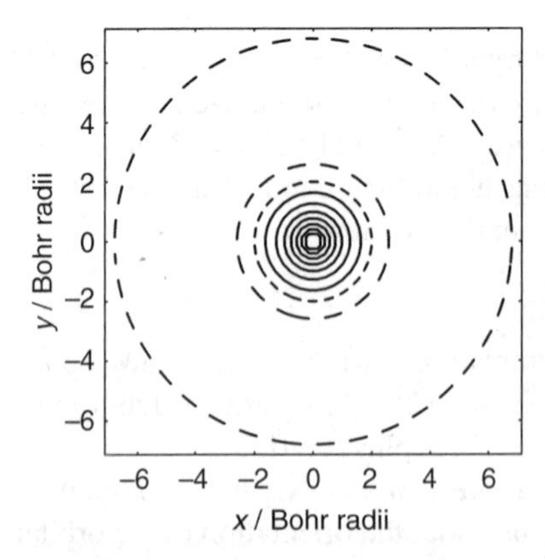

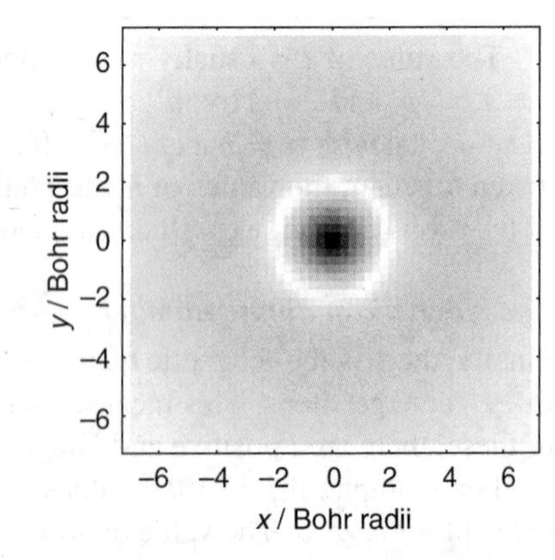

Fig. 4.18 Contour and shaded plots of the $2s$ wavefunction. In the contour plot positive values of the wavefunction are indicated by solid lines and negative values by lines with long dashes; the contour corresponding to the wavefunction being zero (the radial node) is indicated by a line with short dashes. In the shaded plot no distinction is made between positive and negative values of the wavefunction; the radial node is clearly visible as the white ring around the intense central core.

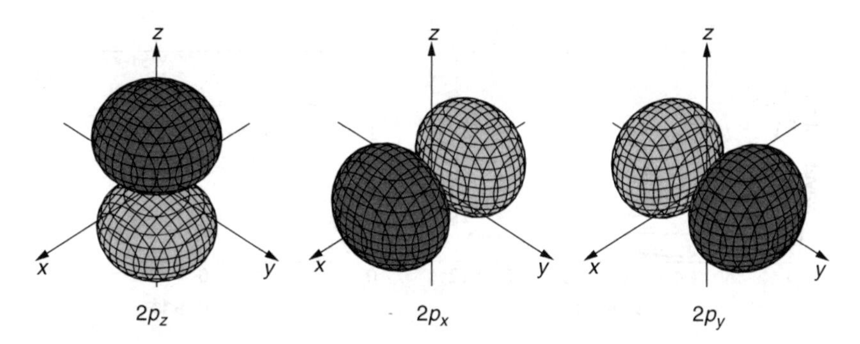

Fig. 4.19 Surface plots of the three $2p$ orbitals. The dark grey shading indicates a positive value for the wavefunction and light grey indicates a negative value. The three orbitals point along the z-, x-, and y-axes, and so are called $2p_z$, $2p_x$ and $2p_y$, respectively. Each has a positive lobe and a negative lobe. The $2p_z$ orbital is zero everywhere in the xy-plane; this is called a *nodal plane* or an *angular node*. Similarly, the other two orbitals each have a nodal plane.

the plane in which the $2p_z$ wavefunction is zero is the one which has θ equal to 90° and ϕ takes any value. This plane can be called an *angular node* as it corresponds to an angle (here θ) at which the wavefunction is zero (compare this to a radial node which is a radial distance, r, at which the wavefunction is zero).

As the electron density depends on the *square* of the wavefunction (the orbital), the sign of the wavefunction is not of any particular significance. However, as we will see in the next chapter, the sign is very important when we start to allow orbitals to interact with one another, which is what happens when bonds are formed.

The mathematical form of the $2p_z$ wavefunction is

$$\Psi_{2p_z} = C\,[\cos\theta]\left[r\,\mathrm{e}^{-Dr}\right] \tag{4.6}$$

where C and D are constants given by quantum mechanics. We can see from this that if $\theta = 90°$ then $\cos\theta = 0$ and so Ψ_{2p_z} is zero; this is the origin of the angular node.

When θ is greater than 0° but less than 90°, $\cos\theta$ is positive; this corresponds to the positive lobe of the $2p_z$ orbital which points along the $+z$-axis. When θ is greater than 90° but less than 180°, $\cos\theta$ is negative; this corresponds to the negative lobe of the $2p_z$ orbital which points along the $-z$-axis.

The wavefunction Ψ_{2p_z} is a product of two parts: the *angular part* (given in the first square bracket in Eq. 4.6) which only depends on the angles θ and ϕ, and the *radial part* (given in the second square bracket) which only depends on the distance r. It turns out that *all* of the orbitals are separable in this way into a radial and an angular part.

The three $2p$ orbitals only differ in their angular parts – they all have the same radial part, which is plotted in Fig. 4.20 along with the radial part of the $2s$ orbital. Whereas the $2s$ shows a radial node at $r = 2$, the $2p$ has no radial nodes. Furthermore, the $2s$ is non-zero at the nucleus ($r = 0$) whereas the $2p$ is zero at this point; note that having the wavefunction equal to zero at the nucleus is not counted as a radial node.

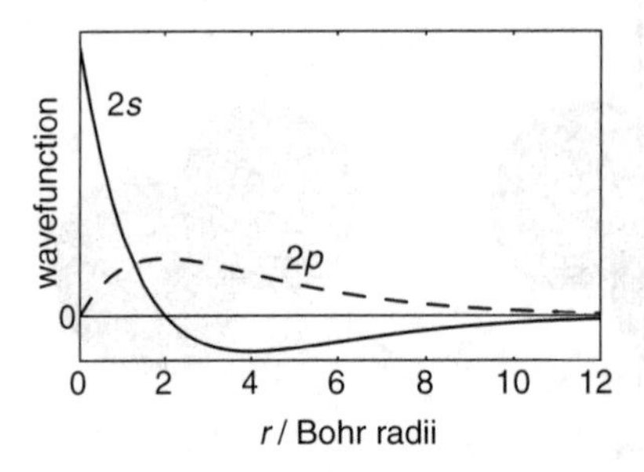

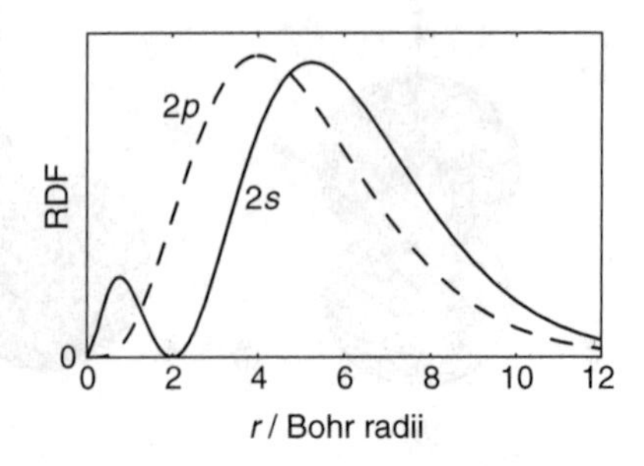

Fig. 4.20 On the left are plotted the radial parts of the $2s$ and $2p$ orbitals; note that whereas the $2s$ orbital shows a radial node at $r = 2$, the $2p$ orbital has no radial node. At $r = 0$ the $2s$ orbital is non-zero but the $2p$ is zero (this is not, however, counted as a radial node). The RDFs are shown on the right; note the presence of two maxima for $2s$ but only one for $2p$. The minimum in the RDF of $2s$ at 2 Bohr radii corresponds to the radial node in the wavefunction.

Figure 4.20 also shows the RDFs for $2s$ and $2p$. Again, there are significant differences. For $2s$, the RDF has its principal maximum at 5.2 Bohr radii and there is a subsidiary maximum at 0.76 Bohr radii; there is also a minimum at 2.0 Bohr radii which corresponds to the radial node in the wavefunction. In contrast, the RDF for $2p$ shows just one maximum at 4.0 Bohr radii.

Recall from Section 4.5 on p. 42 that the RDF gives the probability of finding the electron *anywhere* in a thin shell. This definition means that the probability has to be summed over all possible values for the angles θ and ϕ. We can therefore talk about the RDF of a $2p$ orbital even though such orbitals depend on the angles θ and ϕ, as the way an RDF is calculated involves summation over all angles.

Comparing Fig. 4.20 with the corresponding plot for $1s$ (Fig. 4.15 on p. 43) shows clearly how much larger the $2s$ and $2p$ orbitals are when compared to $1s$. It is also interesting to note that the principal maximum of the $2s$ is at a larger distance from the nucleus than is the maximum for $2p$, which might lead us to conclude (falsely) that if the electron is in the $2s$ it will be less tightly bound than if it is in the $2p$ on the grounds that a $2s$ electron is 'further away' from the nucleus. In fact, *in hydrogen* (and only in hydrogen) the $2s$ and all three $2p$ orbitals have exactly the same energy; as was described on p. 45 the energy depends *only* on the value of the principal quantum number, n.

Distinct orbitals which have the same energy are said to be *degenerate*; so in the L shell there are four degenerate orbitals: $2s$, $2p_x$, $2p_y$ and $2p_z$. This degeneracy of the $2s$ and $2p$ is a result of there only being one electron present in hydrogen; we will see later on that in atoms with more than one electron the $2s$ and $2p$ are not degenerate. However, even in such an atom the three $2p$ orbitals are degenerate.

If we wish to represent the $2p$ orbitals using contour or shaded plots then we have to be careful about which cross-section we choose. For example, if we plotted the xy-plane for the $2p_z$ orbital we would see nothing! Figure 4.21 shows the cross-section taken in the xz-plane (at $y = 0$) through the $2p_z$ orbital. The two lobes are clearly visible, as is the nodal plane.

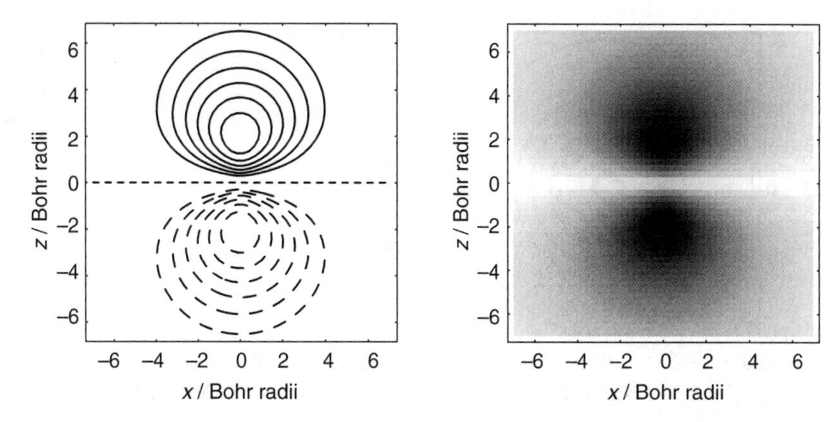

Fig. 4.21 Contour and shaded plots of the $2p_z$ wavefunction taken through the xz-plane (at $y = 0$); the coding of the contours is as in Fig. 4.18. Note the positive and negative lobes, and the nodal plane (indicated by the zero contour with the short dashes). In the shaded plot, no distinction is made between positive and negative parts of the wavefunction.

The $2s$ orbital has one radial node and no angular nodes, whereas the $2p$ orbitals have one angular and no radial nodes. So, all orbitals with $n = 2$ have *one* node of some kind; this is an example of the general rule that the total number of nodes (angular plus radial) is $(n - 1)$.

The M shell ($n = 3$)

For $n = 3$ there are three values of l: 0, 1 and 2 which correspond to the $3s$, $3p$ and $3d$ orbitals respectively. As before, there is one $3s$ orbital and three $3p$ orbitals. For the $3d$ the possible values of m_l are 2, 1, 0, -1 and -2 giving a total of five orbitals. Since the $3s$, three $3p$ and five $3d$ orbitals all have the same principal quantum number they all have the same energy – that is they are degenerate (in the case of the hydrogen atom).

Figure 4.22 compares the radial parts of the $3s$, $3p$ and $3d$ orbitals; comparing these to the plots shown in Fig. 4.20 on p. 48 shows that the orbitals with $n = 3$ are considerably larger than those with $n = 2$. Looking at both these plots,

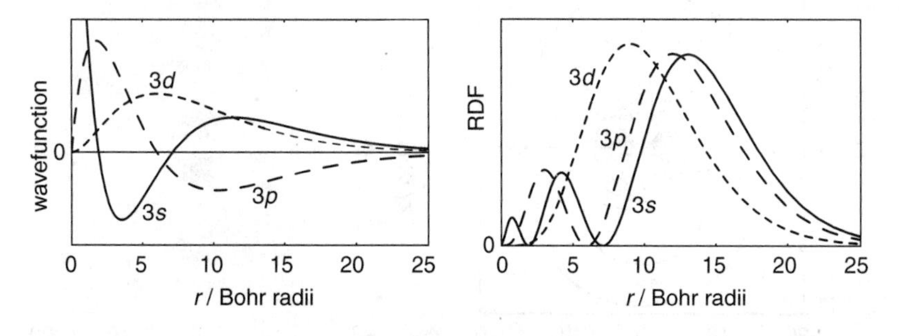

Fig. 4.22 On the left are shown plots of the radial parts of the orbitals with $n = 3$ and on the right the corresponding RDFs; the wavefunction for $3s$ goes to a finite value at $r = 0$, but this is not shown on the plot. Note that the number of radial nodes for $3s$, $3p$ and $3d$ are 2, 1 and 0 respectively. The principal maximum in the RDFs moves in closer to the nucleus as l increases; despite this, in hydrogen all the orbitals have the same energy.

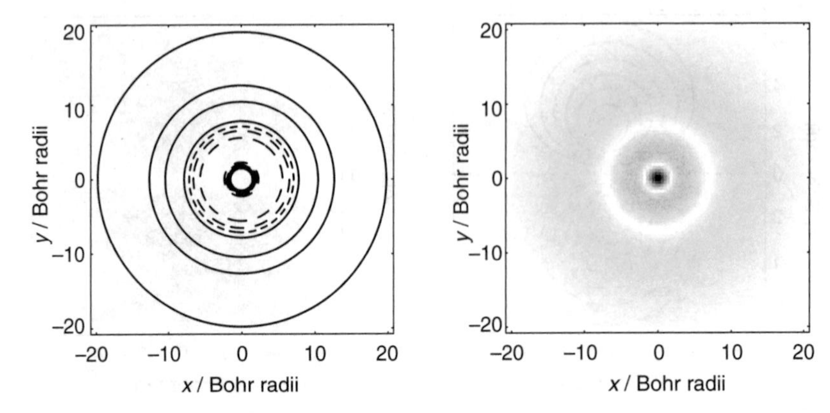

Fig. 4.23 Contour and shaded plots for the 3*s* orbital; the orbital is spherically symmetric and the two radial nodes are clearly visible in the right-hand plot.

a pattern begins to emerge: the 3*s* orbital has two radial nodes, the 3*p* has one and the 3*d* has none. If we compare this to the orbitals in the *L* shell (Fig. 4.20 on p. 48) we see that the number of nodes has increased by one for the *s* and *p*.

For the RDFs a pattern is also discernible: the 3*s* shows three maxima, the 3*p* shows two and there is just one for the 3*d*. Note that the position of the principal maximum moves in towards the nucleus as *l* increases, just as was the case for 2*s* and 2*p*.

As before the minima in the RDF correspond to the values of *r* at which the radial part of the wavefunction goes to zero. In between two such minima there must of course be a maximum in the RDF, a feature which we can clearly see from Fig. 4.22.

Figure 4.23 illustrates clearly the spherical symmetry of the 3*s* orbital; the two radial nodes are clearly visible in the shaded plot. Figure 4.24 shows the form of the $3p_z$ orbital; like $2p_z$ this orbital has an angular node (the *xy*-plane). However, in contrast to 2*p*, 3*p* also has a radial node, clearly visible as the white circle on the shaded plot. Inside this radial node the two lobes appear

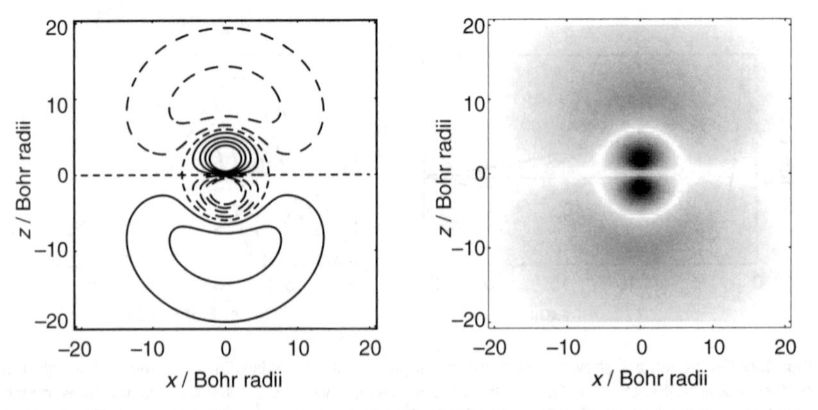

Fig. 4.24 Contour and shaded plots for the $3p_z$ orbital. Like a 2*p* orbital, the 3*p* basically has a positive and a negative lobe, separated by a nodal plane. In addition for a 3*p* orbital there is a radial node closer in to the nucleus; this node is clearly visible in the shaded plot as a white ring.

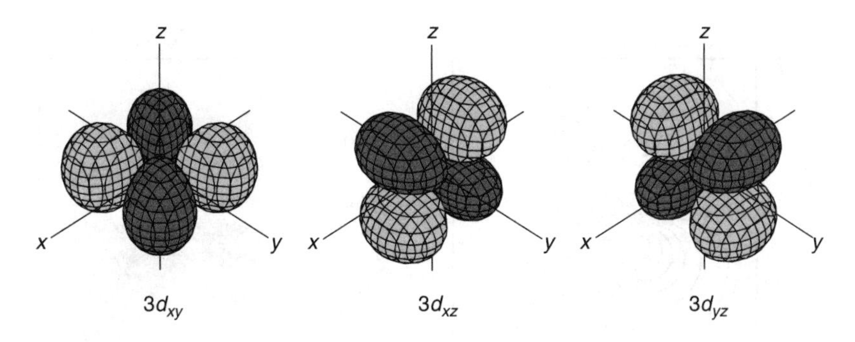

Fig. 4.25 Surface plots of the three 3*d* orbitals whose lobes point *between* the axes. Each orbital has two positive and two negative lobes, and has two angular nodes.

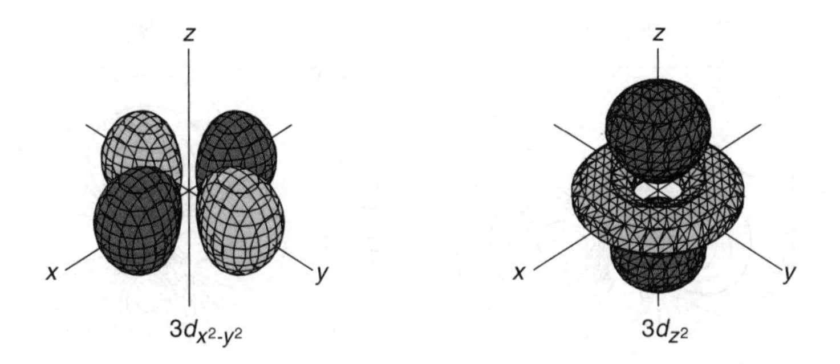

Fig. 4.26 Surface plots of the two 3*d* orbitals whose lobes point *along* the axes.

again. The other two 3*p* orbitals are similar to $3p_z$ except that they point along the *x*- and *y*-axes.

Finally we have the five 3*d* orbitals, which are shown as surface plots in Figs. 4.25 and 4.26. These five orbitals come in two groups. The first group, shown in Fig. 4.25, consists of three orbitals each of which lies in a particular plane and whose lobes point *between* the axes. For example, take the $3d_{xy}$ orbital: this lies in the *xy*-plane and has two nodal planes (or angular nodes): the *xz*-plane and the *yz*-plane. The other two orbitals in this first group are similar to $3d_{xy}$ except that they lie in the *xz*- and *yz*-planes.

The second group, shown in Fig. 4.26, consists of two orbitals whose lobes point *along* the axes. The orbital $3d_{x^2-y^2}$ lies in the *xy*-plane and has its positive lobes pointing along the *x*-axis and its negative lobes along the *y*-axis. Like $3d_{xy}$ this orbital has two nodal planes, but in this case they bisect the *xz*- and *yz*-planes.

The orbital denoted $3d_{z^2}$ points along the *z*-axis. For this orbital the two angular nodes are not planes but take the form of two cones about the *z*-axis defined by the angles $\theta = 54.7°$ and $125.3°$ with ϕ taking any value, as is shown in Fig. 4.27.

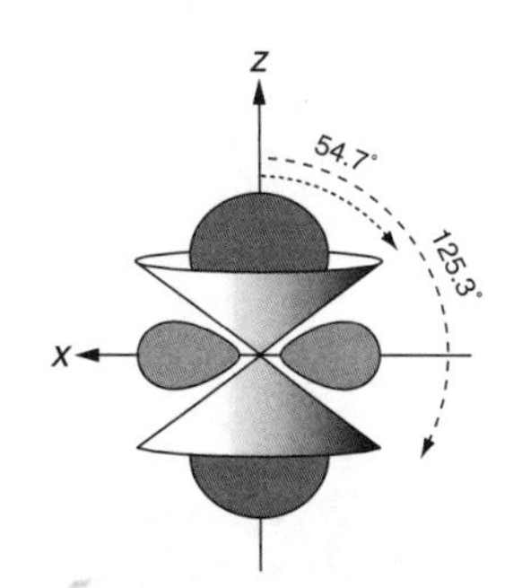

Fig. 4.27 Illustration of the two angular nodes present in the $3d_{z^2}$ orbital. These nodes take the form of two cones with angles of 54.7° and 125.3°; the two lobes and the ring of the orbital are shown in cross section.

We see that, in accordance with the rule that the total number of nodes is $(n - 1)$, all of the orbitals with $n = 3$ each have two nodes of some kind. You should not worry about the names of these *d* orbitals – they are in fact derived from the mathematical forms of these functions when they are expressed in

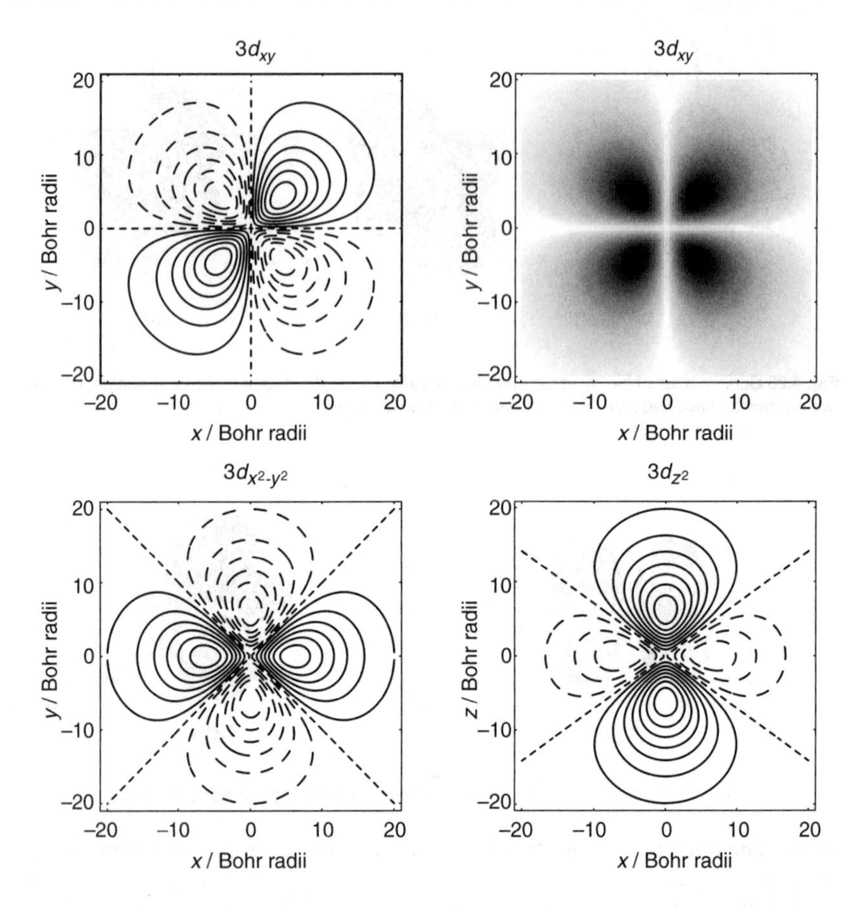

Fig. 4.28 At the top are shown contour and shaded plots of the $3d_{xy}$ orbital. The 'clover leaf' pattern of the four lobes is clearly visible, as are the two nodal planes (here the *xz*- and *yz*-planes). At the bottom are shown contour plots of the $3d_{x^2-y^2}$ and $3d_{z^2}$ orbitals. The former also shows the clover leaf pattern; for the latter the two angular nodes making angles of 54.7° and 125.3° to the *z*-axis are clearly shown by the zero-value contour (short dashed lines).

cartesian coordinates (that is, in x, y and z coordinates).

Finally, Fig. 4.28 shows contour and shaded plots of a selection of the $3d$ orbitals. The 'clover leaf' patterns which the lobes of the $3d_{xy}$ and $3d_{x^2-y^2}$ orbitals form are clearly visible, as are the two nodal planes. As we have already noted, for the $3d_{z^2}$ orbital the two angular nodes take the form of cones rather than planes, and this can clearly be seen.

The general pattern

If we look back over these orbitals we see that there is a pattern to the way in which the radial and angular nodes occur. The number of angular nodes (nodal planes) is l and the number of radial nodes is $(n - l - 1)$; the total number of radial and angular nodes is therefore $(n - 1)$.

Using these general ideas we can have a good guess at what other orbitals might look like. For example, the $4s$ orbital will have three radial and no angular nodes: it is therefore spherically symmetric like the other s orbitals.

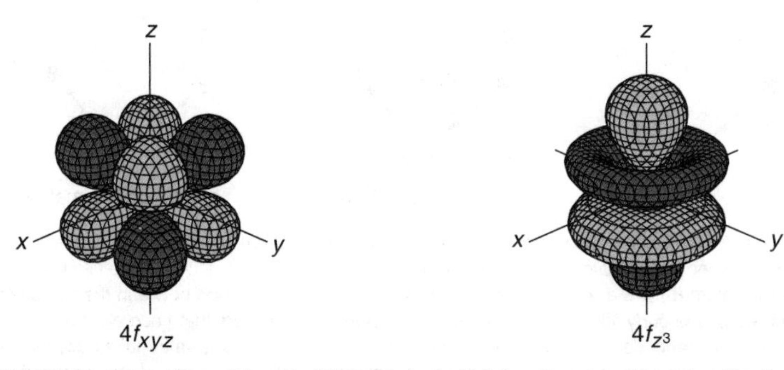

Fig. 4.29 Surface plots of two of the 4*f* orbitals, each of which has three angular nodes. For the orbital shown on the left these angular nodes are the *xy*-, *xz*- and *yz*-planes; the orbital is thus divided into eight lobes. For the orbital on the right, cones at angles of 41° and 139° to the *z*-axis form two of the angular nodes and the *xy*-plane forms the third.

A 4*f* orbital, which has $l = 3$, will have no radial nodes and three angular nodes; these can clearly be seen in the surface plots of two of the 4*f* orbitals shown in Fig. 4.29.

For the $4f_{xyz}$ orbital these three angular nodes are the *xy*-, *xz*- and *yz*-planes; the orbital is thus divided into eight lobes. It is interesting to compare this orbital with the $3d_{xy}$ (shown in Fig. 4.25) in which the angular nodes are the *xz*- and *yz*-planes. These planes divide the $3d_{xy}$ into four lobes; for the $4f_{xyz}$ orbital the *xy*-plane is an additional angular node which further divides these four lobes into eight.

The $4f_{z^3}$ orbital, also shown in Fig. 4.29, is reminiscent of the $3d_{z^2}$ orbital shown in Fig. 4.26. However, for this 4*f* orbital the angles of the two cones are different to those for the $3d_{z^2}$. The additional angular node, the *xy*-plane, effectively 'cuts' the $3d_{z^2}$ orbital in half.

4.7 Atoms with more than one electron

All we have said so far applies to an atom with only one electron, such as hydrogen or ions such as He^+ and Li^{2+}. In a one-electron atom the only interaction we need to consider in the quantum mechanical treatment of the problem is the electrostatic attraction between the nucleus and the electron (Fig. 4.30 (a)); however, if there are two electrons present (as in He) we also need to consider the electrostatic repulsion between the like charges of the two electrons (Fig. 4.30 (b)).

Unfortunately, simply going from one electron to two changes the problem from one which quantum mechanics can deal with exactly to one which cannot be solved 'on paper' (although computer calculations can still solve the problem to very high accuracy). We want to be able to retain the simple picture which the hydrogen orbitals provide rather than having to resort to computer calculations, and this is where the *orbital approximation* comes in.

The orbital approximation

The idea behind the orbital approximation is to make a multi-electron atom 'look like' a one-electron atom. If we achieve this it is then easy to find the

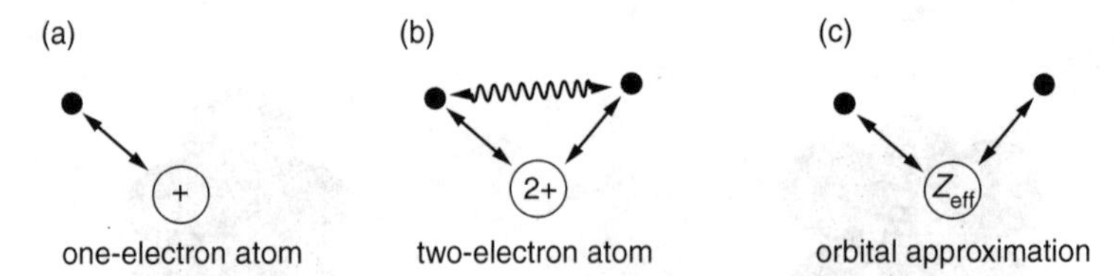

Fig. 4.30 In a one-electron atom (a), the only interaction we need to consider is the electrostatic attraction between the positive nucleus and the negative electron (represented by the filled circle). In a two-electron atom (b), we also need to consider the electrostatic repulsion between the two electrons (represented by the curly line). The orbital approximation (c), considers that each electron in a multi-electron atom experiences only an effective nuclear charge, Z_{eff}, whose value takes account of the effect of electron–electron repulsion. Each electron then effectively forms a one-electron atom as there is no electron–electron repulsion to consider explicitly.

wavefunctions as the problem is then essentially the same as the hydrogen atom. Remember that finding the wavefunctions is crucial as once we know these we can find the probability distribution of the electrons, their energies and so on.

In the orbital approximation we concentrate on just one of the electrons. Then, rather than considering the details of its repulsive interactions with all the other electrons we imagine that the effect of these is simply to alter the nuclear charge seen by the electron we are interested in from its actual value to an 'effective' value.

For example, in the case of helium the actual nuclear charge is two. However, as the electrons repel one another they are not as tightly held as would be expected solely on the basis of a nuclear charge of +2. To account for this reduced binding of the electrons, we imagine that the electrons are held by an *effective nuclear charge* of somewhat less than 2 (Fig. 4.30 (c)).

The orbital approximation allows us to consider each electron as if it were on its own (a one-electron system) experiencing an effective nuclear charge. It turns out that in such a situation the orbitals (wavefunctions) for the electron are of the same form as those in hydrogen, but with their energies and sizes modified in rather a simple way which just depends on Z_{eff}.

We saw on p. 45 that the energy of the orbitals in hydrogen is a function of the principal quantum number, n. If the effective charge is Z_{eff}, this energy becomes

$$E_n = -\frac{Z_{eff}^2 R_H}{n^2}. \tag{4.7}$$

The Rydberg constant, R_H, is conveniently expressed in electron-volts (eV) and takes the value 13.6 eV. We see from Eq. 4.7 that increasing the effective nuclear charge makes the energy more negative, which means that the electron is more tightly held – this is exactly what we would expect.

We can use computer-calculated values of the orbital energies in multi-electron atoms to get a feel for the kinds of values that the effective nuclear charge takes. For example, the energy of the $1s$ orbital in hydrogen is -13.6 eV and so, just as we would expect, the value of Z_{eff} calculated using Eq. 4.7 is 1. In helium, the orbital energy of a $1s$ electron is -25.0 eV, and using Eq. 4.7 we can work out that this corresponds to $Z_{eff} = 1.4$; the value is considerably less than the real nuclear charge of 2 on account of the effect of the electron–electron repulsion.

Changing Z_{eff} also affects the size of the orbitals as well as their energies. Increasing the effective nuclear charge causes the orbitals to contract, which is hardly surprising as we expect that the increased attraction caused by a higher nuclear charge will pull the electrons in. It turns out that the effect is inversely proportional to Z_{eff}. So, for example, as is shown in Fig. 4.31 the maximum in the RDF for a $1s$ orbital occurs at (a_0/Z_{eff}), where a_0 is the Bohr radius. All of the orbitals scale in this simple way.

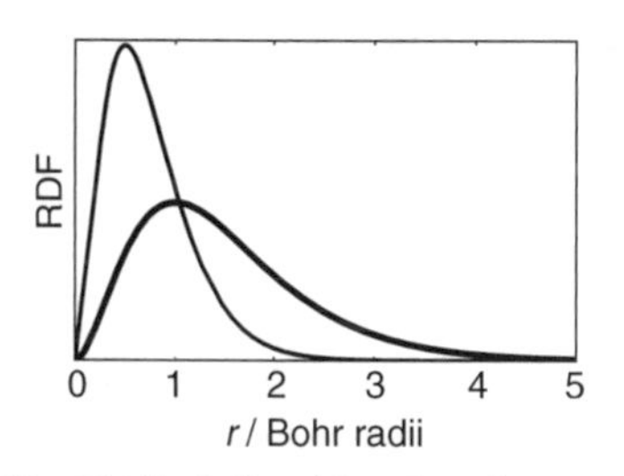

Fig. 4.31 Illustration of the effect of increasing the nuclear charge on the form of the $1s$ orbital. The thick line shows the RDF for a $1s$ orbital with a nuclear charge of 1; the thinner line shows the RDF for the same orbital but with a nuclear charge of 2. Note how the maximum shifts from 1 Bohr radius when the nuclear charge is 1 to 0.5 Bohr radii when the charge is 2.

Electronic configurations

The orbital approximation allows us to imagine that each electron is behaving as if it is in a one-electron atom, and so the wavefunctions of each of the electrons (the orbitals) are the same as those for hydrogen albeit modified to take account of the effective nuclear charge. Each electron can therefore be assigned to a hydrogen-like orbital. This is the basis for something which you have already been doing, which is writing *configurations* for multi-electron atoms.

When we write the electronic configuration of helium as $1s^2$ what we imply is that both electrons are in a hydrogen-like $1s$ orbital. As you know, we can have up to two electrons in one orbital, provided we oppose the spins, i.e. one spin is 'up' and the other is 'down'.

For lithium ($Z = 3$) the configuration is $1s^22s^1$; we cannot put the third electron into the $1s$ orbital as this is already full with two spin-opposed electrons, so it has to go into the next lowest energy orbital, which is the $2s$. We can carry on in this way building up the electronic configuration of any element.

We know that the ground state electronic configuration of lithium is $1s^22s^1$ and not $1s^22p^1$, because we have learnt that the $2s$ orbital is lower in energy than is the $2p$. However, for hydrogen we have already commented (on p. 48) that $2s$ and $2p$ have the *same* energy – what, then, has changed? The crucial point here is that, in contrast to hydrogen, in a multi-electron atom the $2s$ and $2p$ are no longer degenerate. In the next section we will explore why this is so.

Orbital energies in multi-electron systems

From Eq. 4.7 on p. 54 we see that the energy of an orbital depends on the principal quantum number and the effective nuclear charge; we should also recall that this effective nuclear charge is a result of the balance between the actual nuclear charge and the electron–electron repulsion.

We have already seen that the effective nuclear charge in helium is 1.4, a value significantly less than the real nuclear charge. For lithium the $2s$ orbital energy is -5.3 eV; if we assume that this electron has $n = 2$ we can use Eq. 4.7 to compute the effective nuclear charge as 1.3. The first thing to notice is that this is much less than the actual nuclear charge of 3. What this is telling us is that the two $1s$ electrons are repelling the outer electron to a very significant extent.

A convenient way to describe the effect of this repulsion is to use the concept of *screening*. Let us think about the outer electron in lithium – what does it see? There is the nucleus with a charge of 3 and this is surrounded by the two $1s$ electrons, each of which has a single negative charge. If the outer electron is far enough away, then we might reasonably assume that it would 'see' an effective nuclear charge of +1, i.e. +3 from the nucleus and −1 from each 1*s* electron giving +1 overall. The idea is illustrated in Fig. 4.32.

We say that the two 1*s* electrons *screen* the outer electron from the nuclear charge. If they form a perfect screen, the negative charges of the two 1*s* electrons will cancel out two positive charges from the nucleus leaving an effective nuclear charge of +1.

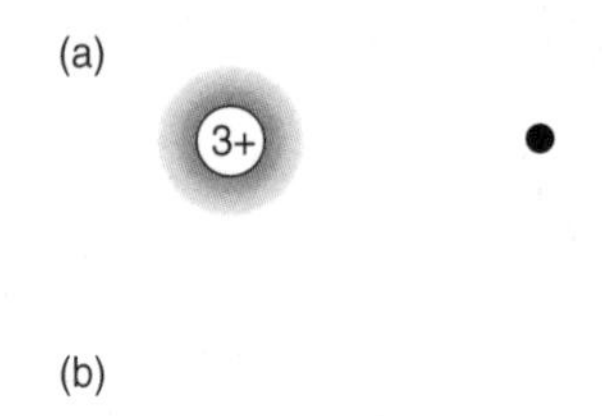

Fig. 4.32 Illustration of the concept of screening. In (a) the lithium atom is represented as an outer electron (the filled circle), a nucleus of charge 3 and the two 1*s* electrons (the shaded area close to the nucleus). If the outer electron is far enough away, it experiences an effective nuclear charge of 1 as shown in (b). We say that the two inner electrons *screen* the outer electron from the nuclear charge. In practice the screening is not perfect and so the effective nuclear charge will be greater than 1.

In fact we know that for lithium Z_{eff} is 1.3, and this tells us that the screening by the two inner 1*s* electrons is not perfect. This should come as no surprise as the representation given in Fig. 4.32 does not fit in with the quantum mechanical view that the electron is not localized to an orbit, but is smeared out over space. We can investigate how this might affect the shielding by comparing the RDFs of the 1*s*, 2*s* and 2*p* orbitals; these are shown in Fig. 4.33.

To make these plots we have assumed that the effective nuclear charge experienced by the 1*s* electrons is 2. What this value of Z_{eff} does is to contract the 1*s* orbitals, and this is clearly visible in Fig. 4.33 where the RDF for the 1*s* forms a compact peak close to the nucleus.

Figure 4.33 shows us clearly that much of the probability density of the 2*s* and 2*p* electrons falls well outside the area occupied by the 1*s* electrons. It is not surprising, therefore, that the 1*s* electrons do form such a good screen of the nuclear charge, pretty close to the rather crude picture of Fig. 4.32.

However, there is some probability of the 2*s* and 2*p* electrons being *inside* the area occupied by 1*s*. When the electrons are in this region they will experience a larger nuclear charge than when they are further out. We describe this effect by saying that the outer electron *penetrates* the screen formed by the 1*s* electrons and so experiences a greater nuclear charge. This is the explanation for why Z_{eff} for the outer electron in lithium is greater than 1.

Figure 4.33 also provides an explanation of why the 2*s* orbital is lower in energy than the 2*p*. Close inspection of the RDFs shows that the 2*s* has greater amplitude than the 2*p* inside the region occupied by 1*s*; the 2*s* is said to be *more penetrating* than the 2*p*. The effective nuclear charge experienced by the 2*s* is greater than that for 2*p* and this is why the former is lower in energy.

The crucial thing about the 2*s* orbital which makes it more penetrating than

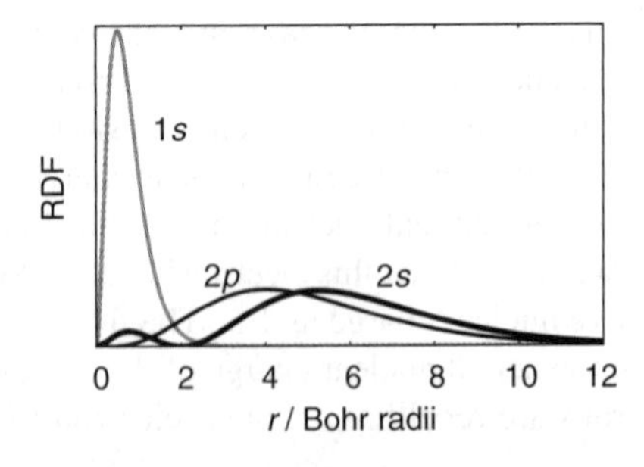

Fig. 4.33 Plots of the RDFs of the 1*s*, 2*s* and 2*p* orbitals, all to the same scale. For the 1*s* the effective nuclear charge has been taken as 2 in order to mimic the likely charge experienced by these electrons in lithium; for the other orbitals the effective nuclear charge has been taken as 1. Most of electron density of the 2*s* and 2*p* falls outside the region occupied by the 1*s*, and so 2*s* and 2*p* are well shielded from the nucleus. However, there is some probability of the 2*s* and 2*p* electrons penetrating inside the 1*s*, and this probability is greater for 2*s* than for 2*p*. This explains why the 2*s* is lower in energy than the 2*p*.

$2p$ is the presence of the small subsidiary maxima close in to the nucleus. The principal maximum for the $2s$ is in fact further out than for $2p$.

If we compare the RDFs for $3s$, $3p$ and $3d$ (Fig. 4.22 on p. 49) we see a similar pattern. The $3s$ has a subsidiary maximum close in to the nucleus, as does the $3p$, but the latter does not come in as close to the nucleus as the former. Not surprisingly, therefore, the $3s$ is more penetrating than the $3p$ which is in turn more penetrating than the $3d$. The $3s$ is thus the lowest in energy, followed by the $3p$ and then the $3d$.

In summary, the energies of the electrons in multi-electron atoms are most strongly affected by the value of the principal quantum number, n. Within a given shell (a given value of n) the ordering of the orbitals depends on the degree of penetration each shows, with those with the lower value of l being the more penetrating and hence the lowest in energy.

4.8 Orbital energies

We can put all we have learnt about multi-electron atoms together to discuss the trends in the orbital energies of the elements H to Ne; the data are shown in Fig. 4.34. Recall that these orbital energies are negative as they are measured downwards from the energy of the ionized electron which is taken as zero (see Fig. 4.16 on p. 45). So the lower – meaning more negative – the energy goes the more tightly the electron is held.

Looking first at the data for the $1s$ orbital, we see that the energy for helium is lower than that for hydrogen. We have already discussed why this is so: the nuclear charge has increased to two and although the electron–electron repulsion results in an effective nuclear charge of 1.4 this is significantly greater than that for hydrogen. Thus the $1s$ orbital is lower in energy in helium than it is in hydrogen.

Going from helium to lithium we see that the energy of the $1s$ drops sharply. We can attribute this to the fact that the third electron in lithium, the $2s$, does *not* shield the $1s$ electrons from the nucleus. As we have noted, the $1s$ electrons have most of their electron density closer to the nucleus than the $2s$ and so the latter is not in a position to provide any shielding. For the remaining elements

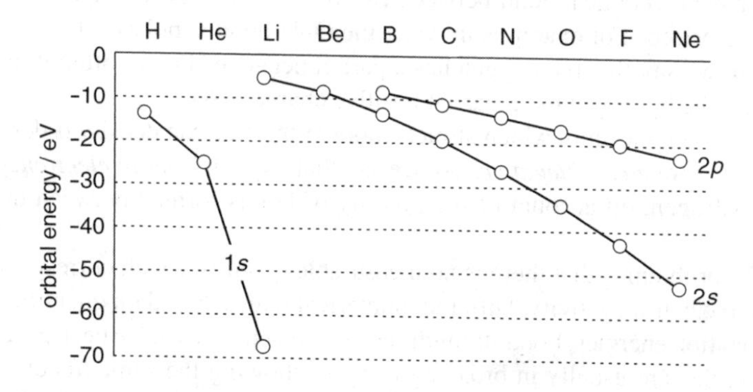

Fig. 4.34 Plot showing the energies of occupied orbitals for the elements H to Ne. Data for the $1s$ electron are only shown for the first three elements, as after that the energies fall very steeply. The lines connecting the data points are of no significance; they just serve to highlight the trends.

the energy of the $1s$ orbital drops even lower and, as we shall see, such electrons are so tightly held that they are hardly involved in bonding.

Turning now to the energies of the $2s$ orbital we see that there is a general fall as we go across the first row of the Periodic Table. The reason for this is that the nuclear charge is increasing, resulting in the electron being more tightly held. Of course, each time the nuclear charge increases an additional electron is added, but as these electrons are all going into the $n = 2$ shell they do not shield one another as well as the $1s$ electrons shield those in the $n = 2$ shell. This is because the electrons in the $n = 2$ shell broadly occupy the same region of space, whereas the $1s$ electrons are well inside the region occupied by the $2s$ and $2p$ electrons.

As we discussed on p. 56 the $2s$ orbital is lower in energy than the $2p$ so the $2s$ is occupied first. It is only by the time that we get to boron that the $2p$ orbital is occupied for the first time. Then, like the $2s$, the $2p$ orbital steadily falls in energy as we go across the first row of the Periodic Table. This fall in energy is for exactly the same reason as we gave for the $2s$.

Looking at Fig. 4.34 it is striking how the energy separation between the $2s$ and $2p$ increases as we go across the first row. The explanation for this is that as the $2s$ is more penetrating than the $2p$, the increase in nuclear charge as we go from one element to the next has a greater effect on the $2s$ than the $2p$. So while both orbitals drop in energy, the $2s$ falls more than the $2p$. We will see in the next chapter that this increasing separation of the $2s$ and $2p$ has some direct consequences when it comes to bonding.

As you might expect, the orbital energy of the $3s$ electron in sodium, -5.0 eV, is much less negative than for the $2p$ in neon; the effective nuclear charge for this electron is 1.8. In sodium the outer electron is now in the $n = 3$ shell and we argue that this is quite well shielded from the nucleus by the electrons in the $n = 1$ and $n = 2$ shells. As we continue across the second row of the Periodic Table, we find a pattern very similar to that shown in Fig. 4.34 for the first row.

4.9 Electronegativity

If we have a chemical bond between two different atoms we expect the bond to be polarized. For example, in water the O–H bond is polarized towards the oxygen; we say that the oxygen has a partial negative charge (often written as $\delta-$) and the hydrogen has a partial positive charge ($\delta+$).

In a bond, the atom which attracts more of the electron density is described as being more *electronegative*. So, we say that oxygen is *more electronegative* than hydrogen, on account of the polarity of bonds formed between the two atoms.

The problem is that there is no measurable quantity which corresponds directly to electronegativity. Different numerical scales, based on quantities such as ionization energies, bond strengths and orbital energies, have been proposed; these scales are usually in broad agreement, showing the same trends. Typically we find that the electronegativity increases as we go across the first row of the Periodic Table, e.g. oxygen is more electronegative than nitrogen which is in turn more electronegative than carbon.

We will see in the next chapter that the polarity of a bond between two atoms depends on the relative energies of their atomic orbitals. Generally, the atom with the lowest energy orbitals will end up with the partial negative charge – in other words, the more electronegative element is the one with the lower energy orbitals. Looking at Fig. 4.34 on p. 57 we see that the orbital energies decrease as we go across the first row, which means that we would expect the electronegativity to increase, which is exactly what it does.

Electronegativity is a consequence of the energies of the orbitals – the lower the orbital energy the more electronegative the element. If we are trying to work out which of two atoms has the lowest energy orbitals it is sometimes convenient to relate this to the familiar concept of electronegativity and note that the more electronegative element has the lower energy orbitals. However, it is important to realize that electronegativity is a consequence of orbital energies, and not the other way round.

5 Electrons in simple molecules

Electrons form the bonds which hold molecules together, and in chemical reactions electrons are rearranged as bonds are made and broken. So, understanding what the electrons are doing in a molecule is essential for understanding both chemical structures and reactions.

In this chapter we will introduce the powerful *molecular orbital* (MO) description of bonding, first applying it to simple diatomics and then, in the following chapter, extending it to larger molecules. Using the MO approach we will be able to rationalize many observations about the bonding in molecules, such as why H_2 readily forms but He_2 is unknown, why O_2 is paramagnetic (meaning that it is drawn into a magnetic field) but N_2 is not and why the bond in N_2 is stronger than that in O_2 which in turn is stronger than that in F_2. Simpler models of bonding are quite unable to explain these observations in any satisfactory way – we need the more subtle MO approach.

5.1 The energy curve for a diatomic

Figure 5.1 illustrates how the energy of two hydrogen atoms varies as we bring them together from a large separation.

When the atoms are far apart there will be no interaction between them and so the total energy is just the sum of the energies of the two atoms. As we bring the atoms closer together they start to interact in a favourable way and the energy falls, eventually reaching a minimum; this distance corresponds to

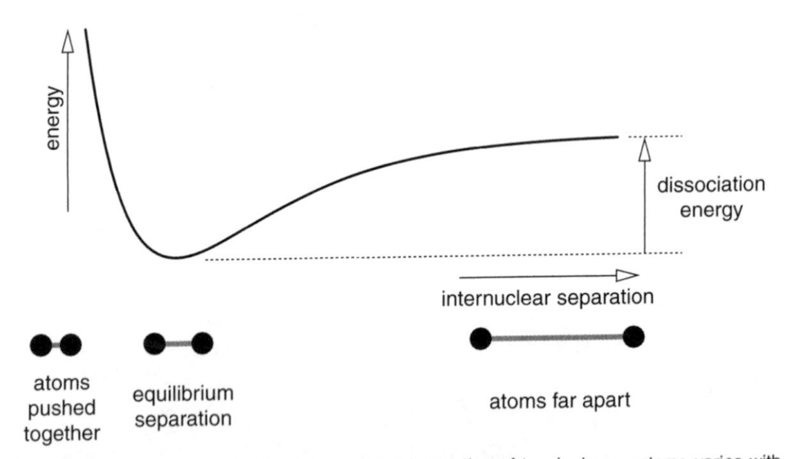

Fig. 5.1 Illustration of the way in which the energy of interaction of two hydrogen atoms varies with the distance between them. At large distances there is little interaction and the energy is just the sum of the energies of the two atoms. As the distance decreases there is a favourable interaction and eventually the energy reaches a minimum; this is the equilibrium separation. Finally, as the separation decreases further the atoms start to repel and the energy rises steeply. The energy change between the equilibrium position and the separated atoms is the bond dissociation energy.

the equilibrium bond length of H_2. Finally, if the atoms are brought still closer together, the energy rises, slowly at first and then more rapidly as the two nuclei begin to repel one another.

The difference in energy between the bottom of the potential energy curve and the separated atoms is the *dissociation energy*; this is the energy needed to separate the two atoms starting from the equilibrium position. What we have to do is to explain why it is that the energy of two hydrogen atoms decreases as they come together – in other words why a bond is formed; the MO approach provides the explanation we need.

5.2 Molecular orbitals for H_2

In Chapter 4 we saw that atomic orbitals (AOs) are used to describe the behaviour of electrons in atoms; in an analogous way we use molecular orbitals (MOs) to describe electrons in molecules. As with AOs, we can talk about the shape and energy of an MO. Also, electrons fill the MOs in molecules in the same way as AOs are filled in atoms: we start with the lowest energy MO and work upwards, placing up to two (spin-paired) electrons in each orbital. Our task is to determine the shapes and energies of these MOs.

In quantum mechanics a common way of solving a complex problem is to 'add together' the solutions to a simpler, but related, problem. We will use this approach to construct MOs by combining AOs of the atoms which are forming the bond. The attractive feature of this method is that there is a smooth transition from the molecule, where the MOs are a combination of AOs, to the separated atoms, where the AOs are uncombined. Also, we already know a lot about the shapes and energies of the AOs, and will be able to use this knowledge to good effect.

This method of constructing MOs is called the *linear combination of atomic orbitals* (LCAO) method; the name comes from the fact that the method simply involves adding together AOs. In the case of H_2 we write the MOs as

$$\text{MO} = c_1 \times (\text{AO on atom 1}) + c_2 \times (\text{AO on atom 2})$$

where c_1 and c_2 are just some numerical coefficients whose values are determined by quantum mechanics.

For hydrogen we need only consider MOs formed from the $1s$ AOs on the two atoms. It turns out that when *two* AOs on separate atoms combine the result is *two* MOs. In one MO the two coefficients have the *same* sign and in the other they have *opposite* signs. For a homonuclear diatomic, such as H_2, the coefficients are either $+1$ or -1, and so the two MOs are:

$$\text{bonding MO} = 1s_1 + 1s_2$$
$$\text{anti-bonding MO} = 1s_1 - 1s_2$$

where $1s_1$ and $1s_2$ are $1s$ AOs on atoms 1 and 2, respectively. We see that in the bonding MO the two coefficients have the same sign whereas in the anti-bonding MO they have the opposite sign. Just why these orbitals are called *bonding* and *anti-bonding*, we will consider next.

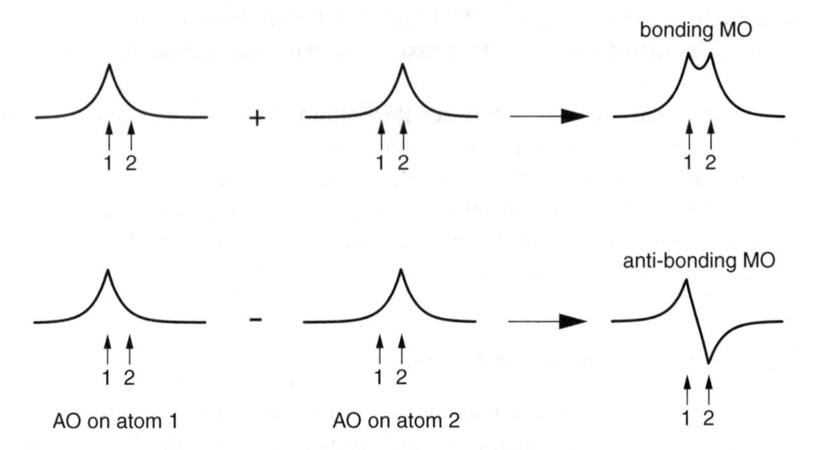

Fig. 5.2 Illustration of the formation of a bonding and an anti-bonding MO from two 1*s* orbitals. What is plotted is a cross-section through the orbitals along the line of the bond; the arrows marked 1 and 2 indicate the location of the two nuclei. The top part of the diagram illustrates the case where both coefficients are positive; this leads to *constructive* interference of the two AOs and the formation of the *bonding* MO, which has a build-up of electron density between the two atoms. The lower part of the diagram illustrates the case where the coefficients are of opposite sign; this leads to *destructive* interference and the formation of the *anti-bonding* MO. Note that this MO has a node at the mid-point between the two atoms.

Bonding and anti-bonding MOs

Figure 5.2 illustrates the formation of a bonding MO and an anti-bonding MO. When the two coefficients are both positive the two AOs reinforce one another and form an MO which extends across both atoms; this is the bonding MO. The overlap between the orbitals is described as being *constructive* or *in phase*. Remember that the square of the wavefunction (the orbital) gives the electron density, so we conclude that this MO places electron density *between* the two nuclei.

In contrast, when the two coefficients have opposite sign, the two AOs detract from one another in what is called *destructive* or *out of phase* overlap. This leads to the formation of the anti-bonding MO which has a node at the mid-point along the bond. Figures 5.3 and 5.4 show surface and contour plots of these two MOs; from these we can see clearly the way in which the bonding MO envelops the whole molecule, whereas the anti-bonding MO has a nodal plane between the two atoms.

It turns out that the bonding MO is *lower* in energy than the two AOs from which it is formed; this can be rationalized by noting that the bonding MO concentrates electron density in the region between the two nuclei. When the electrons are in this region, they are attracted by *both* nuclei, and this contributes to the lowering of the energy of the orbital when compared to the constituent AOs. The anti-bonding MO is *higher* in energy than the AOs from which it is formed. We can rationalize this by noting that in this orbital, electron density is pushed away from the favourable internuclear region, as evidenced by the nodal plane between the two atoms.

The relative energies of the AOs and MOs are usually illustrated with an *MO diagram* of the type shown in Fig. 5.5. This diagram shows the lowering of energy of the bonding MO, and the raising of energy of the anti-bonding

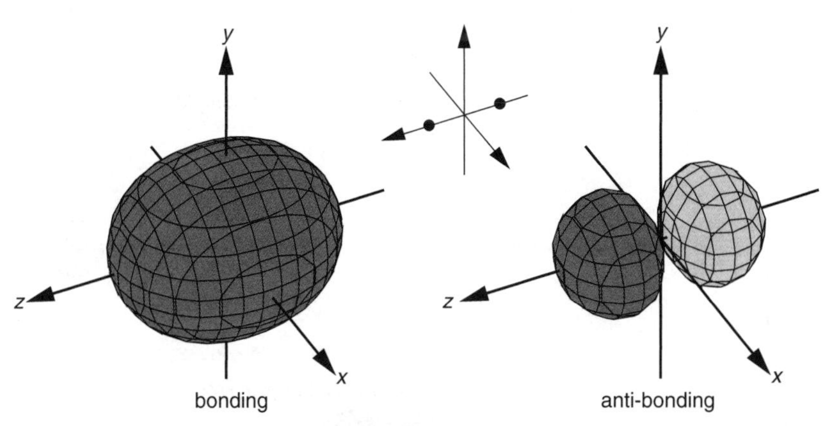

Fig. 5.3 Surface plots of the bonding and anti-bonding MOs formed from the overlap of two 1s orbitals; the two nuclei are placed along the z-axis, symmetrically about $z = 0$, as shown by the dots on the set of axes at the top of the diagram. Note that the bonding orbital has the same sign throughout and encompasses the whole molecule. In contrast, the anti-bonding orbital has a nodal plane (the xy-plane) between the two atoms. The dark grey indicates a positive value of the wavefunction and the light grey a negative value.

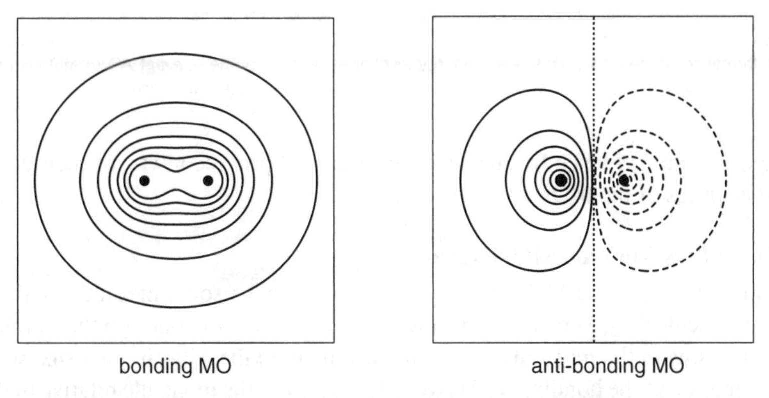

Fig. 5.4 Contour plots (taken in the xz-plane) of the bonding and anti-bonding MOs shown in Fig. 5.3. Positive values of the wavefunction are indicated by solid lines and negative values by lines with long dashes; the contour corresponding to the wavefunction being zero (the nodal plane) is indicated by a line with short dashes. The way in which the bonding MO encompasses the whole molecule is clearly seen, as is the nodal plane which bisects the anti-bonding orbital. The black dots indicate the location of the nuclei.

MO, when compared to the energies of the AOs. Also shown are 'cartoons' of the MOs which should be compared to Figs. 5.3 and 5.4; these cartoons are not supposed to be very precise but simply illustrate the relative signs of different parts of the orbitals.

It is important to realize that this description of the MOs as resulting from the combination of AOs is just a way of calculating the form of the MOs. We do not mean to imply that the AOs really do interfere constructively and destructively to form MOs – rather, this approach is just a way of predicting and thinking about the MOs.

Just as we can probe the energy levels of atoms using atomic emission spectroscopy (Section 4.1 on p. 35) or photoelectron spectroscopy (Section 4.2 on p. 37), we can also do the same for molecules. Such experiments provide direct

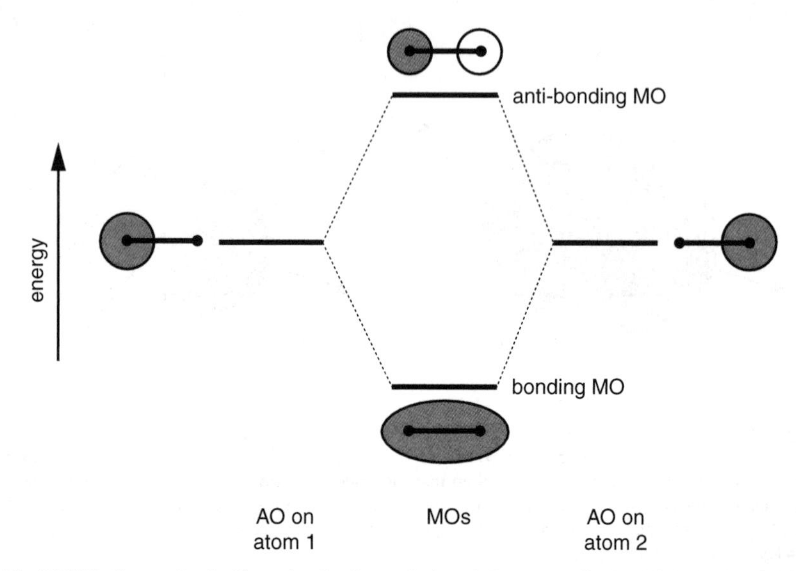

Fig. 5.5 MO diagram for the H_2 molecule; the vertical scale is energy. On the left the horizontal line shows the energy of the AO from atom 1; the line on the right shows the same for atom 2. In the middle are shown the energies of the MOs, with the dashed lines showing which AOs are involved in the formation of the MO. Beside each energy level is a 'cartoon' of the relevant AO or MO with the positive parts of the orbitals shown shaded. The small dumb-bell is there to indicate where the nuclei and bond are.

experimental evidence for the MO approach and provide a way of measuring the energies of the MOs.

Predictions from the MO diagram

We can now use the MO diagram (Fig. 5.5) to make some predictions about the molecules H_2, He_2 and their ions. What we do is to count up the number of electrons in the molecule and then place them in the MOs in the usual way. Occupation of the bonding MO lowers the energy of the molecule relative to the separated atoms and so *favours* the formation of a bond. In contrast, occupation of the anti-bonding MO raises the energy of the molecule relative to the atoms and so *disfavours* the formation of the molecule.

Figure 5.6 shows how the MOs are occupied for a series of molecules with increasing numbers of electrons. The molecular ion H_2^+ has just one electron which goes into the bonding MO; we therefore predict that the energy will be lowered when the molecule is formed and so expect H_2^+ to be stable with respect to dissociation into $H + H^+$. This prediction is borne out by experiments which have been able to detect and measure the properties of this molecular ion.

H_2 has two electrons, both of which go into the bonding MO and so we predict that the molecule is stable with respect to dissociation into atoms. Whereas H_2 has two bonding electrons, H_2^+ has only one, so we might expect the bond in the latter to be weaker than in the former. Experiment bears this out: the bond dissociation energy of H_2^+ is 256 kJ mol^{-1} whereas that of H_2 is 432 kJ mol^{-1}.

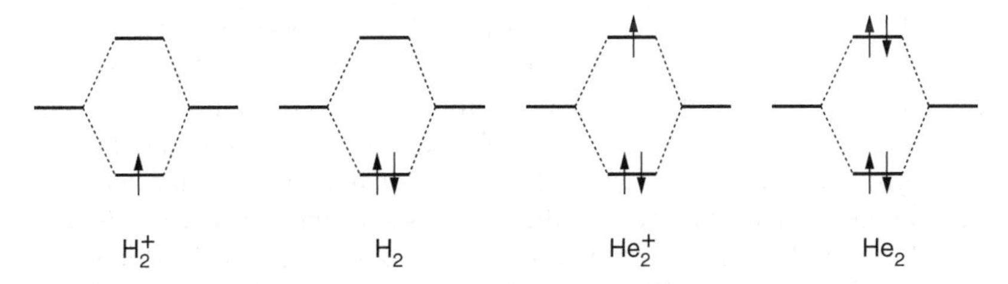

Fig. 5.6 MO diagrams (as in Fig. 5.5) for a series of molecules with increasing numbers of electrons; the occupation of the orbitals by the electrons is indicated by the arrows. Occupation of the bonding MO favours the formation of a bond as it lowers the energy relative to the AOs; occupation of the anti-bonding MO has the opposite effect. These diagrams predict that H_2^+, H_2 and He_2^+ should be stable with respect to dissociation into atoms or ions whereas He_2 should not form.

He_2 has four electrons, so we have to put two in the bonding MO and two in the anti-bonding MO. Occupation of the bonding MO favours formation of the bond, but occupation of the anti-bonding MO disfavours formation of the bond. Overall, the two effects cancel one another out and so there is no tendency for the molecule to form as the energy is not lowered as the atoms come together. Detailed calculations show that the anti-bonding MO is raised in energy by slightly more than the bonding MO is lowered, so if both are occupied equally the overall effect is to disfavour the formation of a bond. In agreement with our prediction, no He_2 molecules have ever been observed experimentally.

Finally, He_2^+ represents an interesting case. Here there are three electrons, two in the bonding and one in the anti-bonding MO. So, overall there is net bonding as the number of bonding electrons is greater than the number of anti-bonding electrons. We therefore predict that the molecule He_2^+ should be stable with respect to dissociation into He + He^+. Experiment confirms this prediction: He_2^+ has been observed and found to have a dissociation energy of 290 kJ mol^{-1}.

This first application of the MO approach is very nice indeed – it gives us a simple and straightforward explanation of the bonding and occurrence of the molecules H_2, He_2 and their ions. A 'dot and cross' picture of these molecules simply cannot explain what is going on – we need the MO approach.

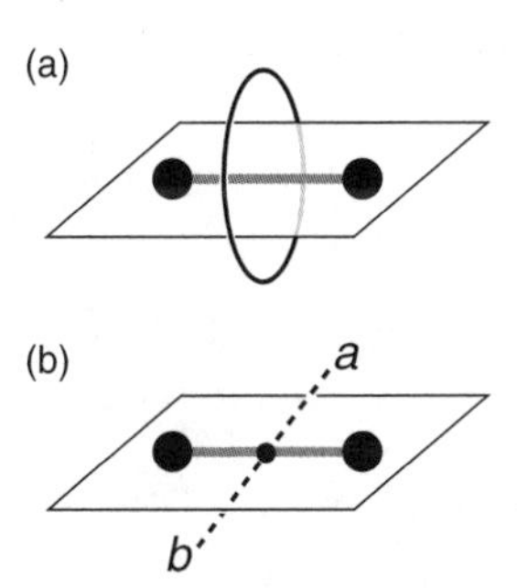

Fig. 5.7 Illustration of the two symmetry labels which are used to describe MOs in diatomic molecules. The first symmetry label refers to traversing a circular path around the internuclear axis, as shown in (a). If the wavefunction does not change sign along this path the orbital is given the label σ. If the circular path crosses a nodal plane the orbital is given the label π. The second label, shown in (b), refers to what happens when we start at any point *a*, go in a line to the centre of symmetry of the molecule (marked with a dot) and then carry on an equal distance to point *b*. If the orbital has the same sign at *a* and *b* the symmetry label is *g*; if the sign has changed, the label is *u*. Note that the second symmetry label only applies to *homonuclear* diatomics.

5.3 Symmetry labels

MOs are given labels which tell us about their symmetry with respect to the molecule. For a diatomic, the first of these symmetry labels refers to what happens to the orbital when we traverse a circular path around the internuclear axis; this is illustrated in Fig. 5.7 (a). The circular path is centred on the bond and is in a plane perpendicular to the internuclear axis.

If, when we make a complete circuit of this path, the orbital does not change sign it is given the label σ ('sigma'). Looking at Fig. 5.3 on p. 63 we see that neither the bonding nor the anti-bonding MO has a sign change along such a path, so they are both given the label σ. Sometimes we distinguish the anti-bonding orbital by adding a superscript star: $\sigma^\star$.

If during a complete revolution we cross one nodal plane (i.e. a plane where the wavefunction is zero) the symmetry label is π ('pi'). We will not encounter orbitals which have other than σ or π symmetry.

A homonuclear diatomic possesses a kind of symmetry called a *centre of inversion*, which is the point mid-way between the two atoms; the MOs can be classified according to their behaviour with respect to this kind of symmetry. The process is shown in Fig. 5.7 (b). We start at any point, a, and move in a straight line towards the centre of inversion; having reached the centre we carry on in the same direction for the same distance until we reach point b. If the orbital has the same sign at points a and b it is given the label g (for *gerade*, 'even' in German). From Fig. 5.3 on p. 63 we can see that the bonding MO has this g symmetry; this label is added to the σ as a subscript: σ_g.

If, on going from point a to b, the orbital changes sign we attach the label u (for *ungerade*, 'odd' in German). We see from Fig. 5.3 that the anti-bonding MO has this symmetry, and so it is given the label σ_u or $\sigma_u^\star$.

The σ and π labels can still be applied to MOs in heteronuclear diatomics, but the *g/u* symmetry labels cannot as heteronuclear diatomics do not possess a centre of inversion. We sometimes also label the MOs with the AOs from which they are derived, so for example the bonding MO in H_2 can be labelled $1s\,\sigma_g$.

5.4 General rules for forming molecular orbitals

We now want to move on to construct the MOs for molecules more complex than H_2, so we need to consider what happens when there are more AOs present than just the two $1s$ AOs in H_2. It turns out that there are some simple rules about how MOs are constructed from AOs; we will look at each of these rules in turn.

Only AOs with similar energies interact to a significant extent

When two AOs interact to form MOs, the extent of the interaction is reflected in the amount by which the bonding MO goes down in energy, or equivalently the amount by which the anti-bonding MO goes up in energy. So, a large interaction results in the MOs differing significantly in energy from the AOs, whereas if there is little interaction the energies of the MOs are not much different from those of the original AOs.

When two AOs interact, it is always the case that the bonding MO is *lower* in energy than the lowest energy AO, and that the anti-bonding MO is *higher* in energy than the highest energy AO. Figure 5.8 illustrates the effect on the MO energies of increasing the energy separation of the AOs.

In (a) the AOs are matched exactly and the interaction is strong, as evidenced by the large drop in energy of the bonding MO when compared to the AOs. In (b) the match is not quite as good and we see that the drop in energy of the bonding MO is less; in (c) the mismatch is even greater and the fall in energy of the bonding MO is correspondingly smaller. Finally in (d) there is hardly any interaction and the bonding MO is more or less at the same energy as the lowest energy AO.

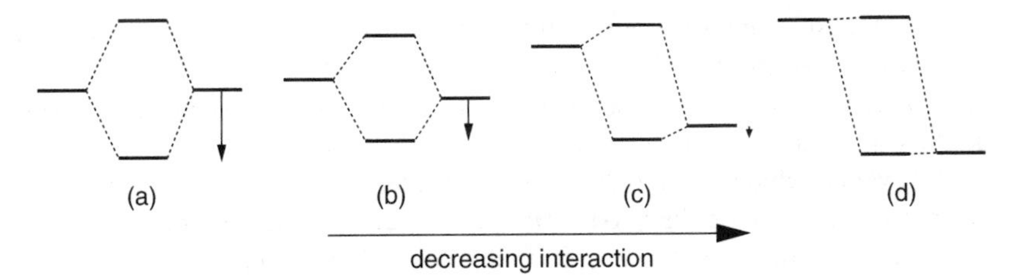

Fig. 5.8 Illustration of how the interaction between two AOs decreases as the energy match between them gets poorer; the downward pointing arrow gives the amount by which the bonding MO is lower in energy than the lower energy AO.

The size of the AOs is important

The two AOs interact in the region where their wavefunctions overlap; the extent of overlap will depend on the size of the orbitals and how close together they can come. The size of an AO is determined by the principal quantum number and the effective nuclear charge, whereas how close two AOs can approach depends on the bond length.

Generally what is found is that when considering the interaction between identical orbitals increasing their size decreases the degree of overlap. So, for example, two $3s$ orbitals overlap to a lesser extent than two $2s$ orbitals. Similarly, we find that the overlap between a $2s$ and a $3s$ orbital is less than between two $2s$ orbitals.

AOs must be of the correct symmetry to interact

This condition is best illustrated by giving an example of a case where the two AOs are not of the correct symmetry. Consider the case shown in Fig. 5.9 in which the overlap along the z-axis (the axis of the bond) of a $2p_x$ orbital with a $1s$ orbital is depicted; recall that the $2p_x$ orbital points along the x-axis, so it will point perpendicular to the z-axis. In (a) the orbitals are far apart so that we can see clearly the positive and negative lobe of the p orbital and the single positive lobe of the $1s$ orbital.

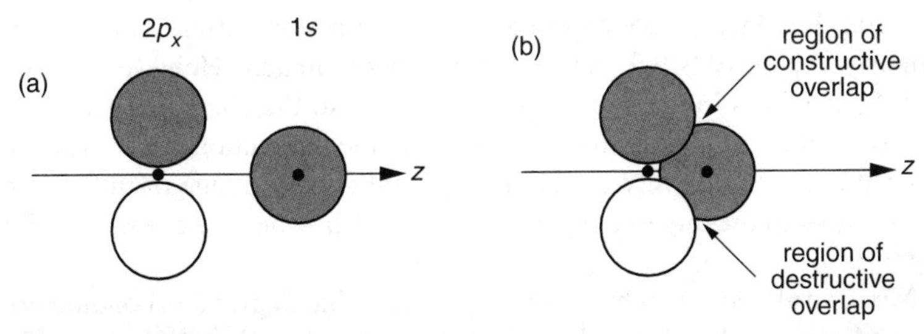

Fig. 5.9 Illustration of the symmetry requirement for AOs to overlap. (a) Shows a $2p_x$ and a $1s$ AO about to overlap along the z-axis; the positive parts of the orbitals are shaded grey. In (b) the overlap has begun, but we see that there is a region of constructive overlap (where the two orbitals have the same sign) and a region of destructive overlap (where the orbitals have opposite sign). These two regions cancel one another out leading to no net overlap. The black dots indicate the location of the nuclei.

As the orbitals come together, shown in (b), the positive lobe of the $2p$ orbital overlaps with the positive $1s$ orbital; this is the kind of constructive overlap which leads to the formation of a bonding MO. However, the negative lobe of the $2p$ orbital overlaps with the positive $1s$ orbital; this is the kind of destructive overlap which leads to the formation of an anti-bonding MO. In the case shown here the constructive and destructive overlaps exactly cancel one another out, leading to no net overlap.

So, we conclude that the $2p_x$ and $1s$ orbitals do not have the correct symmetry to overlap along the z-axis. The same is true for the $2p_y$ and $1s$ orbitals. However, as is illustrated in Fig. 5.10, the $2p_z$ and $1s$ orbitals do have the correct symmetry to overlap.

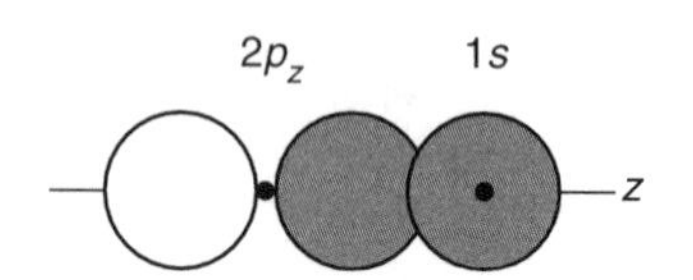

Fig. 5.10 In contrast to the case shown in Fig. 5.9, a $2p_z$ orbital has the correct symmetry to overlap with a $1s$ AO along the z-axis.

When n AOs interact the same number of MOs is formed

We have already seen an example of this when two AOs in H_2 gave two MOs. In Chapters 6 and 10 we will see some examples of the interaction of more than two AOs.

5.5 Molecular orbitals for homonuclear diatomics

We will now use these rules to construct the MO diagram for the series of diatomic molecules Li_2, Be_2, ..., Ne_2 formed by the elements of the first row of the Periodic Table. The first thing we need to think about is which AOs might overlap with one another; from our knowledge of the electronic configurations of these atoms we know that the relevant orbitals are the $1s$, $2s$ and $2p$. In fact we need not consider the $1s$ as they are so contracted towards the nucleus (due to the high effective nuclear charge they experience) that, as illustrated in Fig. 5.11, they do not overlap to a significant extent. So, the only orbitals we need to consider are $2s$ and $2p$.

(a)

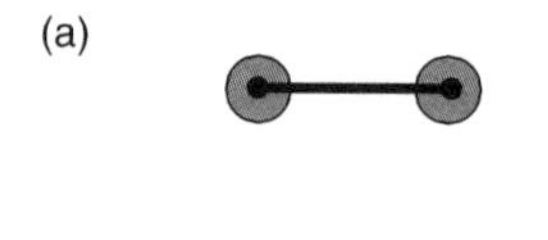

(b)

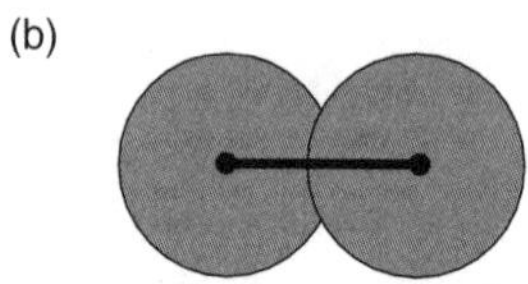

Fig. 5.11 Illustration of why the $1s$ orbitals are not involved in bonding in diatomic molecules formed from elements of the first row of the Periodic Table. On account of the high effective nuclear charge experienced by the $1s$ electrons the orbitals are highly contracted and so, for typical bond lengths, they do not overlap to a significant extent, as shown in (a). In contrast, the $2s$ orbitals are much larger and so overlap well as shown in (b).

The $2s$ orbital is spherical, just like the $1s$ orbital, but as we have seen (Fig. 4.18 on p. 46) the $2s$ has a radial node close in to the nucleus. However, this does not really affect the way in which the $2s$ AO overlaps with other orbitals as the region of overlap is much further out than is this node. So, to all practical intents and purposes, we can simply treat the $2s$ as a spherical orbital.

There will be a perfect energy match between the $2s$ AO on one atom and the $2s$ on the other, so we expect a strong interaction between these orbitals. Similarly, the $2p$ AOs will have a perfect energy match. However, as we saw on p. 56 the $2s$ is lower in energy than the $2p$, so the energy match between these two orbitals is not perfect and therefore the interaction is not as strong as between $2s$ and $2s$ or between $2p$ and $2p$. Taking advantage of this effect we will for now ignore any overlap between $2s$ and $2p$ when we construct our MO diagram.

We have already seen how two $1s$ AOs combine to give a σ bonding orbital (σ_g) and a σ anti-bonding orbital (σ_u); the same thing happens for the $2s$ orbitals and we will denote the resulting MOs $2s\ \sigma_g$ and $2s\ \sigma_u$. The $2p$ orbitals can overlap to give MOs of both σ and π symmetry, as described in the next section.

MOs from the overlap of *p* orbitals

If we take the internuclear axis to be the z-direction, then two $2p_z$ orbitals can overlap 'head on' to give σ MOs as shown in Fig. 5.12 (a). The bonding MO

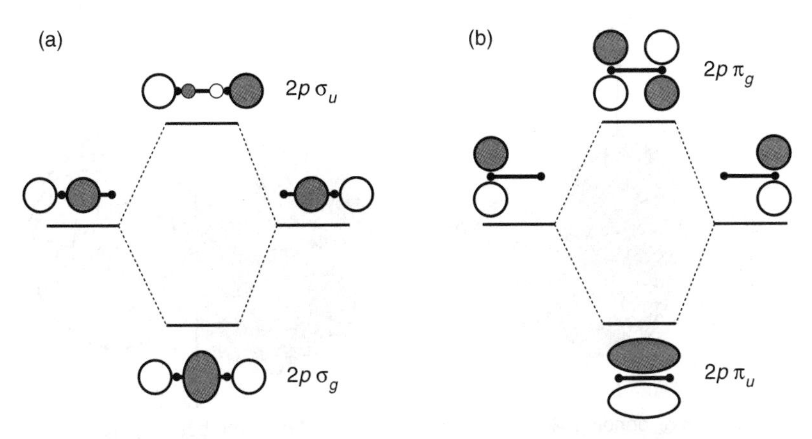

Fig. 5.12 MO diagrams illustrating the formation of σ and π type MOs from the overlap of $2p$ AOs. The 'head-on' overlap, shown in (a), leads to σ MOs, whereas the 'side-on' overlap (b) leads to π MOs. The $2p$ AOs are drawn so that adding them together leads to constructive interference. The dots indicate the location of the nuclei.

is formed when there is constructive interference between the two lobes which face toward one another along the z-axis; as with the $2s$ σ MO the concentration of electron density between the two nuclei is largely responsible for the lowering in energy of the MO.

In Fig. 5.12 (a) the $2p$ orbitals are drawn so that simply adding them together gives constructive interference. The other combination we need to consider comes from subtracting the orbitals; this leads to destructive interference and the formation of the anti-bonding MO in which electron density is pushed away from the internuclear region.

It is important to realize that the way we choose to represent the $2p$ AOs is arbitrary; we could just as well have made the left-hand lobe on both AOs positive. With this choice, adding together the two AOs would lead to destructive interference and the formation of the anti-bonding MO, whereas subtracting the two AOs would lead to the formation of the bonding MO. Either way, we still end up with a bonding and an anti-bonding MO.

Figure 5.13 shows surface plots of the two MOs and Fig. 5.14 shows contours plots of the same orbitals. The two MOs are both σ; with respect to the centre of inversion the bonding MO has symmetry g and the anti-bonding MO has symmetry u. We therefore denote the orbitals $2p\,\sigma_g$ and $2p\,\sigma_u$ to indicate their symmetries and that they are derived from $2p$ orbitals.

The $2p_x$ and $2p_y$ AOs point in directions perpendicular to the internuclear axis and, as was discussed on p. 67, these orbitals do not have the correct symmetry to overlap with the $2s$ orbitals. A similar line of reasoning, illustrated in Fig. 5.15, shows us that a $2p_x$ on one atom cannot overlap with a $2p_z$ on the other.

However, the two $2p_x$ AOs can overlap 'side on' in the way illustrated in Fig. 5.12 (b); the shapes of the resulting orbitals are best appreciated from the surface and contour plots shown in Figs. 5.16 and 5.17.

Constructive interference leads to a bonding MO in which there is electron density above and below the internuclear axis; note the contrast to a σ bonding MO which has electron density directly between the two nuclei.

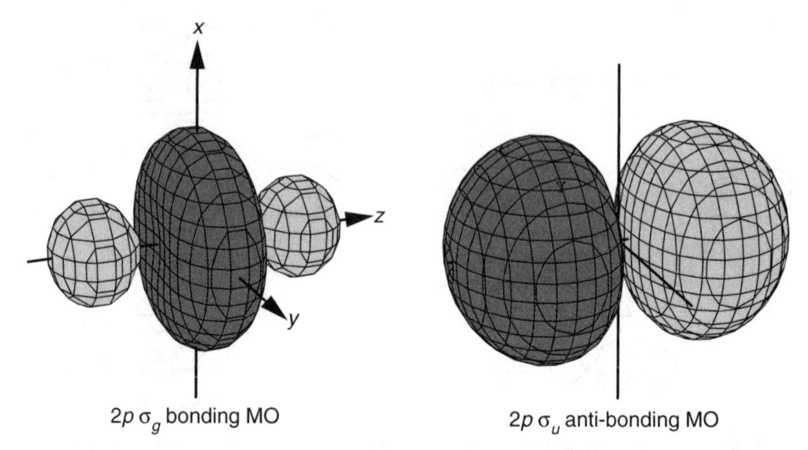

Fig. 5.13 Surface plots of the $2p\ \sigma_g$ (bonding) MO and the $2p\ \sigma_u$ (anti-bonding) MO formed from the head-on overlap of two $2p$ orbitals. Note how the bonding MO concentrates electron density in the internuclear region whereas the opposite is true for the anti-bonding MO. The two nuclei are placed along the z-axis, symmetrically about $z = 0$.

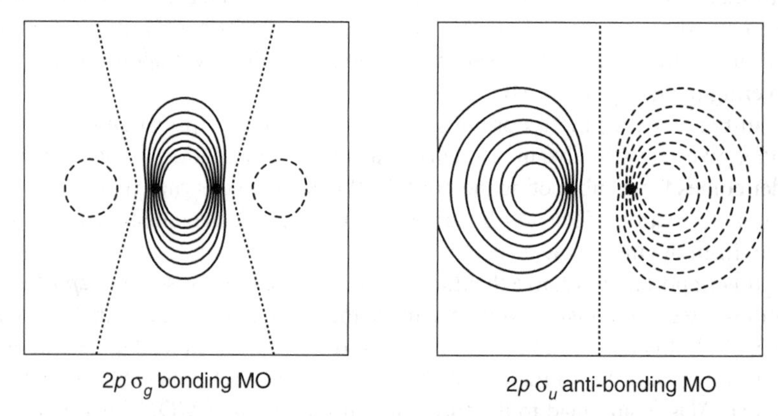

Fig. 5.14 Contour plots of the $2p\ \sigma_g$ (bonding) MO and the $2p\ \sigma_u$ (anti-bonding) MO. The cross-section shown is the xz-plane (with z running horizontally); the positions of the two nuclei are indicated by the small black dots.

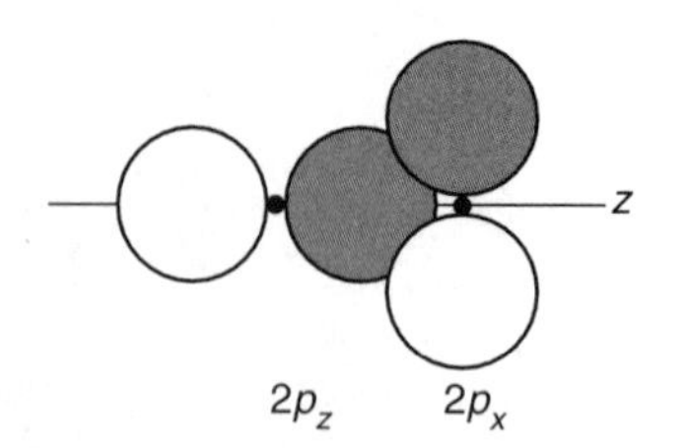

Fig. 5.15 Illustration of why a $2p_z$ AO and a $2p_x$ AO have no net overlap; the region of constructive interference is balanced out by the region of destructive interference, just as in Fig. 5.9 on p. 67.

This difference in the electron distribution explains why the MO formed by the side-on overlap is less strongly bonding than the σ bonding MO. Side-on destructive interference leads to an anti-bonding MO in which the electron density is pushed away from the region above and below the internuclear axis.

Both the bonding and anti-bonding MOs have a nodal plane which contains the two nuclei. In addition, the anti-bonding MO has a second nodal plane which lies between the two nuclei, perpendicular to the internuclear axis.

Inspection of Fig. 5.16 shows that the symmetry is π as when traversing a circular path perpendicular to the internuclear axis (see Fig. 5.7 (a) on p. 65) we cross a nodal plane. The bonding orbital has u symmetry and so is given the full label $2p\ \pi_u$ and the anti-bonding orbital is g so is given the label $2p\ \pi_g$.

The two $2p_y$ AOs can also overlap with one another to give two more π MOs, one bonding and one anti-bonding. So, there is a total of two bonding π orbitals and two anti-bonding π orbitals: one bonding and anti-bonding pair

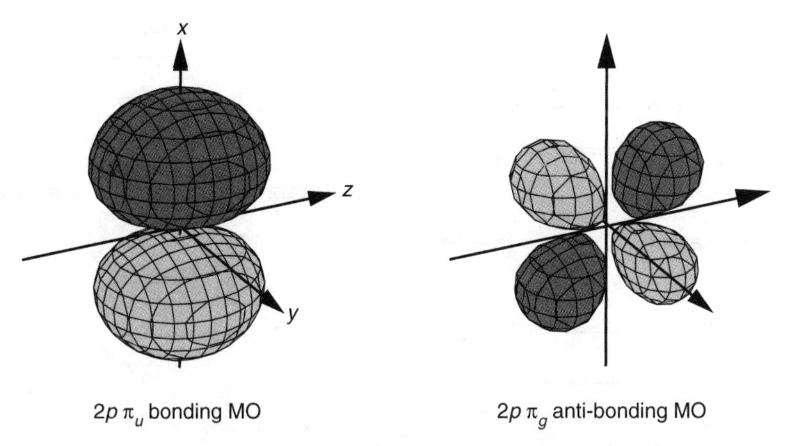

Fig. 5.16 Surface plots of the $2p\ \pi_u$ (bonding) MO and the $2p\ \pi_g$ (anti-bonding) MO formed from the side-on overlap of two $2p_x$ orbitals. Both MOs have a nodal plane containing the two nuclei, and the anti-bonding MO also has a nodal plane between the two nuclei, perpendicular to the internuclear axis. Note that for the bonding MO there is a concentration of electron density in the regions above and below the internuclear axis. The two nuclei are placed along the z-axis, symmetrically about $z = 0$.

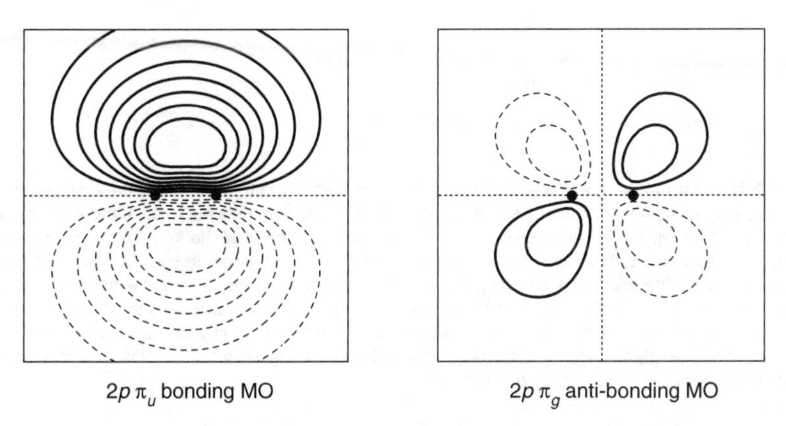

Fig. 5.17 Contour plots of the $2p\ \pi_u$ (bonding) MO and the $2p\ \pi_g$ (anti-bonding) MO; the nodal planes are clearly visible in both plots. The cross-section shown is the xz-plane (with z running horizontally); the positions of the two nuclei are indicated by the small black dots.

lie in the xz-plane and the other pair lie in the yz-plane. The two bonding MOs are degenerate, as are the two anti-bonding MOs.

Two MO diagrams for the A_2 molecules

If we consider only the $2s$-$2s$ and $2p$-$2p$ overlap it is quite easy to assemble the complete MO diagram, which is shown in Fig. 5.18 (a). On the left and right are shown the AOs, with the $2s$ lower in energy than the $2p$. The three degenerate $2p$ orbitals are represented by the three closely spaced lines labelled $2p$ on the diagram.

The $2s$ orbitals overlap to form a bonding MO, $2s\ \sigma_g$, and an anti-bonding MO, $2s\ \sigma_u$. The $2p$ orbitals which point along the bond (the z-direction) also overlap to form σ orbitals: $2p\ \sigma_g$ (bonding) and $2p\ \sigma_u$ (anti-bonding).

The two $2p_x$ orbitals overlap to give a π bonding MO ($2p\ \pi_u$) and a π anti-

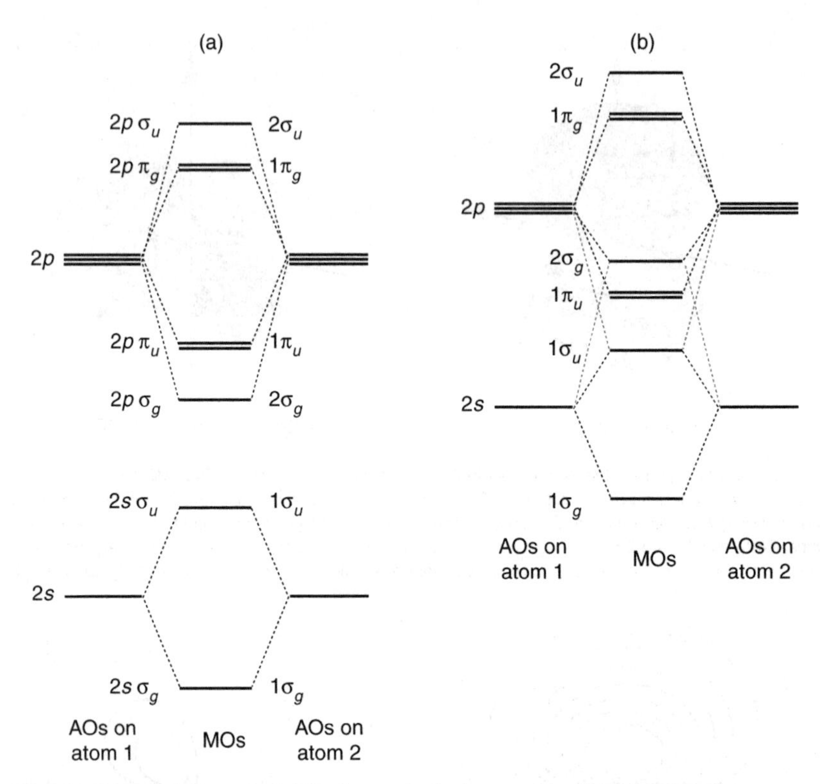

Fig. 5.18 Shown in (a) is a simple MO diagram for the homonuclear diatomics of the first row, constructed by considering only 2*s*-2*s* and 2*p*-2*p* overlap; this diagram turns out to be appropriate for O_2 and F_2. However, for the molecules $Li_2, \ldots, N_2$ the MO diagram is of the form shown in (b). The difference between (b) and (a) is that the *s* and *p* orbitals are much closer in energy in (b) so that we need to include the effect of *s*-*p* mixing; this results in a re-ordering of the π and σ bonding orbitals derived mainly from the 2*p* AOs. The MOs in diagram (a) are labelled according to which AOs each MO is derived from; diagram (b) is labelled using a different scheme in which orbitals of the same symmetry are simply distinguished from one another by numbering them 1, 2, ...; the MOs in diagram (a) are also labelled using this second scheme.

bonding MO ($2p\ \pi_g$). The same is true of the two $2p_y$ orbitals, so there are two degenerate $2p\ \pi_u$ orbitals, indicated by the two closely spaced lines, and two degenerate $2p\ \pi_g$ orbitals indicated in the same way. Generally σ interactions are stronger than π, so we have shown the $2p\ \sigma_g$ as being lower in energy than the $2p\ \pi_u$.

Right at the start of this section (p. 68) we made the assumption that the 2*s* and 2*p* orbitals are far enough apart in energy that we do not need to consider the possibility of σ overlap between them. Therefore, the MO diagram shown in Fig. 5.18 (a) is only suitable for cases where this assumption is appropriate.

We saw in Fig 4.34 on p. 57 that the energy separation between the 2*s* and 2*p* increases as we go across the first row of the Periodic Table. It turns out that this separation is sufficient in oxygen and fluorine for the MO diagram of Fig. 5.18 (a) to be correct. However, for the other elements (Li, ..., N), the separation of the 2*s* and 2*p* is such that we cannot ignore the effect of σ overlap between these AOs.

The most significant effect of these *s-p* interactions (usually called *s-p mixing*) is to make the $2p\ \sigma_g$ MO less bonding (i.e. to raise its energy) and the $2s\ \sigma_u$ MO less anti-bonding (i.e. to lower its energy). For the elements Li, ..., N the shift of the $2p\ \sigma_g$ is large enough to move it above the $2p\ \pi_u$ MO, as is shown in Fig. 5.18 (b).

When there is significant *s-p* mixing we cannot really label an orbital $2p\ \sigma_g$ as it is not correct to identify it as being derived solely from $2p$; the symmetry labels σ and g are still applicable, though. Under these circumstances we drop the $2p$ prefix and simply number the orbitals to distinguish ones with the same symmetry, so the first σ_g MO is labelled $1\sigma_g$ and the second $2\sigma_g$. The other orbitals are labelled using a similar scheme, as is shown in Fig. 5.18 (b); the same scheme can be used to label the orbitals in MO diagram (a).

Using the MO diagram

We can now use the MO diagrams of Fig. 5.18 to make some predictions about the homonuclear diatomics of the first row. All we have to do is to slot the appropriate number of electrons into the MOs and then work out what are the consequences of the resulting electronic configuration.

For example, lithium has the configuration $1s^2 2s^1$, but as we commented on before, the $1s$ electrons are not really involved in bonding, so we need only be concerned with the single $2s$ electron. The electrons which are involved in bonding are called the *valence electrons*, and there are two of these in Li_2.

For this molecule the appropriate MO diagram is Fig. 5.18 (b) and so the two valence electrons go spin paired into the $1\sigma_g$ orbital giving the electronic configuration $1\sigma_g^2$. We can carry on in the same way for the diatomics Li_2, ..., N_2, giving the results shown in the table below (a star has been added to indicate the anti-bonding MOs). The table also shows the bond order, the meaning of which we will come on to shortly.

	$1\sigma_g$	$1\sigma_u$	$1\pi_u$	$2\sigma_g$	$1\pi_g$	bond order
Li_2	↑↓					1
Be_2	↑↓	↑↓				0
B_2	↑↓	↑↓	↑ ↑			1
C_2	↑↓	↑↓	↑↓ ↑↓			2
N_2	↑↓	↑↓	↑↓ ↑↓	↑↓		3

In Li_2 both electrons are in a bonding orbital and there are no electrons in anti-bonding orbitals, so we predict that the molecule is stable with respect to dissociation into atoms. Experiment bears this out: molecular Li_2 has been observed in the gas phase and it is found to have a bond dissociation energy of 101 kJ mol^{-1}. The two electrons in a σ bonding MO create a σ bond between the two atoms, so the molecule is described as having a single σ bond.

In Be_2 there are two electrons in a bonding MO and two in an anti-bonding MO; the simple expectation is that the molecule has no net bonding and so will not form (just as was the case for He_2). In fact, because of the *s-p* mixing, the $1\sigma_u$ orbital is somewhat less anti-bonding than the $1\sigma_g$ is bonding, so overall Be_2 is expected to be a weakly bound species. The molecule has been observed in the gas phase and is found to have a rather low dissociation energy of 59 kJ mol^{-1}.

B_2 has six valence electrons; four of these are spin paired in σ orbitals and the last two go into different $1\pi_u$ orbitals with their spins parallel. Remember that when there are degenerate orbitals the lowest energy arrangement for two electrons is to put them into separate orbitals with their spins parallel.

There is a special feature of B_2 which we can explain using the MO diagram, which is that the molecule is found to be *paramagnetic*. Paramagnetic substances are drawn into a magnetic field – an effect which can be measured experimentally. Paramagnetism is associated with the presence of *unpaired* electrons, which, according to the MO diagram, is precisely what we have for B_2.

The bonding effect of having two electrons in the $1\sigma_g$ is roughly cancelled out by having two electrons in the anti-bonding $1\sigma_u$, so the bonding in B_2 is mainly due to the two $1\pi_u$ electrons. A useful way of expressing the degree of bonding is to calculate the *bond order*, which is given by

$$\text{bond order} = \tfrac{1}{2}\left[\text{number of bonding electrons} - \text{number of anti-bonding electrons}\right].$$

The factor of one-half is included so that a pair of bonding electrons, which we think of as comprising a bond, gives a bond order of one.

In the case of Li_2 the bond order is $\tfrac{1}{2}(2-0) = 1$, whereas for Be_2 it is $\tfrac{1}{2}(2-2) = 0$. These values fit in with the description of Li_2 having a single bond and Be_2 having (to a first approximation) no bond. Similarly the bond order in B_2 is $\tfrac{1}{2}(2-2+2) = 1$, i.e. a single bond. The table on p. 73 gives the values of the bond order for each molecule.

The bond order is a rather rough measure of the degree of bonding as it takes no account of the relative contribution to the bonding of different bonding orbitals (and similarly for anti-bonding MOs). However, there is quite a good correlation between the bond dissociation energy (a measure of the bond strength) of the molecules A_2 and the bond order, as shown in Fig. 5.19.

The next diatomic is C_2 which has eight valence electrons. These fill the $1\sigma_g$, the $1\sigma_u$ and both of the $1\pi_u$ orbitals with all spins paired; the result is two π bonds. N_2 has 10 valence electrons and these occupy the MOs in such a way that there are two π bonds and one σ bond; this is consistent with the simple view that N_2 has a triple bond and indeed the bond order is 3. This molecule has the largest bond order of all of the diatomics of the first row and so is expected to have the strongest bond, which is indeed the case.

Once we get to oxygen and fluorine the appropriate MO diagram is Fig. 5.18 (a) which has a different ordering of the orbitals. The electronic configurations for these last two diatomics are therefore:

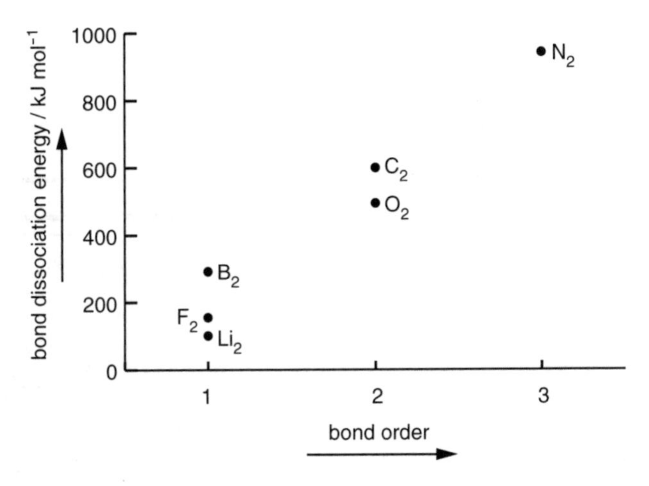

Fig. 5.19 Plot showing the correlation between bond dissociation energy (bond strength) and bond order for the homonuclear diatomics of the first row. There is a useful correlation between these two quantities.

	$1\sigma_g$	$1\sigma_u$	$2\sigma_g$	$1\pi_u$	$1\pi_g$	bond order
O_2	↑↓	↑↓	↑↓	↑↓ ↑↓	↑ ↑	2
F_2	↑↓	↑↓	↑↓	↑↓ ↑↓	↑↓ ↑↓	1

In O_2 all of the bonding MOs have been occupied and so the final two electrons have to go into the π anti-bonding MOs. As is the case for B_2 the last two electrons are split between two degenerate orbitals with their spins parallel, making the molecule paramagnetic – which is what is found experimentally.

In F_2 the π anti-bonding orbitals are now full, so the net bonding is simply from the $2\sigma_g$ orbital. Finally, the hypothetical molecule Ne_2 has all of the bonding and anti-bonding orbitals filled and so is not expected to form.

The MO approach is remarkably successful at explaining the properties of the A_2 molecules. Using it, we have been able to rationalize the variation in bond strengths, explain why Be_2 and Ne_2 are not expected to form, and predict the observed paramagnetism of B_2 and O_2. We can be well satisfied with this outcome!

Figure 5.20 shows a plot of the energies of the occupied MOs for the homonuclear diatomics of the first period. We see that in general the energies of the MOs fall steadily across the period; this has the same origin as the fall in the orbital energies of the atoms, i.e. it is due to the increasing nuclear charge – see Section 4.8 on p. 57.

For B_2 and C_2 the highest occupied MO is the $1\pi_u$, showing that this lies lower in energy than the $2\sigma_g$. In contrast, for O_2 and F_2 the $2\sigma_g$ lies lower than the $1\pi_u$. For N_2 these two MOs are rather close in energy and there is some uncertainty over their ordering.

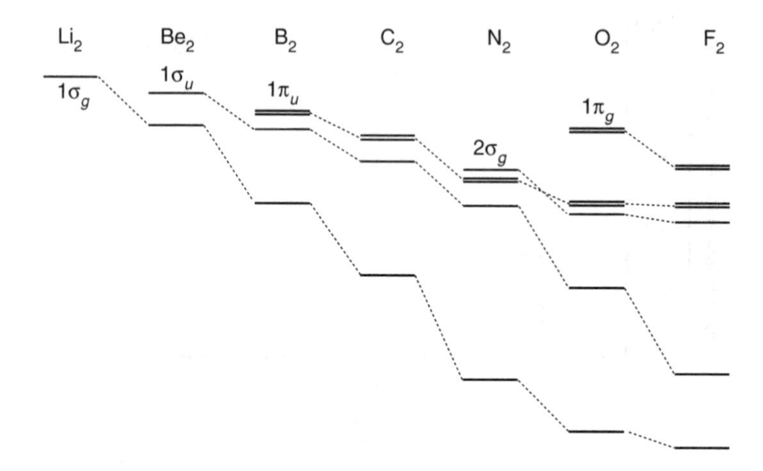

Fig. 5.20 Plot showing the calculated energies (to scale) of the occupied MOs for the homonuclear diatomics of the first row; the dashed lines connect similar MOs. Note that there is a general fall of the energies of the MOs as we go across the period and that for O_2 and F_2 the $2\sigma_g$ lies below the $1\pi_u$.

5.6 Heteronuclear diatomics

The bonding in heteronuclear diatomics (ones in which the two atoms are different) can be described in a similar way to that used for homonuclear diatomics. There is one significant difference which is that there will no longer be an exact match between the energies of the AOs of the two atoms; this makes it more difficult to construct the MO diagram as we have to decide which AOs will overlap. We will also see that this energy mismatch between AOs leads to polarized bonds.

On p. 66 and in Fig. 5.8 it was explained that when the energy separation between two AOs becomes larger the interaction between them decreases. A further consequence of this energy mismatch is that the contribution of the two AOs to the MOs is not the same: this is illustrated in Fig. 5.21.

Figure 5.21 (a) shows the situation in which the two AOs have the same energy. The bonding MO has equal contributions from the two AOs, and the same is true of the anti-bonding MO except that the orbitals are combined with opposite coefficients. When the two AOs have different energies what we find is that the AO which is *closer* in energy to the MO is the major contributor to that MO. This is illustrated in (b) in which the bonding MO is closest in energy to the AO of atom 2 and so this AO is the major contributor to the MO. In contrast, the anti-bonding MO is closest in energy to the AO from atom 1 and so this is the major contributor to the MO.

As the energy mismatch increases further the relative contributions of the two AOs to a given MO become even more unequal, as is shown in Fig. 5.21 (c). Now the bonding MO has a much larger contribution from the AO on atom 2 than from that on atom 1, and vice versa for the anti-bonding MO.

If the bonding MO of Fig. 5.21 (c) is occupied the electron distribution will be uneven across the molecule, simply because the orbital (wavefunction) is greater in the region of atom 2 (recall that the electron density is proportional to

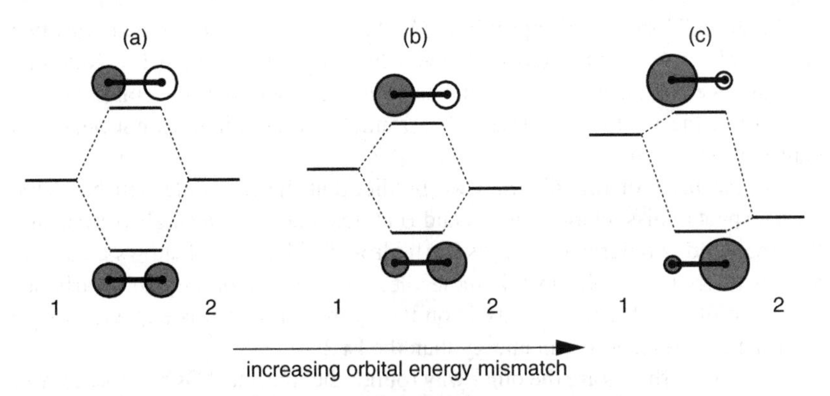

Fig. 5.21 Illustration of the effect on the MOs of increasing the energy mismatch between the two interacting AOs. In (a) the two AOs have identical energies, and so contribute equally to the bonding and anti-bonding MOs; this is shown in the cartoons of these orbitals where the MO is drawn such that the size of the constituent AO indicates its relative contribution. If the energy of the AO on atom 2 is lower than that of atom 1, as shown in (b), the AO on atom 2 is a more significant contributor to the bonding MO whereas the AO on atom 1 is the more significant contributor to the anti-bonding MO. As the energy mismatch between the AOs becomes greater, the effect is more pronounced, as is shown in (c).

the square of the wavefunction). As a consequence, the bond will be *polarized* towards atom 2, with a partial negative charge on atom 2 and a partial positive charge on atom 1. The bond will have a *dipole moment*, as shown in Fig. 5.22.

It is clear from this discussion that the relative energies of the AOs are important as this determines which will overlap most effectively and the polarity of any resulting bonds. In Section 4.8 on p. 57 it was described how the energies of the AOs decrease as we go across the first row of the Periodic Table. We can therefore use this observation to work out which of any two atoms has the lowest energy orbitals and hence draw up an appropriate MO diagram.

In Fig. 5.22 the AO from atom 2 is lower in energy and so the bond is polarized towards this atom. If we were describing this polarization in terms of electronegativities we would say that atom 2 is more electronegative than atom 1 *because* the bond is polarized towards atom 2. We can now see that the reason for the greater electronegativity of atom 2 is that its AOs are lower in energy than those of atom 1. This connection between orbital energies and electronegativity has been discussed earlier on p. 58.

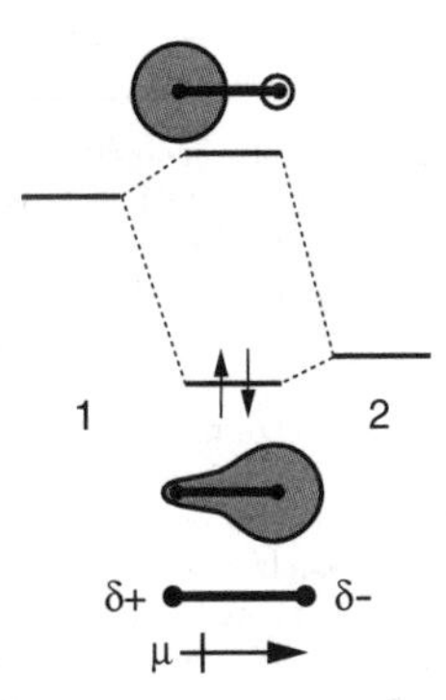

Fig. 5.22 The occupation of an MO which has unequal contributions from the two AOs leads to a polarized bond. Here, the bonding MO is closest in energy to the AO from atom 2 and so this AO is the major contributor to the MO. As a result, the electron density is skewed towards atom 2, which acquires a partial negative charge. The bond is therefore polarized and has a dipole moment.

Two simple examples: LiH and HF

The molecules LiH and HF provide a good illustration of the concepts introduced in the previous section. They also illustrate two extremes, as in LiH the H has the lower energy AOs (see Fig. 4.34 on p. 57) and so the bond is polarized towards hydrogen, whereas in HF it is the fluorine which has the lower energy AOs resulting in a bond which is polarized the other way round.

In LiH the valence orbitals are the 2*s* on lithium and the 1*s* on hydrogen. In Section 4.8 on p. 57 we saw that the orbital energy of the 1*s* in hydrogen is considerably lower than that of the 2*s* in lithium, so in constructing our MO diagram we must take this into account.

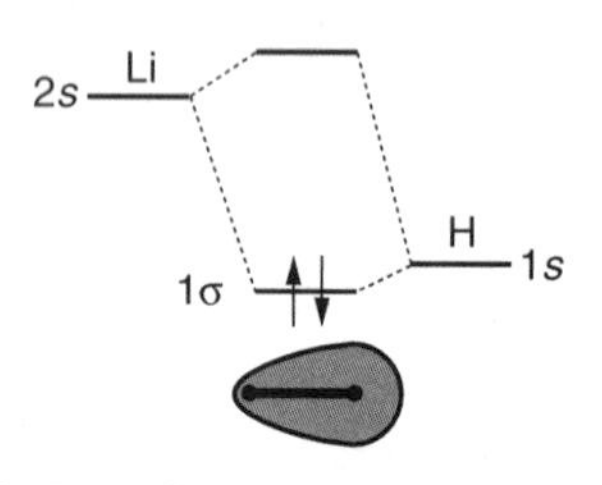

Fig. 5.23 MO diagram for LiH; note that the 1*s* orbital on H is lower in energy than the 2*s* orbital on Li (see Fig. 4.34 on p. 57). There are two valence electrons and these occupy the 1σ bonding orbital; the result is a single bond, polarized towards the hydrogen.

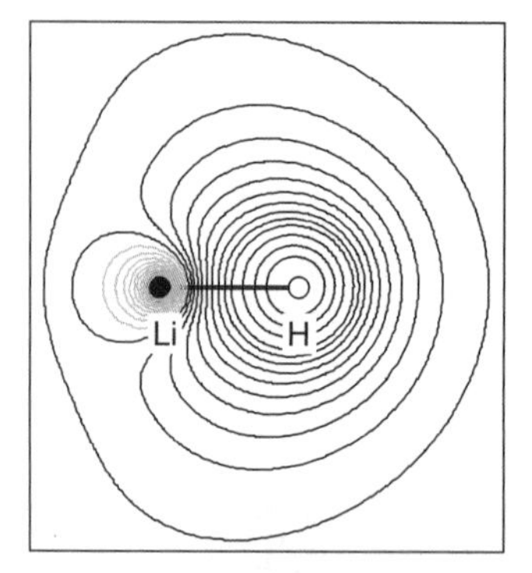

Fig. 5.24 Computer generated contour plot of the σ bonding MO of LiH which corresponds to the 2σ MO shown schematically in Fig. 5.23. Positive and negative contours are shown in black and grey, respectively; the atoms and Li–H bond are also shown. The Li AO involved in this orbital is the 2*s* which has a radial node close-in to the nucleus; this can clearly be seen by the change in colour of the contours around the Li. As expected, the electron density for this MO is mainly on the hydrogen on account of the lower energy of the hydrogen AO.

Figure 5.23 shows an approximate MO diagram for LiH; there are just two valence electrons and these occupy the σ bonding MO, labelled 1σ. Note that although we can classify this orbital as σ (according to the rules on p. 65) we cannot use the *g* and *u* labels as a heteronuclear diatomic does not possess a centre of inversion.

On the basis of this diagram we predict that the molecule will be stable with respect to dissociation into Li and H atoms, that it has a single σ bond and that the bond is polarized towards the hydrogen. Figure 5.24 shows a contour plot of the bonding MO in this molecule. From this plot we can clearly see that most of the electron density is on the hydrogen, which is expected as the hydrogen 1*s* AO is lower in energy than the Li 2*s*.

In HF the orbitals are the other way round: the fluorine AOs are much lower in energy that those of hydrogen (see Fig. 4.34 on p. 57). We also know that the 2*s* lies at lower energy than the 2*p*, so to a first approximation we will ignore the 2*s* and just assume that the only overlap is between the 2*p* on fluorine and the 1*s* on hydrogen.

There are, of course, three 2*p* orbitals, but as we saw in Fig. 5.10 on p. 68 only the $2p_z$ orbital (the one which points along the bond) has the correct symmetry to overlap with the *s* orbital, to give a σ MO. The major contributor to this σ bonding orbital (labelled 2σ) is the fluorine $2p_z$ as this is closest to it in energy. The other 2*p* orbitals have no interaction with the 1*s* and so remain non-bonding. The resulting approximate MO diagram is shown in Fig. 5.25.

HF has a total of eight valence electrons: two are in the fluorine 2*s* (labelled 1σ in Fig. 5.25) and the remainder occupy the 2σ bonding MO and the two non-bonding fluorine AOs, $2p_x$ and $2p_y$. We predict, therefore, that there will be a single σ bond between the H and F, and that there is a concentration of electron density towards the fluorine in this bond. In addition, there are further electrons localized on fluorine in the 2*s*, $2p_x$ and $2p_y$ AOs. The electron density in this molecule is therefore very biased toward the fluorine. We can attribute

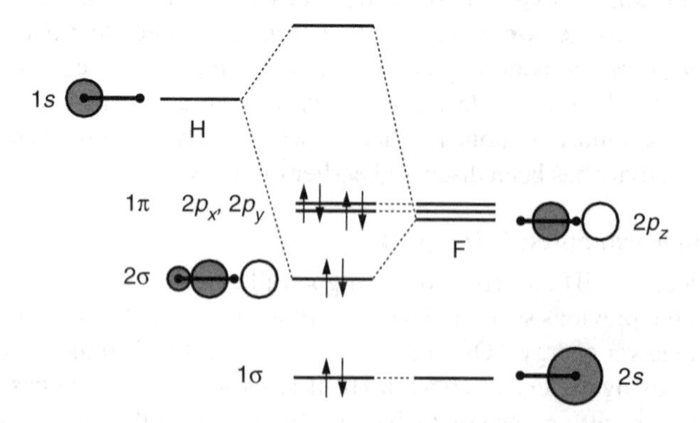

Fig. 5.25 Approximate MO diagram for HF assuming that only the fluorine 2*p* and hydrogen 1*s* AOs are involved in bonding. The fluorine AOs are much lower in energy than the hydrogen 1*s* AO (see Fig. 4.34 on p. 57). Only the $2p_z$ orbital has the correct symmetry to overlap with the hydrogen 1*s*, giving rise to the 2σ orbital which is polarized towards fluorine. The fluorine 2*s* is assumed to be too low in energy to overlap with the 1*s* and so remains non-bonding. The MOs labelled 1σ and 1π are really the 2*s*, $2p_x$ and $2p_y$ fluorine AOs, but the labels are included to facilitate comparison with Fig. 5.26.

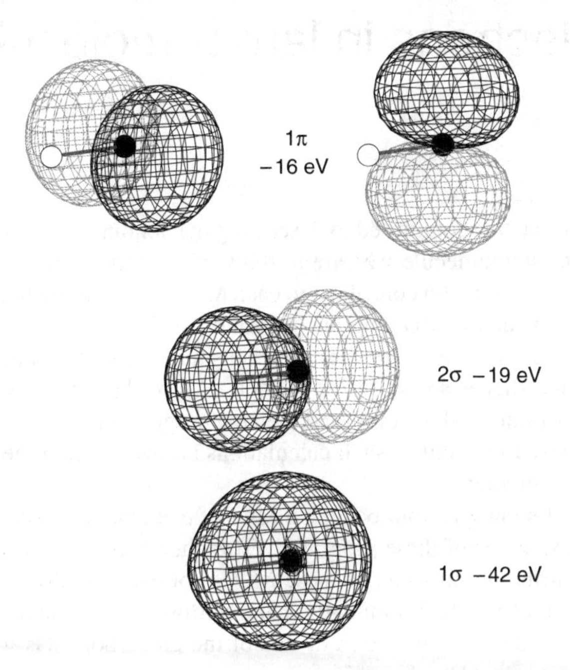

Fig. 5.26 Surface plots of the four highest energy occupied MOs of HF; the MOs have been calculated using a computer program. The surface is shown as a 'net' so that we can see through it to the balls which represent the atoms; the line joining them is drawn in to help identify the bond. The fluorine is shown in black and the hydrogen, which is coming toward us, in white. The energy of each MO is also given. These MOs should be compared with the approximate picture shown in Fig. 5.25.

this imbalance to the fact that the fluorine orbitals are much lower in energy than the hydrogen AOs and that there are three non-bonding electron pairs on the fluorine.

It is interesting to compare our cartoons of the MOs with the computer calculated MOs shown in Fig. 5.26. In the figure, we recognize the MO labelled 1σ as arising from the overlap of the fluorine $2s$ with the hydrogen $1s$. In drawing up the simple MO diagram we discounted this interaction on the grounds of the large energy separation between these two AOs, but the more complete calculation shows that they do have some interaction. Note how much lower in energy the 1σ MO is than all the other MOs; this reflects the low energy of the fluorine $2s$.

The next highest MO, 2σ, is exactly what we expected; it is formed from the overlap of a $2p$ orbital on fluorine with the hydrogen $1s$. At only slightly higher energies are the two degenerate 1π MOs, which can clearly be identified as the two $2p$ orbitals from fluorine which remain non-bonding.

Moving on

So far we have used the molecular orbital approach to describe the bonding in diatomic molecules – in the next chapter we will extend this to rather more complex molecules. We will also see how the use of hybrid atomic orbitals greatly simplifies the description of bonding in these larger molecules and enables us to focus on the parts which are really important when it comes to thinking about reactions.

6 Electrons in larger molecules

The MO approach is not limited to describing the bonding in diatomics but can be applied to any molecule we care to think of. In more complex molecules, AOs on several atoms can contribute to each MO, and it rapidly becomes rather difficult to draw up the MO diagram 'by hand'.

Luckily for us, many computer programs have been developed for calculating both the shapes and energies of MOs. All we have to do is specify the types and positions of the atoms, and then the program does the rest. For small to medium sized molecules, such calculations are well within the capabilities of desk-top computers.

Figure 6.1 shows the four occupied MOs of methane (CH_4) computed (and displayed) using one of these programs. From their form, we can guess which AOs are contributing to some of these MOs. For example, MO (a) is clearly a combination of the $2s$ AO on carbon and all four $1s$ AOs on the hydrogens; MO (b) looks like a combination of one of the $2p$ carbon AOs with just two of the hydrogen $1s$ AOs. The thing to notice is that *several* AOs contribute to each MO.

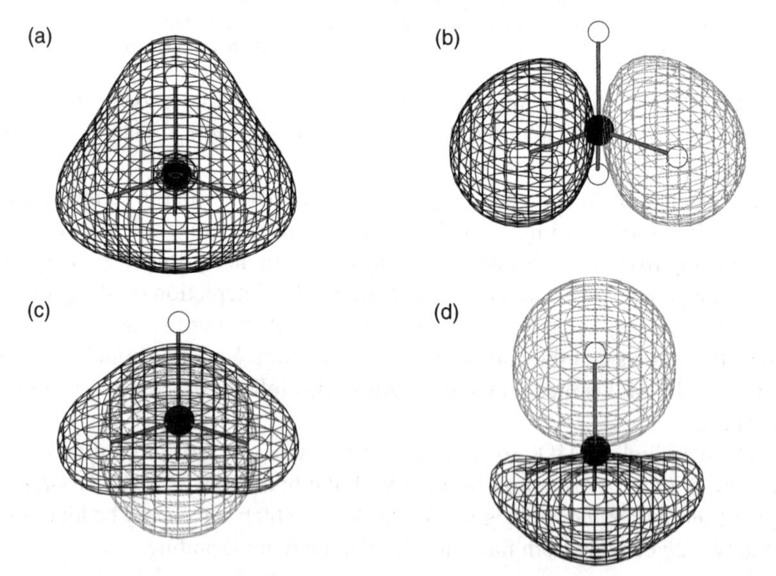

Fig. 6.1 Surface plots of the four occupied MOs of methane. The surface is shown as a transparent 'net' so that we can see through it to the model of the molecule which shows the positions of the atoms and bonds; the carbon atom is shown in black and the hydrogen atoms in white. Positive parts of the orbital are shown by a black net and negative by a grey net. The lowest energy MO is (a); (b), (c) and (d) are all degenerate.

Energy minimization

Experimentally, it is found that methane has a tetrahedral geometry with a bond angle of 109.5° and a C–H bond length of 1.089 Å; this geometry must therefore be the lowest energy arrangement of one carbon and four hydrogen atoms.

1 Å (Ångström) is 10^{-10} m or 100 pm.

The computer programs used to calculate MOs usually have an option for finding this lowest energy arrangement (or *equilibrium geometry*, as it is often called) for a given molecule. All that the program has to do is to compute the MOs for a given arrangement of atoms, assign the electrons to the MOs and hence find the total energy of the molecule. Then the atoms are moved around and the calculation repeated until the minimum energy arrangement is found – a process called *energy minimization*.

Computers are very good at this kind of repetitive calculation and clever procedures have been developed to enable the program to locate the energy minimum in an efficient way. So, even if we do not know the geometry of our molecule, all we need to do is to make a reasonable initial guess and then ask the program to determine the equilibrium geometry.

Working from what we already know

Sophisticated and convenient though these computer programs are, we really need to describe the bonding in a simpler way which we can use on a day to day basis without resorting to computer programs. The key point which helps us here is that for most molecules we already have a pretty good idea of their shape, so we do not need to work out the equilibrium geometry – all we need to do is find a description of the bonding for a molecule of known shape.

For example, we know that the four bonds around a saturated carbon are arranged in a tetrahedral geometry, that ethene (C_2H_4) is flat with bond angles of 120° and that the bond angle in water is 104.5°. We also know that N_2 has a triple bond, and that there is a double bond joining together the two carbons in ethene. What we need is a simple 'back of the envelope' method for determining the MOs in molecules like these whose geometry we already know. We can then use these MOs to think about the reactions of the molecules, which is ultimately what we are aiming at.

6.1 Two-centre, two-electron bonds

When we first learn about the formation of covalent bonds it is common to use 'dot and cross' diagrams such as those for methane and water shown in Fig. 6.2. The diagram for methane shows how the four electrons from carbon are each paired up with one electron from hydrogen, resulting in four *bonding pairs* of electrons. As their name implies, each pair is associated with a bond between carbon and hydrogen which we represent by a line joining the two atoms. Such a bond is called a *two-centre, two-electron* (2c-2e) bond, as it involves two atoms and two electrons. For water, there are just two bonded pairs, and hence two 2c-2e bonds. The remaining four electrons form two *lone* or *non-bonding* pairs on the oxygen.

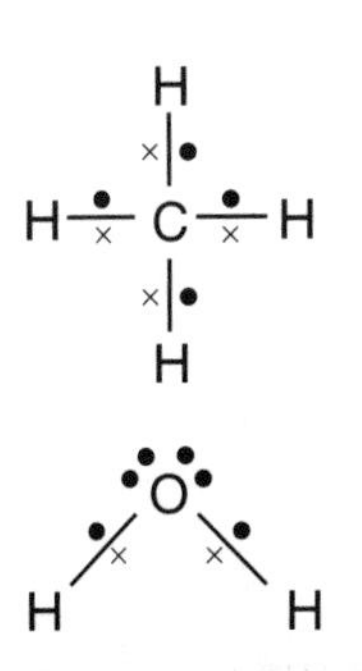

Fig. 6.2 'Dot and cross' pictures of the bonding in methane and water. The dots indicate the electrons from carbon (or oxygen) and the crosses indicate electrons from hydrogen. The lines indicate the presence of two-centre two-electron bonds; water also has two lone pairs of electrons. Methane is, of course, tetrahedral rather than flat.

Our problem is how to reconcile this simple (but very useful) picture of the bonding in methane with the MO description. The key difference between the two models is that the MO picture is a *delocalized* one in the sense that the MOs (and hence the electrons) are spread over several atoms, whereas the picture in Fig. 6.2 is a *localized* one in which electrons are either localized between two atoms forming 2c-2e bonds or are localized on the atoms as lone pairs.

In the MO picture, two orbitals overlap to form a bonding MO and an anti-bonding MO, and if there are just two electrons in the bonding MO then we say that there is a bond between the two atoms. This overlap of just two orbitals is essentially the MO picture of a 2c-2e bond as shown in Fig. 6.3.

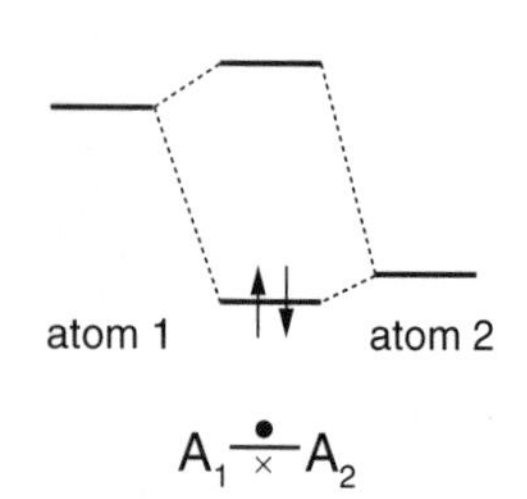

Fig. 6.3 The overlap of one orbital from atom 1 with one from atom 2 gives a bonding MO and an anti-bonding MO. Occupation of the bonding MO by two electrons gives the MO description of a 2c-2e bond between atoms 1 and 2.

Suppose that we try to apply this MO picture of 2c-2e bonds to methane. We have four valence orbitals on carbon (the 2*s* and the three 2*p*) so each could overlap with one of the hydrogen 1*s* orbitals. This would give us four 2c-2e bonds each of which can be described by an MO diagram just as in Fig. 6.3.

The problem with this approach is that the carbon AOs do not point in the right directions to make good overlap with the hydrogen AOs. Remember that the geometry of the molecule is tetrahedral, but that the three 2*p* orbitals point along the *x*-, *y*- and *z*-axes; the 2*s* orbital is spherical so does not 'point' in any particular direction at all. With such a set of orbitals it is hard to see how we can sensibly form four 2c-2e bonds.

The way out of this problem is to combine the carbon AOs into new orbitals called *hybrid atomic orbitals* (HAOs). These hybrids are designed to point in the appropriate directions to overlap with the hydrogen AOs. Each HAO overlaps with one hydrogen 1*s* AO to give a bonding and an anti-bonding MO; two electrons occupy the bonding MO thus giving a 2c-2e bond.

How we go about constructing these HAOs is the subject of the next section, but before we move on to that we need to give a few words of caution about how we represent molecules.

Not all lines are bonds

In Fig. 6.2 the lines joining the atoms certainly imply the presence of 2c-2e bonds, but we have to be careful, as when we draw lines on chemical structures we do not always mean to imply that a 2c-2e bond is present. Consider, for example, the two structures shown in Fig. 6.4.

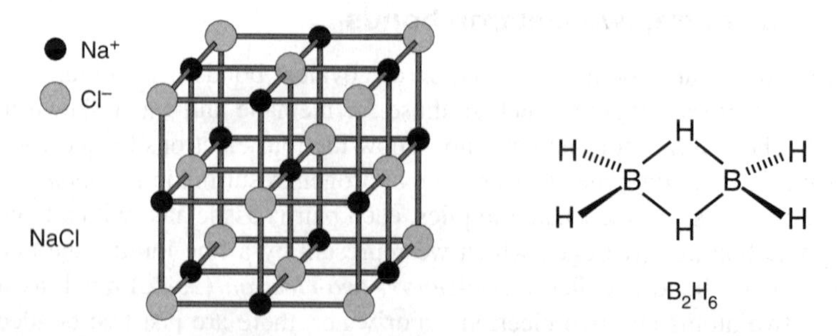

Fig. 6.4 On the left is shown part of the lattice of the ionic solid NaCl and on the right is shown the structure of gaseous diborane (B_2H_6). In diborane the two central hydrogens and the two borons lie in the same plane, with the two hydrogens joined by the dashed lines lying below this plane and those joined by the wedges lying above it. For both structures, the lines joining the atoms are *not* 2c-2e bonds.

Solid sodium chloride is a giant lattice structure consisting of Na^+ and Cl^- ions. We often draw it, as in Fig. 6.4, with lines joining the ions together; however these lines *do not* represent 2c-2e bonds. Rather, the lines are just there to guide our eye and indicate the spatial relationship between the ions. In fact, as we discussed in Chapter 3, it is electrostatic forces between all of the ions (not just adjacent ones) which are responsible for the binding energy of this lattice.

Diborane, B_2H_6, provides another example; this molecule has 12 valence electrons, yet the structure shown in Fig. 6.4 has eight lines and if each represented a 2c-2e bond we would be four electrons short. In fact, as with NaCl, the lines are there just to help us understand the spatial arrangement of the atoms – they do not indicate 2c-2e bonds.

So, we have to be careful: not every line drawn on a chemical structure represents a 2c-2e bond. However, for many simple compounds, especially those involving carbon and other first row elements, it is often the case that the lines do represent 2c-2e bonds. It is to the description of the bonding in such compounds that we now turn our attention.

6.2 Hybrid atomic orbitals

Recall that what we are trying to do is to find an orbital description of the bonding in methane which is consistent with the view that there are four 2c-2e bonds. The problem we have is that the carbon AOs do not 'point' in the right directions to overlap with the hydrogen AOs; we will overcome this by combining the carbon AOs to form hybrid atomic orbitals, HAOs, which do point towards the hydrogens.

HAOs are formed by making a linear combination of AOs just as we do when we form MOs. However, the important distinction is that to form HAOs we combine AOs *on the same atom*. This is in contrast to MOs which are formed by combining AOs on *different* atoms.

Each HAO is constructed by adding together AOs:

$$\mathrm{HAO} = c_1 \times \mathrm{AO}_1 + c_2 \times \mathrm{AO}_2 + c_3 \times \mathrm{AO}_3 + \ \ldots$$

where c_1, c_2, etc. are simply numerical coefficients which determine the directional properties of the HAO; AO_1 is the first AO and so on. As with MOs, the number of HAOs is equal to the number of AOs being combined. So, for example, if we combine the $2s$ and the three $2p$ AOs we will obtain four HAOs.

The shapes and energies of the HAOs are determined by the values of the coefficients $c_1, c_2, \ldots$. Generally, we will not be able to guess at what these coefficients are so instead we will look at a few particular cases where the resulting HAOs are useful for describing the bonding in commonly encountered molecules.

sp^3 hybrids

If we take the $2s$ and the three $2p$ AOs we can form a set of four equivalent HAOs which point towards the corners of a tetrahedron; these HAOs, depicted in Fig. 6.5, are called sp^3 hybrids; the name indicates which AOs were used to

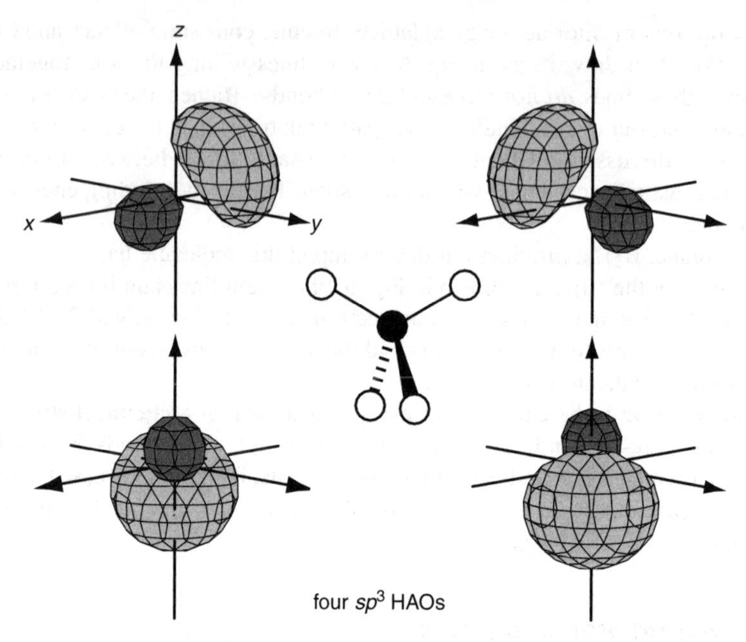

Fig. 6.5 Surface plots of the four sp^3 HAOs; each hybrid points toward one of the corners of a tetrahedron. The conventional representation of a tetrahedral carbon is also shown to indicate the directions in which the HAOs point.

construct the hybrids. Often it is said that HAOs are 'more directional' than AOs. We can understand what this means by thinking about the directional properties of the AOs which are used to construct these HAOs: the $2s$ does not point in any particular direction and each $2p$ orbital points in two directions (e.g. along x and $-x$ for the $2p_x$). In contrast, as we can see from Fig. 6.5, the major lobe of each of the hybrids points in just one direction.

Using these HAOs, the description of the bonding in methane is straightforward and is shown in Fig. 6.6. Each sp^3 HAO overlaps with *one* of the hydrogen $1s$ orbitals, forming a bonding MO and an anti-bonding MO; two

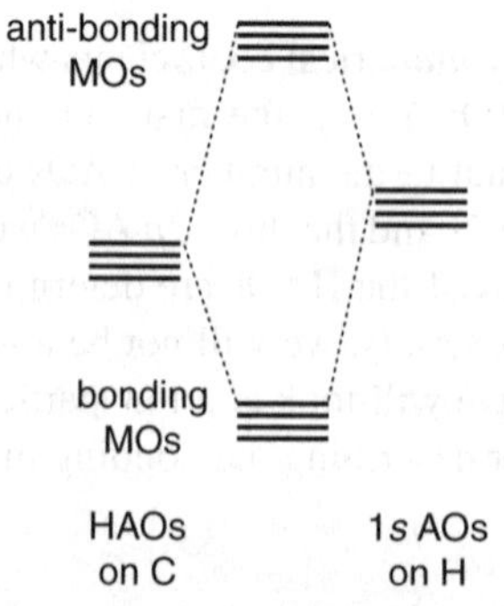

Fig. 6.6 MO diagram for methane. On the left are shown the four equivalent sp^3 HAOs on carbon, and on the right are the four $1s$ orbitals on hydrogen. One HAO overlaps with one hydrogen orbital to give a bonding and an anti-bonding MO; two electrons occupy the bonding MO, giving a 2c-2e bond. The other HAOs overlap, one-on-one, with the other hydrogen orbitals; it is important to understand that each HAO only overlaps with a *single* hydrogen orbital.

electrons are placed in each bonding MO thus generating four 2c-2e bonds and using up all eight valence electrons. The use of HAOs has allowed us to develop a simple MO description of the bonding in methane which matches up with the 2c-2e description.

Just as the energies of MOs are different from those of the AOs from which they are formed, the energies of the HAOs are different from the AOs which form them. The energy of the four equivalent sp^3 HAOs lies between that of the 2s and 2p but closer to 2p than 2s, as shown in Fig. 6.7.

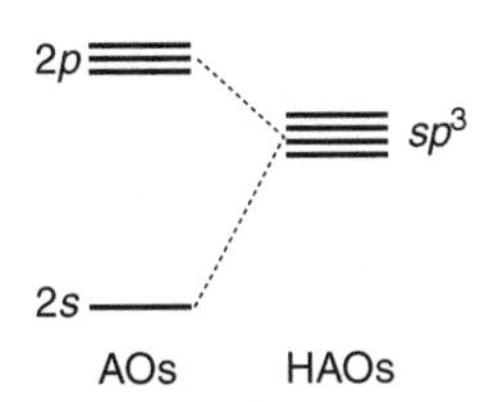

Fig. 6.7 Illustration of the relative energies of the sp^3 HAOs and the AOs from which they are formed. Note that the three 2p orbitals are degenerate, but are shown slightly separated on this diagram; the same is true of the four sp^3 HAOs.

Ammonia and water

The sp^3 hybrids can also be used to give an approximate description of the bonding in ammonia (NH_3) and water (H_2O). Ammonia is best described as being *trigonal pyramidal*, a shape which we can imagine as being derived from tetrahedral methane simply by 'removing' one of the hydrogens, as shown in Fig. 6.8. This gives us the clue as to how to describe the bonding in ammonia: we form four sp^3 hybrids on the nitrogen and let three of these overlap with hydrogen 1s AOs to form three 2c-2e bonds. Ammonia has eight valence electrons, so we assign the final pair not used for the N–H bonds to the fourth sp^3 hybrid, classifying it as a lone pair.

Of course there is a problem with this description, which is that the H–N–H bond angle in ammonia is 106.7° – not the same as the tetrahedral angle of 109.5° between the sp^3 hybrids. The difference is small, however, so our description is not far from the truth and we will see later on how to refine the model to produce the correct bond angle.

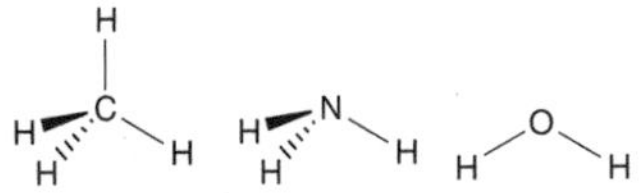

The same approach can be used for water; we imagine its structure as being derived from methane with two of the hydrogens removed. As before, we form four sp^3 hybrids on the oxygen, and overlap two of these with the hydrogens to form two 2c-2e bonds. The remaining four electrons are assigned to the other two sp^3 hybrids, forming two lone pairs. Once again, the experimental bond angle of 104.5° does not quite match the angle between the sp^3 hybrids, but it is pretty close.

Fig. 6.8 The structures of ammonia and water can be thought of as being based on the tetrahedral shape of methane with one and two of the bonds to hydrogen being removed, respectively.

These sp^3 hybrids are not only useful for describing the bonding in methane, but also give us a start in describing the bonding in other molecules. In these we not only use the hybrids to form bonds but also use them to accommodate lone pairs. We will now turn to how other hybrids can be formed which point at different angles to the tetrahedral arrangement of sp^3 hybrids.

Other kinds of hybrid atomic orbitals

The molecule BH_3 has a trigonal planar shape in which the three hydrogens lie in a plane and at the corners of an equilateral triangle with the boron at its centre (Fig. 6.9). Clearly the sp^3 hybrids are not going to be useful in describing the bonding in this molecule – in fact what we need are the three equivalent sp^2 hybrids which are formed from the 2s and *two* of the 2p AOs. These hybrids are depicted in Fig. 6.10 along with the remaining $2p_z$ orbital which is not involved in forming the hybrids (the choice of the remaining orbital as pointing along z is arbitrary).

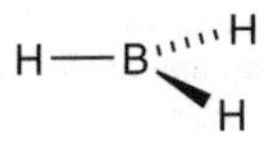

Fig. 6.9 BH_3 is a trigonal planar molecule; all the atoms lie in a plane, with the hydrogens at the corners of an equilateral triangle.

Each sp^2 hybrid overlaps with a hydrogen 1s AO to form a bonding and an anti-bonding MO; a pair of electrons is assigned to each bonding MO, creating

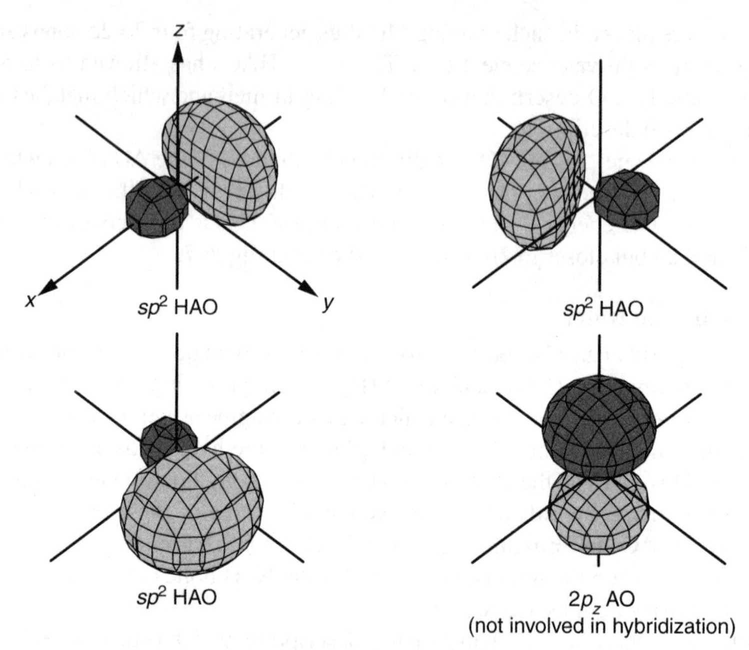

Fig. 6.10 Surface plots of the three equivalent sp^2 hybrids formed from the 2*s*, $2p_x$ and $2p_y$ AOs; also shown is the $2p_z$ AO which is not involved in the hybridization. The three hybrids point at 120° to one another and all lie in the *xy*-plane.

three 2c-2e bonds. BH_3 has six valence electrons, so there are just enough for these three 2c-2e bonds. The remaining out-of-plane $2p_z$ orbital is not occupied, but we will see later on that this orbital plays an important role in the chemical reactions of BH_3. The complete MO diagram is shown in Fig. 6.11.

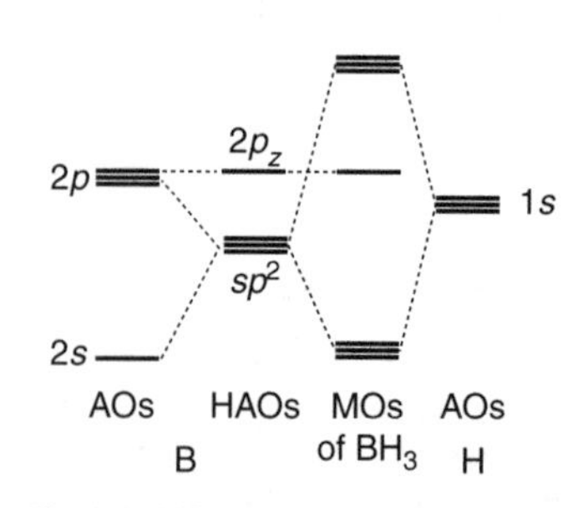

Fig. 6.11 MO diagram for BH_3; on the left are shown the AOs of boron combining to give three sp^2 hybrids and leaving the $2p_z$ orbital unhybridized. The three HAOs overlap one-on-one with the three hydrogen 1*s* AOs shown on the right. The molecule has six valence electrons which occupy the three bonding MOs; the $2p_z$ is empty.

We can also form hybrids of 2*s* with just *one* 2*p* (arbitrarily chosen to be the $2p_z$) to give two equivalent *sp* hybrids; these are depicted in Fig. 6.12 along with the two unhybridized 2*p* orbitals. The *sp* hybrids point at 180° to one another.

A pattern now emerges: the angle between the sp^3 hybrids is 109.5°, between the sp^2 hybrids it is 120° and between the *sp* hybrids it is 180°. The greater the proportion of *s*-character in the HAOs the *larger* the angle between them. Put the other way, the greater the *p*-character the *smaller* the angle between the hybrids. So, by altering the ratio of *s* to *p* in our HAOs we can alter the angle between them.

When we combined the 2*s* with the three 2*p* AOs we described the resulting sp^3 hybrids as *equivalent*, which means that each has the same ratio of *s*- and *p*-character; this is why they all have the same angle between them. However, if we increase the *s*-character in one of the hybrids and proportionately decrease the amount of *s*-character in the other three, we expect to *decrease* the angle between the orbitals which have lower *s*-character (and hence greater *p*-character). There are still four HAOs, but they are no longer equivalent.

This set of inequivalent hybrids is just what we need to describe the bonding in ammonia. The three hybrids with increased *p*-character are used to form the bonds to the hydrogens and the fourth hybrid (the one with greater *s*-character) accommodates the lone pair. Similarly, to describe the bonding in

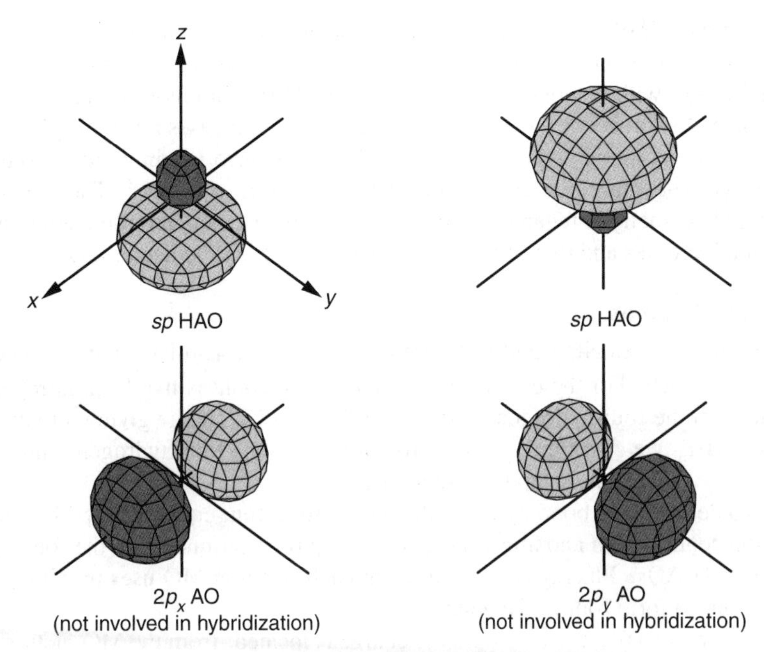

Fig. 6.12 Surface plots of the two equivalent *sp* hybrids formed from the 2*s* and $2p_z$ AOs; also shown are the $2p_x$ and $2p_y$ AOs which are not involved in hybridization. The two hybrids point at 180° to one another, and the unhybridized 2*p* orbitals point in perpendicular directions to the two hybrids.

water, we increase the *p*-character in two of the hybrids until the angle between them is the required 104.5° and use these to form the 2c-2e bonds. The remaining two hybrids (which have increased *s*-character) are used to accommodate the two lone pairs.

The idea that the proportion of *s* and *p* AOs in the hybrid atomic orbitals can be varied in order to account for different bond angles is nicely illustrated by the inversion of ammonia.

The inversion of ammonia

Although the equilibrium geometry of ammonia is trigonal pyramidal it turns out that this molecule constantly undergoes a process called *inversion* by which it is pulled 'inside out' like an umbrella on a windy day (Fig. 6.13). The molecule starts in its lowest energy geometry (a trigonal pyramid) and then the NH bond angles start to open out until the molecule becomes planar. One way of describing what has happened is to say that the three hydrogens have moved from a plane below the nitrogen into a plane containing the nitrogen. The process then carries on with the three hydrogens moving to a plane above the nitrogen as the bond angles return to their equilibrium values. We regain the original geometry but with the hydrogens on the other side of the molecule compared to where they started from.

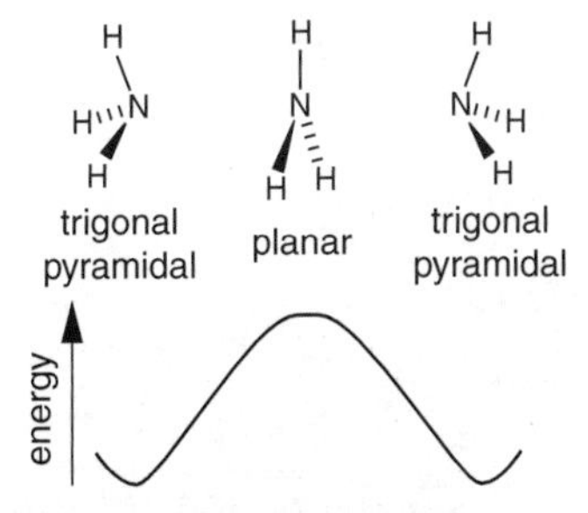

Fig. 6.13 Ammonia undergoes an inversion in which it turns 'inside out' by passing from the equilibrium trigonal pyramidal geometry to a planar geometry and then continuing to a trigonal pyramidal geometry which is the mirror image of the original. The energy is a maximum at the planar geometry and at this point is 24 kJ mol^{-1} above the energy at the equilibrium geometry.

We have already described the bonding in ammonia as being based on sp^3 hybridization, but with the lone pair occupying a hybrid of greater *s*-character. The planar form of NH_3 can best be described as having sp^2 hybridization at the nitrogen. These three hybrids overlap with the three hydrogen 1*s* orbitals, and a total of six electrons are assigned to the resulting bonding MOs. The remaining pair of electrons is assigned to the out-of-plane $2p_z$ AO.

So, as the inversion proceeds we have an interesting progression of bonding. To start with the lone pair is in an sp^3 type HAO with more s-character; as the inversion proceeds the p-character of this HAO increases reaching 100% at the planar geometry. At the same time, the HAOs involved in bonding to the hydrogens decrease in p-character as they move from sp^3 to sp^2. Then, as the trigonal geometry is regained on the other side, the s-character of the lone-pair orbital increases and that of the HAOs involved in bonding decreases.

Tetrahedral ions

The ionic species NH_4^+ and BH_4^- are both tetrahedral and have eight valence electrons each. For the case of NH_4^+ the electron count is five from nitrogen, four from the four hydrogens and -1 for the positive charge giving a total of 8. For BH_4^- we count three from boron, four from the four hydrogens and $+1$ for the negative charge, giving a total of eight.

To describe the bonding for both ions we form four equivalent sp^3 hybrids on the central atom and allow these to overlap (one-on-one) with the four hydrogen $1s$ AOs. Placing two electrons in each bonding MO uses up all eight electrons to form four 2c-2e bonds.

We might ask on which atom the charge is located: from the MO picture it is clear that the charge cannot be associated with any one atom, but is shared around all of the bonds. This is an important point which we will return to when considering the reactions of these two species.

6.3 Using hybrid atomic orbitals to describe diatomics

Hybrid atomic orbitals can be used to describe the bonding in diatomics and sometimes this results in a more convenient description than the full MO treatment which we gave in Chapter 5. We will describe just one example: N_2.

We start out by supposing that the nitrogen is sp hybridized, with the hybrids pointing along the direction of the internuclear axis (z). The two sp HAOs which point towards one another overlap to form σ bonding and anti-bonding MOs; we place two electrons in the bonding MO. The other two HAOs which point away from the bond are occupied by two lone pairs.

There are two sets of out-of-plane p orbitals: the two $2p_x$ orbitals overlap to give a π bonding MO and a corresponding anti-bonding MO; the two $2p_y$ AOs overlap in the same way. Two electrons are placed in each of the π bonding MOs, using up all 10 valence electrons.

Our description of the bonding in N_2 is that there is a σ bond and two π bonds between the nitrogen atoms; in addition each atom has a lone pair, which is directed away from the internuclear axis. The picture is summarized in Fig. 6.14.

In fact, this picture is not too dissimilar to the MO description of the bonding in N_2 given on p. 74 (see the MO diagram of Fig. 5.18 (b) on p. 72). In the MO picture the s-p mixing causes the $1\sigma_u$ MO to become less anti-bonding and the $2\sigma_g$ orbital to become less bonding; it turns out that occupation of these orbitals contributes rather little to the net bonding so they can be described as

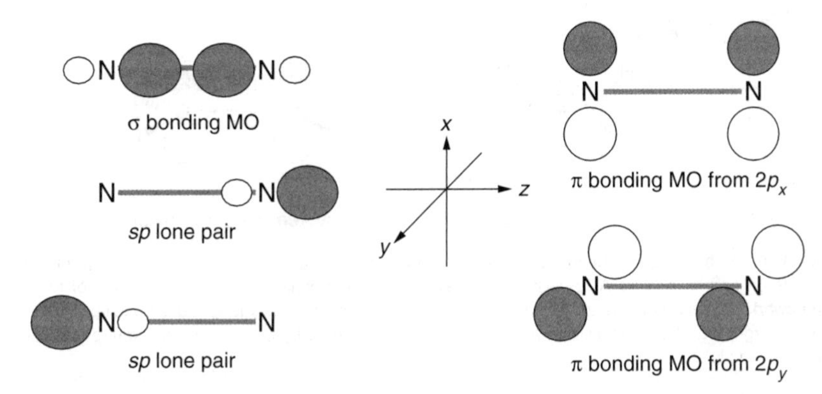

Fig. 6.14 Cartoons showing a description of the bonding in N_2 using *sp* hybrids. Two of the hybrids overlap to form a σ bonding MO and the remaining two hybrids become lone pairs. A π bonding MO is formed from the overlap of the two $2p_x$ AOs, and similarly another π bonding MO is formed from the overlap of two $2p_y$ AOs. In these terms, the description of N_2 is of a σ bond, two π bonds and two lone pairs. Note that for the σ overlap of the two *sp* hybrids the lobes of the HAOs are shown with the signs that will lead to the bonding MO; do not forget, though, that an anti-bonding MO will also be formed. The same applies to the π overlap which gives a bonding and an anti-bonding MO.

approximately non-bonding. It is the occupation of the $1\sigma_g$ MO (two electrons) and $2\pi_u$ MOs (four electrons) which is responsible for most of the bonding in N_2; in other words there is a σ and two π bonds just as in the description given in Fig. 6.14.

6.4 Simple organic molecules

Hybrid atomic orbitals are a very convenient way of describing the bonding in simple organic molecules so in the following chapters we will use this approach very often. Generally, it is clear from the structure what kind of hybrids we should use for a particular carbon: if the coordination is tetrahedral we need sp^3 hybrids, if the carbon is doubly bonded we need sp^2 hybrids and if it is triply bonded we need *sp* hybrids. For other atoms, such as nitrogen and oxygen, the choice is sometimes less clear and we will discuss this later on in this section. First, however, we will look at the description of the bonding in three simple hydrocarbons.

Ethane, ethene and ethyne

In ethane, C_2H_6, both carbons are tetrahedral and so it is clear that we should use sp^3 hybrids to describe the bonding. Of the four hybrids on each carbon, one overlaps with a hybrid on the other carbon and the other three overlap with hydrogen 1*s* orbitals. Each overlap is just between two orbitals and results in a σ bonding and a σ anti-bonding MO. We put two electrons in each bonding MO thus forming a 2c-2e bond; the overall picture is summarized in Fig. 6.15. There are 14 valence electrons in ethane (four from each carbon and six from the hydrogens), and these fill all of the σ bonding MOs.

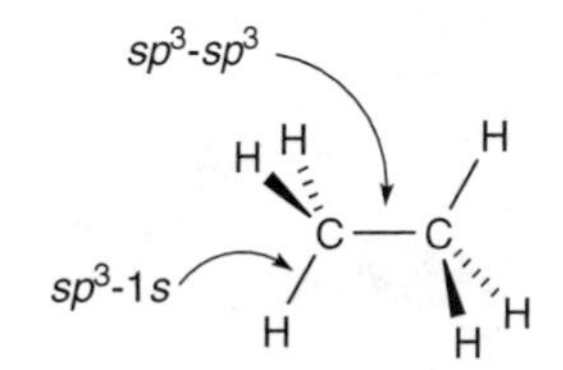

Fig. 6.15 Schematic structure of ethane in which each line represents a 2c-2e bond. Each bond can be described as being the result of the overlap between either two sp^3 hybrids (for the C–C bond) or an sp^3 hybrid and a 1*s* AO on hydrogen (for the C–H bonds).

Ethene, C_2H_4, is flat with 120° H–C–H and H–C–C bond angles; the choice of sp^2 hybrids for the carbons is therefore clear. The bonding in this molecule separates into two parts: firstly, a σ *framework* which involves the sp^2 hybrids

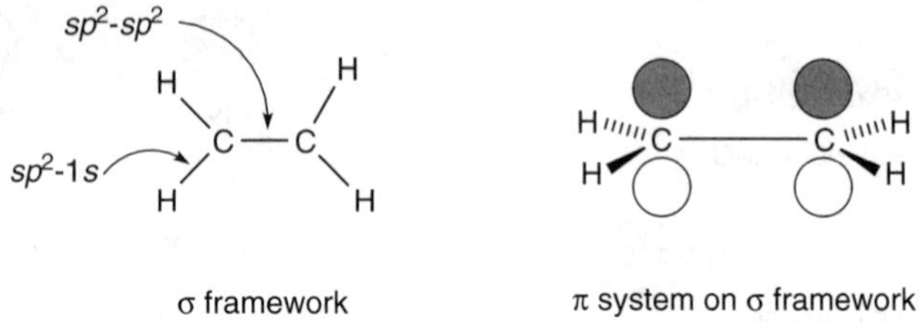

σ framework π system on σ framework

Fig. 6.16 The bonding in ethene can be separated into a σ framework, involving the overlap of sp^2 hybrids on carbon and $1s$ AOs on hydrogen, and a π system which lies out of the plane of the σ framework. The out-of-plane $2p$ orbitals are shown with the signs that lead to the π bonding MO, but do not forget that there will also be a π anti-bonding MO formed by the destructive overlap of these two $2p$ orbitals.

and the $1s$ AOs; secondly, the π bonding between the two carbons.

The σ framework is much the same as that for ethane with the single exception that sp^2, rather than sp^3, hybrids are involved. For the C–C bond, two sp^2 hybrids overlap to give σ bonding and anti-bonding MOs; for the C–H bonds the overlap is between an sp^2 hybrid and a hydrogen $1s$ orbital. Filling all of these σ bonding MOs accounts for 10 electrons.

Pointing out of the plane of the molecule are the two $2p_z$ carbon AOs which were not involved in the hybrids (see Fig. 6.10 on p. 86). These two orbitals overlap to give a π bonding and a π anti-bonding MO. Assigning two electrons to the bonding MO accounts, together with those in the σ framework, for all of the 12 valence electrons in ethene (four from each carbon and four from the hydrogens). Figure 6.16 summarizes the picture of the bonding which we have developed: the two carbons are joined by a σ bond and a π bond, forming the *double bond* which we expect for ethene.

This diagram also explains a marked difference between ethane and ethene. Whereas in ethane there is quite a small energy barrier for rotation about the C–C bond, in ethene a great deal of energy is needed to rotate about the double bond – so much that we can safely regard it as being fixed. The easy rotation about the single bond in ethane comes about because the bond is formed by a σ MO in which the orbitals overlap head-on. Rotation about the bond axis makes no difference to the overlap, and so does not affect the bonding.

In contrast, forming the π bond in ethene requires that the two $2p$ AOs lie in the same plane, as shown in Fig. 6.16. Rotating about the C–C axis will result in the loss of this overlap, thus breaking the π bond; this is why rotation about this bond needs so much energy.

Finally, we come to ethyne (acetylene), C_2H_2, which is a linear molecule; *sp* hybridization at the carbons is therefore appropriate, giving the bonding scheme illustrated in Fig. 6.17. As with ethene, we can separate the bonding into a σ framework and a π system. This time each carbon has two out-of-plane $2p$ orbitals (see Fig. 6.12 on p. 87) which overlap in two pairs to give two π bonding MOs and two π anti-bonding MOs. Occupation of all of the σ and π bonding MOs accounts for all 10 valence electrons.

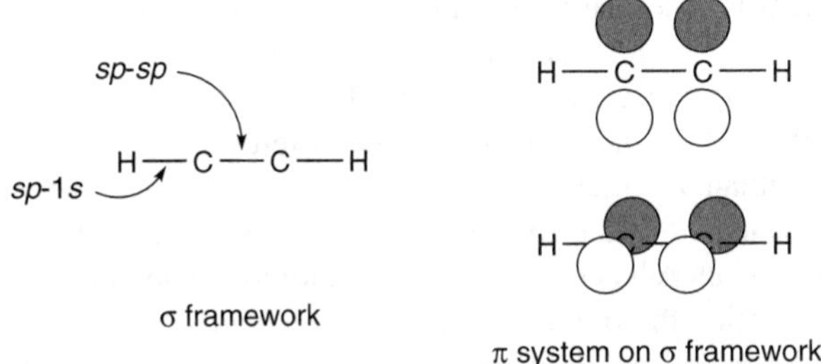

Fig. 6.17 The bonding in ethyne (acetylene) is similar to that in ethene except that the carbons are now *sp* hybridized and there are two sets of π interactions.

Carbonyl compounds

The simplest carbonyl compound is methanal (formaldehyde), CH_2O, shown in Fig. 6.18. The molecule is flat, with bond angles at the carbon of approximately 120°, so the natural choice of hybridization for this atom is sp^2. For the oxygen, the choice is less clear; we know that there is a double bond between the C and the O, so we need to have at least one unhybridized $2p$ orbital on the oxygen which we can use to form a π bond. So the hybridization needs to be sp^2 or sp; for the moment we will choose sp^2.

The σ framework is simple to describe. One of the carbon sp^2 hybrids overlaps with a hydrogen $1s$ to give a σ bonding and anti-bonding MO; two electrons are placed in the bonding MO forming the C–H σ bond. The other C–H bond is formed in the same way. The third carbon sp^2 hybrid overlaps with one of the oxygen sp^2 hybrids to give a bonding and an anti-bonding σ MO; two electrons occupy the bonding MO to form the C–O σ bond.

As the oxygen AOs are lower in energy than those on carbon, the HAOs on oxygen will also be lower in energy than those on carbon. Therefore the major contributor to the σ bonding MO will be the oxygen based orbital; we therefore expect the σ bond to be polarized towards the oxygen. The remaining two sp^2 hybrids on oxygen are not needed for bonding; they are each occupied by two electrons and so form two lone pairs.

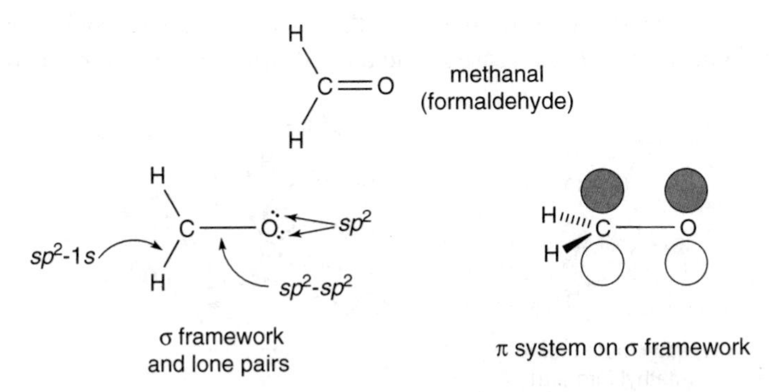

Fig. 6.18 Methanal (formaldehyde) has a carbonyl group, that is a C–O double bond. One description of the bonding involves sp^2 hybridization of both the carbon and oxygen. One of the hybrids on oxygen forms a σ bond to the carbon, and the other two are occupied by lone pairs.

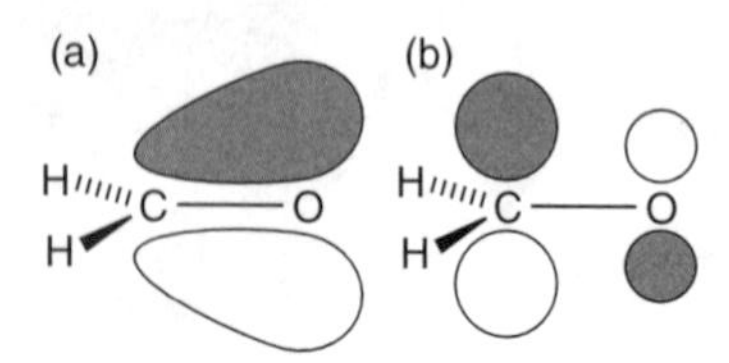

Fig. 6.19 Cartoons of (a) the π bonding and (b) the π anti-bonding MOs in a carbonyl group. The AO on oxygen is lower in energy than that on carbon, and so the oxygen AO is the major contributor to the bonding MO; for the anti-bonding MO, it is the carbon AO which is the major contributor.

The π system is essentially the same as in ethene, with the out-of-plane $2p$ orbitals from oxygen and carbon overlapping to give a π bonding and a π anti-bonding MO; two electrons occupy the bonding MO, forming the π bond between the carbon and the oxygen. The difference between the π bonds in ethene and methanal is that whereas in ethene the two AOs have the same energy, in methanal the oxygen $2p$ AO is lower in energy than that from carbon. As a result, the oxygen AO is the major contributor to the bonding MO and the resulting π bond (like the σ bond) is polarized towards oxygen. The major contributor to the π anti-bonding MO is the carbon, and we will see later on that this is crucial in explaining a great deal of the chemistry of the carbonyl group. Cartoons of these MOs are shown in Fig. 6.19.

We should check that all the electrons are accounted for: methanal has 12 valence electrons (two from the hydrogens, four from carbon and six from oxygen). In our description there are three pairs in σ bonding MOs, one pair in a π bonding MO and two lone pairs in sp^2 hybrids on the oxygen; all the electrons are therefore accounted for.

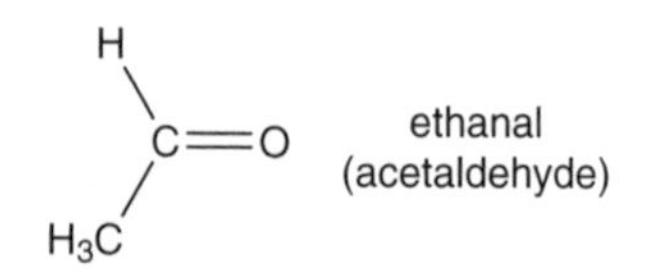

Fig. 6.20 The structure of ethanal (acetaldehyde).

The description of the bonding in ethanal (acetaldehyde, Fig. 6.20), CH_3CHO, is trivially different from methanal. The methyl carbon is sp^3 hybridized and uses these HAOs to form the σ bonds to the carbonyl carbon and the three hydrogens. The description of the bonding in the carbonyl group is the same as before.

We could have chosen the carbonyl oxygen to be *sp* hybridized, in which case one of the lone pairs will occupy an *sp* HAO and one will occupy the $2p$ oxygen AO which is at right angles to the one forming the π bond. It makes little practical difference which hybridization state we choose for oxygen as the key features of the bonding in the carbonyl group are not really affected by this choice.

The simplest ester is methyl methanoate (methyl formate), shown in Fig. 6.21. As in methanal the carbonyl carbon and the carbonyl oxygen are sp^2 hybridized, leading to the same description of the C–O double bond. The C–O–C bond angle at the ester oxygen is 127 ° so sp^2 hybridization is probably the most appropriate choice for this atom. The methyl carbon is sp^3 hybridized, as before.

The σ framework therefore consists of the C–O bond in the carbonyl group, the C–H bond to the carbonyl carbon and the following additional interactions:

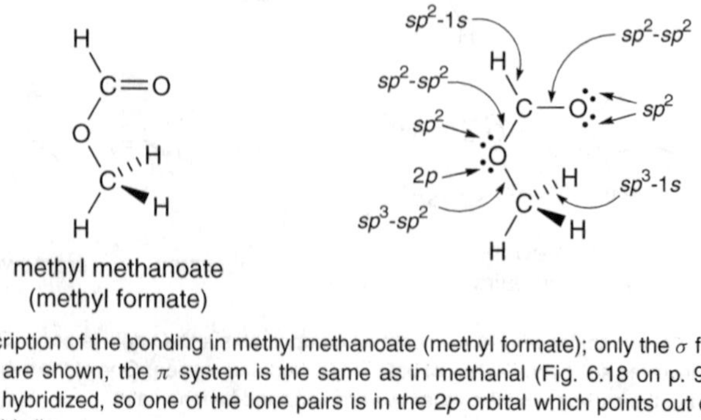

methyl methanoate
(methyl formate)

sp^2-1s
sp^2-sp^2
sp^2-sp^2
sp^2
sp^2
$2p$
sp^3-1s
sp^3-sp^2

Fig. 6.21 Description of the bonding in methyl methanoate (methyl formate); only the σ framework and the lone pairs are shown, the π system is the same as in methanal (Fig. 6.18 on p. 91). The ester oxygen is sp^2 hybridized, so one of the lone pairs is in the $2p$ orbital which points out of the plane in which the hybrids lie.

three C–H bonds in the methyl group, each formed by the overlap of an sp^3 hybrid and a hydrogen 1*s*; a bond between the methyl carbon and the ester oxygen formed by the overlap of a carbon sp^3 and an oxygen sp^2 hybrid; and a bond between the carbonyl carbon and the ester oxygen, formed from sp^2 hybrids on each. The remaining sp^2 hybrid and the unhybridized 2*p* AO on the ester oxygen are occupied by lone pairs. Figure 6.21 summarizes the description.

The description of amides, such as methanamide (Fig. 6.22), is a little more complex as it turns out that the lone pair on the nitrogen interacts with the carbonyl π system. We will discuss how this comes about and the consequences it has for both the structure and reactions of amides later on in Section 10.2 on p. 160.

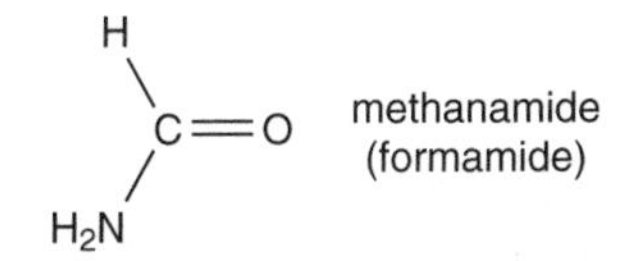

Fig. 6.22 The structure of methanamide (formamide).

Nitriles

Nitriles, such as ethanenitrile (acetonitrile), CH_3CN, shown in Fig. 6.23, have a triple bond between carbon and nitrogen; as with ethyne (p. 90) both the carbon and the nitrogen are *sp* hybridized. The two hybrids which point towards one another along the C–N axis overlap to form σ bonding and anti-bonding MOs; two electrons occupy the bonding MO to form the C–N σ bond. The two pairs of out-of-plane 2*p* orbitals overlap to form two π bonding and two π anti-bonding MOs. Together, occupation of the σ and the two π bonding MOs forms the triple bond.

The remaining *sp* hybrid on nitrogen is occupied by two electrons forming a lone pair; the remaining *sp* hybrid on carbon is involved in forming a σ bond to the methyl carbon.

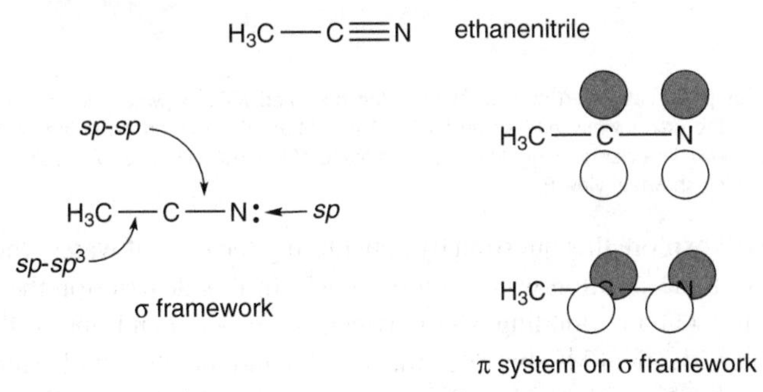

Fig. 6.23 The C–N σ bond in ethanenitrile is formed using *sp* hybrids on carbon and nitrogen, just as in ethyne (Fig. 6.17 on p. 91); the second *sp* hybrid on the nitrogen is occupied by a lone pair of electrons. The out-of-plane 2*p* orbitals form the two π bonds.

6.5 Reconciling the delocalized and hybrid atomic orbital approaches

The MO and HAO approaches take a very different view of the bonding in a molecule. Whereas the MO picture combines AOs from several atoms, the HAO picture is strictly a localized one in which bonding is between pairs of atoms; electrons not involved in bonding (lone pairs) are assigned to orbitals which are localized on a single atom. The question is, do the two approaches give us the same picture of the bonding in a molecule?

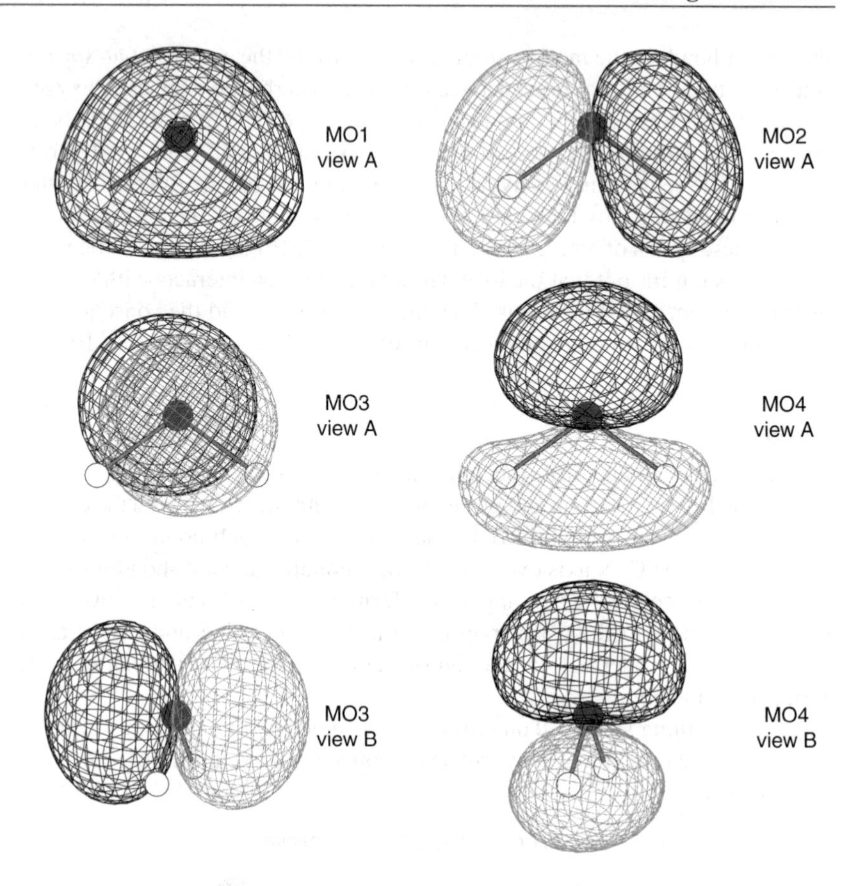

Fig. 6.24 Computer calculated surface plots of the occupied MOs of water. MO1 is the lowest in energy, MO2 the next lowest, and so on for MO3 and MO4; this last orbital is the HOMO. All four orbitals are shown in a view looking down onto the plane of the molecule (view A); a side view of MO3 and MO4 is also shown in view B.

We will explore this question by considering the case of water whose localized bonding description was given on p. 85. In this description the occupied orbitals are: (1) a σ bonding MO between the oxygen and one of the hydrogens; (2) an identical MO between the oxygen and the other hydrogen; (3) two sp^3 type orbitals on oxygen, neither of which is involved in bonding.

A computer calculation gives the MOs shown in Fig. 6.24; at first sight these do not appear to have much connection with the localized picture, but on closer inspection it turns out that there are similarities.

Consider first the two lowest energy MOs, MO1 and MO2. If we form the combination (MO1 + MO2) there will be reinforcement on the right-hand side, where both orbitals are positive, but on the left-hand side the negative part of MO2 will cancel with the positive MO1; the process is shown in cartoon form in Fig. 6.25. The result will be an orbital which is mainly between the oxygen and the right-hand hydrogen; this is the localized orbital between these two atoms. In a similar way if we form the combination (MO1 − MO2) we end up with an orbital which is mainly between the oxygen and the left-hand hydrogen. So, MO1 and MO2 can be thought of as representing between them the two localized σ bonds between the oxygen and the two hydrogens.

This leaves the other two orbitals MO3 and MO4 which, in the localized picture, ought to correspond to our two lone pairs. MO3 is essentially an out-of-plane $2p$ orbital which is clearly non-bonding. However, MO4 is a bonding MO as it has some electron density shared between the hydrogens and the oxygen. Clearly, then, MO3 and MO4 do not correspond simply to the two sp^3 hybrids from the HAO description.

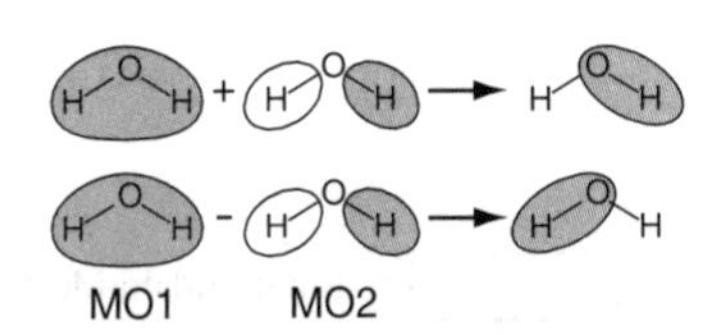

Fig. 6.25 Cartoons showing the way in which combinations of MO1 and MO2 from water (shown in Fig. 6.24) give rise to localized σ bonding orbitals. The combination MO1 + MO2 results in cancellation on the left-hand side and so gives an orbital which is mostly between the oxygen and the right-hand hydrogen. Similarly, the combination MO1 − MO2 gives an orbital which is mostly between the oxygen and the left-hand hydrogen.

In the hybridization approach we made the four sp^3 hybrids equivalent, so it is inevitable that when we use these HAOs to describe the bonding in water the two orbitals which are occupied by the lone pairs will have the same shape and energy. The MO description imposes no such restrictions and so we should not be surprised to find that the orbitals which are occupied by the lone pairs are not the same. Experiments confirm that this prediction from the MO calculation is indeed correct.

The computer generated MOs are certainly a more precise description of the electronic structure of water than is our simple HAO approach. We could hardly expect it to be otherwise as the HAO approach is purely qualitative and based on simple approximations such as only permitting the interaction between pairs of orbitals. Nevertheless the HAO approach does capture the essence of the bonding in water, namely that there are σ bonds between the oxygen and the hydrogens, together with extra electron density on the oxygen. If we want to think about the shape of the water molecule or the way in which it reacts, the HAO picture is a perfectly good starting point.

In summary, although we recognize the deficiencies of the HAO approach, it does allow us to make a reasonable start at describing how a molecule is bonded together, and the form of the orbitals involved, *without* having to resort to computer calculations. A further advantage of the HAO approach is that it clearly associates MOs with particular bonds – an idea which fits in well with the way we think about reactions.

When it comes to describing the bonding in larger molecules we will find that a useful approach is to use HAOs to describe the σ-framework and then use the fully delocalized MO approach to describe the rest of the bonding. Typically this means using the MO approach to describe the π-system. Often the σ-framework is not involved in reactions, so it does not matter too much if our description of the framework is approximate. The more precise MO approach is reserved for dealing with the part of the molecule which is actually involved in the reaction. We will use this approach extensively in Chapters 10 and 11.

6.6 Orbital energies: identifying the HOMO and LUMO

When it comes to chemical reactions it is very important to identify the highest occupied MO (the HOMO) and the lowest unoccupied MO (the LUMO). This is because, as we shall see in the next chapter, these orbitals are the ones which are most directly involved in reactions. Of course, we can always use a computer program to compute the MOs and so identify the HOMO and the LUMO, but in many cases it is not really necessary to go to this level of sophistication.

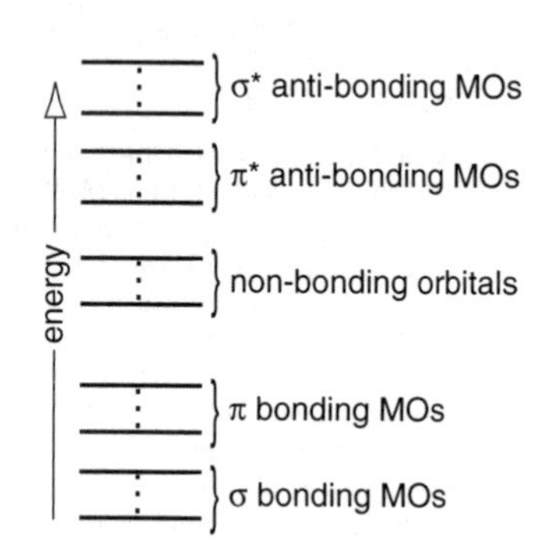

Fig. 6.26 The approximate ordering of orbital energies found in molecules; anti-bonding MOs are indicated by a superscript *.

For our purposes, it will be sufficient to use the HAO approach to draw up a picture of the bonding. As we have seen in this chapter, for simple molecules this approach allows us to identify electron pairs as being involved in 2c-2e bonds (in σ or π bonding MOs) or localized on atoms as lone pairs. All we then need to do is to have a method of ordering these orbitals in energy so that we can determine the HOMO and LUMO.

Typically the ordering of energies is that shown in Fig. 6.26. The σ bonding MOs come lowest, followed by π bonding MOs; as we commented on above, the sideways-on overlap which leads to π MOs is not generally as effective as the head-on overlap which leads to σ MOs.

Not surprisingly, next after the bonding orbitals come the non-bonding orbitals, which can be AOs, HAOs or, as we shall see, MOs. Finally, higher still are the anti-bonding MOs (indicated by a superscript *). For the same reason that the σ bonding MO lies below the π bonding MO, the $\sigma^\star$ lies above the $\pi^\star$ MO.

This order of energies is not immutable, but it is a good starting point. So, for example, we can easily identify the HOMO in methanal and ethanal as a lone pair on the oxygen and the LUMO as the $\pi^\star$ anti-bonding MO between the carbon and the oxygen.

7 Reactions

In Chapter 6 we looked at the molecular orbitals in some simple compounds. We now want to see how these are involved in chemical reactions. The energies of the orbitals help us to understand *why* the reaction occurs and the shapes are crucial in understanding *how* the reaction takes place. As a model, we shall first consider the hypothetical reaction between a single proton, H^+, and a hydride ion, H^-, to produce H_2:

$$H^+ + H^- \longrightarrow H_2.$$

7.1 The formation of H_2 from H^+ and H^-

It might be tempting to say that this reaction occurs because the reactants are oppositely charged; however there is more to it than this. The reaction does not lead to an ion pair held together by electrostatic attraction, as shown in Fig. 7.1 (a), but to the formation of a molecule where the electrons are shared between the two nuclei, as shown in (b). It is true that to start with, due to their opposite charges, the ions are attracted to one another but as the separation between them decreases the electrons are redistributed. Initially both electrons are associated with one nucleus, but in the product they are located in a molecular orbital and shared between the two nuclei. This redistribution is shown in Fig. 7.2 which compares the electron density of the separate ions and the product, H_2.

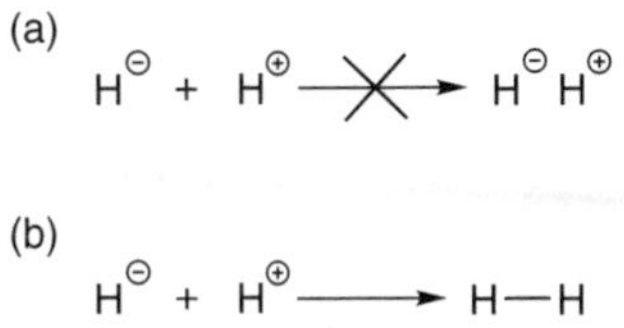

(a) $H^{\ominus} + H^{\oplus} \nrightarrow H^{\ominus} H^{\oplus}$

(b) $H^{\ominus} + H^{\oplus} \longrightarrow$ H—H

Fig. 7.1 The reaction between H^- and H^+ does not give an ion pair, as shown in (a), but a molecule of H_2, as shown in (b).

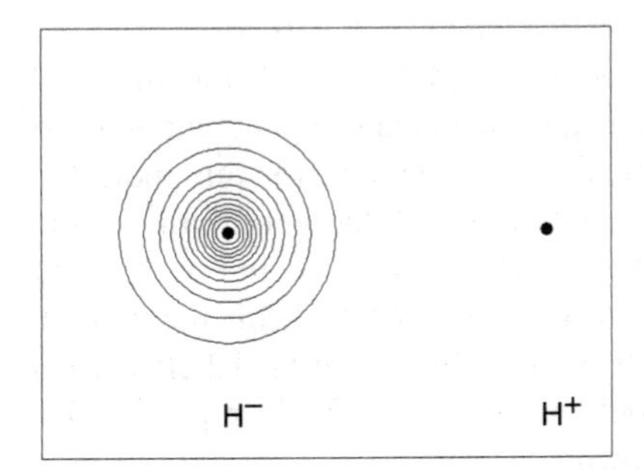

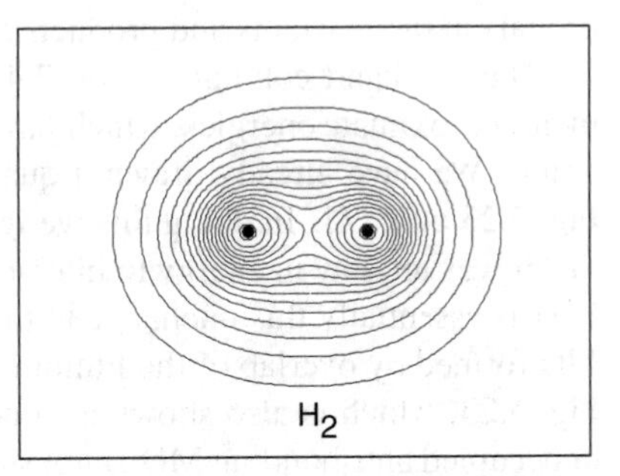

Fig. 7.2 Contour maps showing the electron density associated with (a) the H^- ion and a separate proton (H^+), and (b) with the H_2 molecule.

The sharing of the electrons has led to a product lower in energy than the starting materials – it is an exothermic process. Even though the entropy of the system must have decreased as the two ions combined to form one molecule, the entropy of the Universe has increased due to the heat given out to the surroundings during this exothermic process.

Chemists illustrate this redistribution of electrons by using *curly arrows*; Fig. 7.3 shows how we can represent the reaction between H^+ and H^- using this approach. A curly arrow represents the movement of a *pair* of electrons.

$H^{\ominus}$ $H^{\oplus}$ $\longrightarrow$ H—H

Fig. 7.3 A curly arrow mechanism for the reaction between a hydride ion and a proton.

The arrow starts where the electrons are in the reactant (here in the H^- ion) and shows where this pair of electrons ends up in the product (in this case in between the two nuclei in the new bond).

This curly arrow representation of the reaction between H^+ and H^- is probably not a realistic description of what actually happens. Rather than the two electrons being shared, what probably happens in this case is that one electron is transferred from H^- to H^+ to give two H atoms which then come together to form H_2.

Nonetheless, hypothetical though the reaction is, it is a good starting point for understanding reaction mechanisms. In the next section we will use this approach to discuss a real reaction in which a bond is formed by the sharing of electrons.

7.2 Formation of lithium borohydride

Lithium hydride, LiH, and borane, BH_3, react to form lithium borohydride, $LiBH_4$:

$$LiH + BH_3 \longrightarrow LiBH_4.$$

This reaction occurs very readily at room temperature when borane (which actually exists as a dimer, B_2H_6 – see Fig. 6.4 on p. 82) is mixed with the lithium hydride in ethoxyethane (diethyl ether) as a solvent.

In this reaction there is a decrease in the entropy of the system since two molecules go to one. Just as in the case of $H^+ + H^-$ the reaction is exothermic so there is an increase in the entropy of the surroundings which, it turns out, more than compensates for the decrease in the entropy of the system. The reason why the reaction is exothermic is because the energy of the electrons in the product is lower than in the reactants. We need to look at the molecular orbitals in the reactants and products to understand why this is so.

The left-hand column of Fig. 7.4 shows the MOs of LiH, together with their approximate energies, which have been calculated using a computer program. We have already drawn a qualitative MO diagram for this molecule, Fig. 5.23 on p. 78. In doing this we ignored the $1s$ orbital on lithium, arguing that it was too low in energy to interact with the hydrogen $1s$. In Fig. 7.4 MO-LH1 is essentially this unchanged lithium $1s$ orbital; MO-LH2 is the bonding MO formed by overlap of the lithium $2s$ with the hydrogen $1s$ (labelled 1σ in Fig. 5.23, which is also shown as a contour plot in Fig. 5.24 on p. 78). The unoccupied anti-bonding MO is not shown.

In Chapter 6 we used sp^2 hybrids to draw up an approximate MO diagram for BH_3 (Fig. 6.11 on p. 86); as for LiH, we ignored the 1s orbital on B. We identified three bonding MOs and one non-bonding MO, the out-of-plane $2p$ orbital. The right-hand column of Fig. 7.4 shows computer calculated MOs for BH_3. As we described in Section 6.5 (p. 93) the MOs calculated in this way are delocalized and so do not compare directly with the simpler localized HAO description. However, we can clearly identify three bonding MOs, MO-B2, MO-B3 and MO-B4, together with the non-bonding MO, MO-B5, which is the LUMO. MO-B1 is essentially the boron $1s$ orbital which is not involved in the bonding.

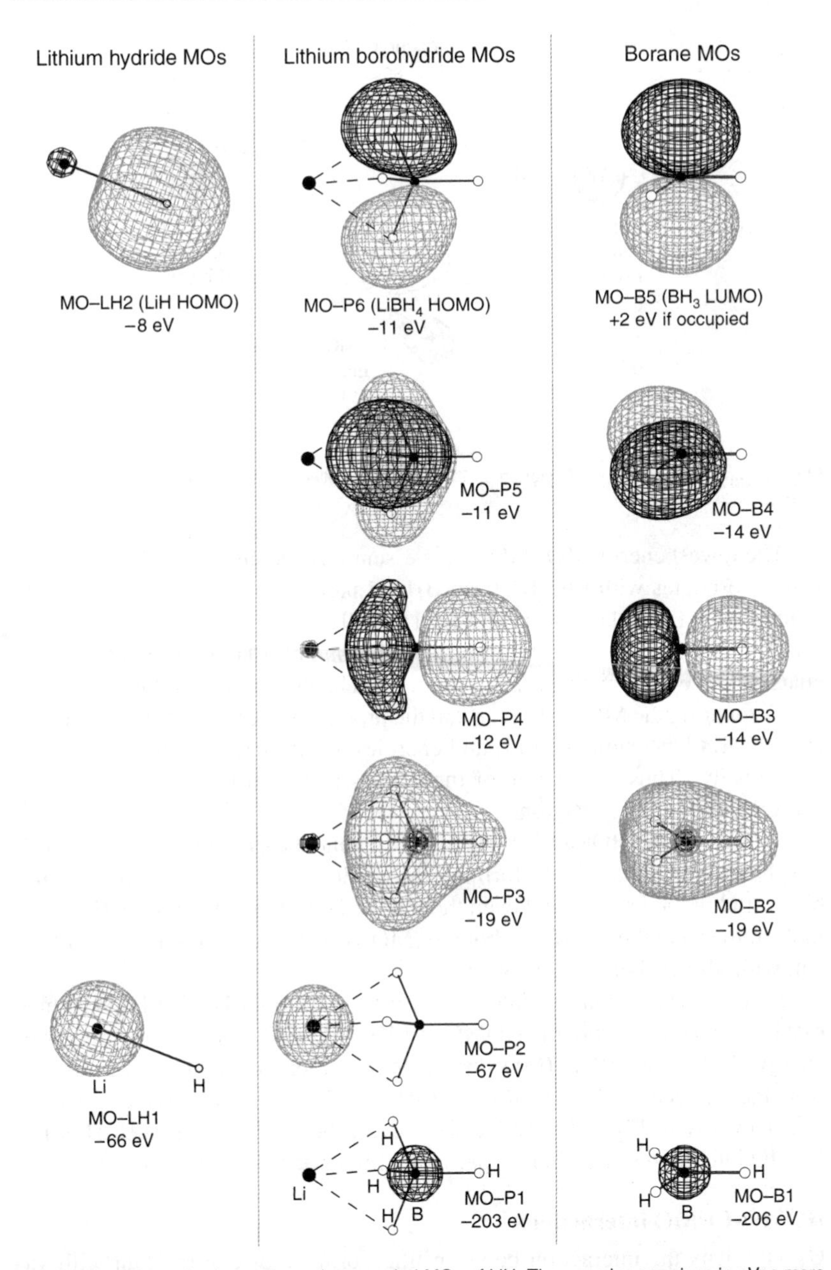

Fig. 7.4 The left-hand column shows the occupied MOs of LiH. The energies are given in eV; a more negative energy means a more tightly held electron. The right-hand column shows the occupied MOs and the LUMO for BH_3. The middle column shows the occupied MOs for the product, $LiBH_4$. The most important orbitals for each molecule are shown at the very top. In the $LiBH_4$ structure the dashed lines are used to illustrate the arrangement of the atoms rather than to imply the existence of bonds.

Just as we can think of a molecular orbital as being formed from two atomic orbitals, we can think of the MOs in the products as arising from the interactions between the MOs in the starting materials. The middle column of Fig. 7.4 shows the occupied molecular orbitals in the product, $LiBH_4$.

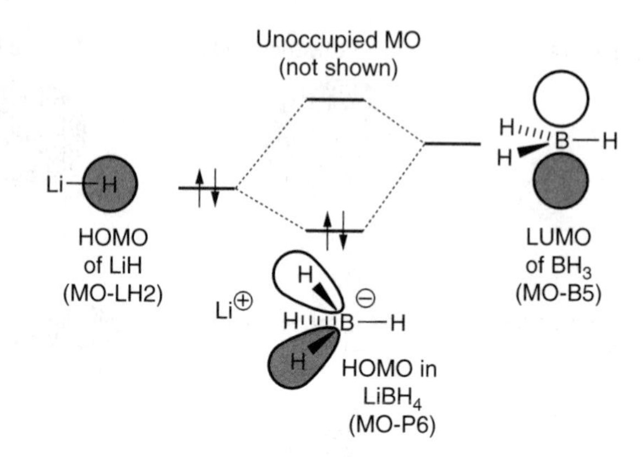

Fig. 7.5 Schematic energy level diagram for the interaction between the HOMO of LiH and the LUMO of BH_3.

The lowest energy MO, MO-P1, is essentially the unaffected boron 1*s* AO which correlates with MO-B1 from BH_3. Likewise, MO-P2 is the 1*s* AO on lithium which correlates with MO-LH1 on LiH. The boron 1*s* orbital is much lower in energy than the lithium because boron has a larger effective nuclear charge. These 1*s* orbitals remain virtually unchanged in the product.

Comparing the MOs in borane and the product we see that MO-B2, MO-B3 and MO-B4 have similar forms and energies to MO-P3, MO-P4 and MO-P5, respectively. Thus occupation of these MOs makes little contribution to the energy change of the reaction.

In contrast, the HOMO of LiH (MO-LH2) and the LUMO of BH_3 (MO-B5) are changed significantly on forming the product. MO-LH2 has no comparable orbital in the product and even though the HOMO in the product (MO-P6) looks rather like the LUMO in borane (MO-B5) their energies are *very* different, with MO-P6 being much lower.

In summary, we can attribute the lowering in energy when $LiBH_4$ is formed to the two electrons from the HOMO of LiH (MO-LH2) moving into the lower energy HOMO of $LiBH_4$ (MO-P6). This orbital can be thought of as arising from the interaction of the HOMO of LiH with the LUMO of BH_3, as is shown schematically in Fig. 7.5. As we will see in the next section, such HOMO-LUMO interactions are often of particular importance in reactions.

HOMO-LUMO interactions

Usually, it is the interaction between filled orbitals of one reactant with vacant orbitals of the other which leads to a favourable lowering of the energy. The interactions between two vacant orbitals is of no consequence as they are unoccupied and so there is no change in the energy of the system. The interactions between two filled orbitals have little net effect on the total energy of the system since the decrease in energy from the newly formed in-phase combination is essentially cancelled out by the increase in energy of the corresponding out-of-phase combination. Only the interaction between a filled orbital and an empty orbital leads to a net lowering of the energy of the electrons, as shown in Fig. 7.6.

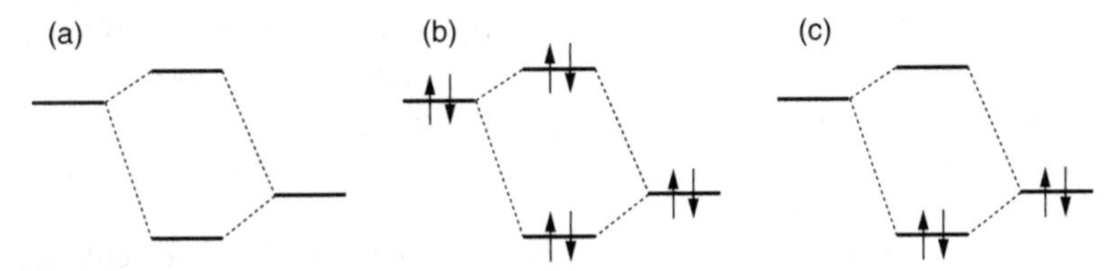

Fig. 7.6 The interaction of two empty orbitals as shown in (a) does not lead to a lowering in energy since there are no electrons whose energy is being lowered. In the interaction between two filled orbitals, shown in (b), two electrons are lowered in energy but this is cancelled out by the two electrons which are raised in energy. Only in (c), the interaction between an empty and a filled orbital, is there a net lowering in the energy since the orbital which is raised in energy is unoccupied.

Of course there are many filled and vacant orbitals which can interact – the question is which interaction is the most important? Recall from the discussion on p. 67 that the strongest interaction is between orbitals of similar energies, so the best candidates for a favourable interaction will be an unoccupied MO and an occupied MO which are close in energy.

If the energies of the MOs of the two reactants are broadly similar, then we can expect the best energy match to be between the LUMO of one with the HOMO of the other, as shown in Fig. 7.7. There is no guarantee that this HOMO-LUMO interaction will be the most favourable, but it turns out that it is often so.

When two species meet, *two* HOMO-LUMO interactions are possible; there is an interaction between the HOMO of reactant A with the LUMO from reactant B and also between the HOMO from B with the LUMO from A. The very best interaction is usually between the occupied MO which has the highest energy of either reactant with the LUMO from the other reactant. In Fig. 7.7, the best interaction (shown with the wavy line) is between the HOMO from reactant A with the LUMO from B since these two MOs are separated by the smallest energy difference. The interaction (shown with the dotted line) between the HOMO of B and the LUMO of A is less significant because the separation in energy is greater.

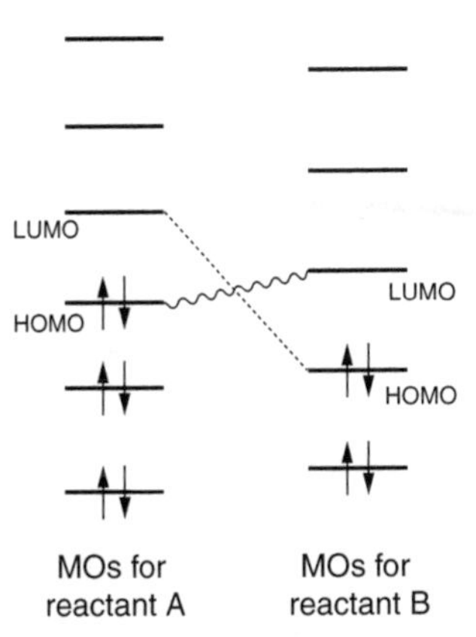

Fig. 7.7 The strongest interactions are between the HOMO of one molecule and the LUMO of another. For the set of MOs shown here, the best interaction, shown by the wavy line, is between the HOMO of reactant A and the LUMO of reactant B. Since the HOMO of A is particularly high in energy, the energy gap between this and the LUMO in B is particularly small.

Frequently we will find that we can understand the outcome of a reaction between two molecules by first identifying which molecule has the very highest occupied MO and then considering the interaction between this and the LUMO of the second molecule. In the reaction between LiH and BH_3, the very highest occupied MO is the HOMO of the LiH (MO-LH2). This is considerably higher in energy than the HOMO of the BH_3 (MO-B4) so the main interaction is between the HOMO from the LiH and the LUMO from the BH_3, MO-B5.

Orbitals and curly arrows

Just as we did for the $H^+ + H^-$ reaction, we can use curly arrows to illustrate, in a concise way, what is going on in the reaction between LiH and BH_3. From studying the orbitals we see that a pair of electrons from the σ-bonding MO of LiH (the HOMO – MO-LH2) end up in an orbital of $LiBH_4$ which contributes to the formation of a new B–H bond. Figure 7.8 illustrates this process using a curly arrow. The arrow starts where the electrons come from – the bond between the Li and H – and finishes where the electrons end up, in a new bond between the H and the B. It tells us that the electrons that were bonding the Li

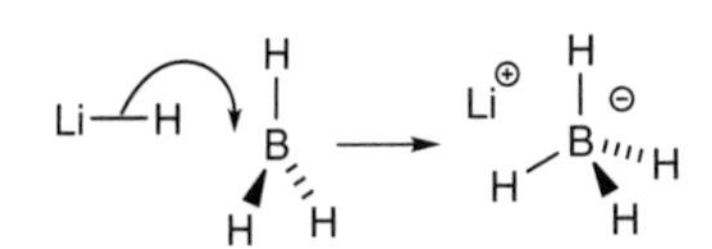

Fig. 7.8 The curly arrow mechanism for the reaction between borane and lithium hydride.

and H are now helping to bond the H to the B. Since the lithium has effectively lost its share of electrons from the Li–H bond, it ends up with a positive charge. The BH_3 molecule initially had no charge; in forming the product it has gained one extra proton and the two electrons that were in the Li–H bond. It therefore now has a negative charge.

In $LiBH_4$, the lithium is mainly attracted to the rest of the molecule electrostatically. Its MOs shown in Fig. 7.4 show little evidence of electrons being shared between the lithium and the rest of the molecule and a calculation confirms that the lithium does indeed bear a substantial positive charge with the corresponding negative charge being spread out over the boron and hydrogen atoms.

The HOMO in lithium borohydride is still fairly high in energy. In the next section we shall see how the interaction between this and a suitable LUMO allows us to understand the use of $LiBH_4$ as a reducing agent.

7.3 Nucleophilic addition to carbonyl

Nucleophilic addition to a carbonyl group is one of the most important reactions in organic chemistry. In this reaction a nucleophile attacks the carbonyl group, forming a new bond to the carbon and simultaneously breaking the C–O π bond; Fig. 7.9 shows the reaction of a general nucleophile, Nu^-. Hydroxide, cyanide (CN^-) and alkyllithium reagents (R–Li) are all examples of nucleophiles which will attack carbonyl groups in this way.

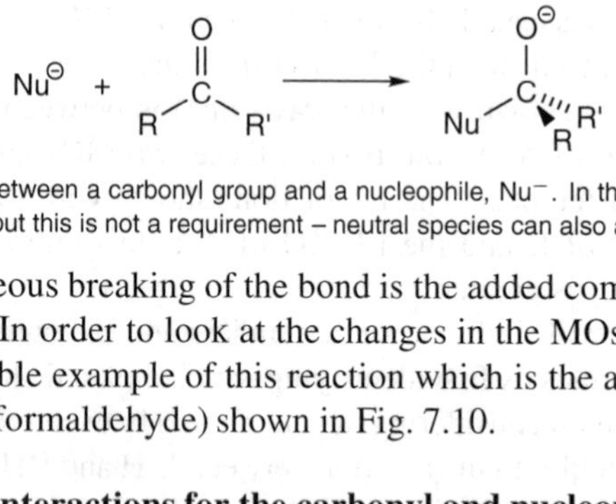

Fig. 7.9 The reaction between a carbonyl group and a nucleophile, Nu^-. In this case the nucleophile is negatively charged, but this is not a requirement – neutral species can also act as nucleophiles.

This simultaneous breaking of the bond is the added complication we now need to consider. In order to look at the changes in the MOs we will consider the simplest possible example of this reaction which is the attack of a hydride ion on methanal (formaldehyde) shown in Fig. 7.10.

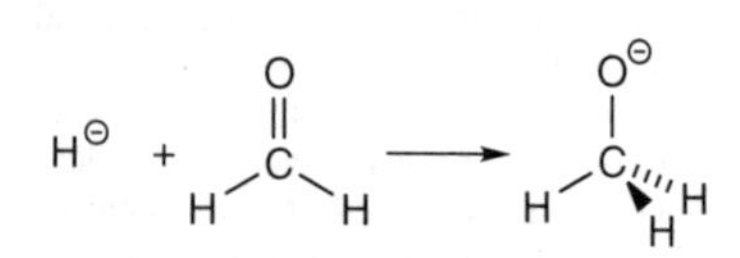

Fig. 7.10 The reaction between methanal and hydride; this is an example of nucleophilic addition.

HOMO-LUMO interactions for the carbonyl and nucleophile

The two reactants are initially attracted towards each other since the carbonyl carbon has a partial positive charge and the hydride ion is negatively charged. However, the two reactants do not just remain attracted towards each other electrostatically; bonds are formed and broken. In other words, electrons are redistributed.

The reaction starts with a negative charge on hydrogen and ends with the charge on oxygen. Since oxygen has a greater effective nuclear charge than hydrogen, having the increased electron density on oxygen is favourable and so contributes to the lowering of the energy of the product.

We can see how this redistribution of charge takes place by looking at the HOMO-LUMO interactions. The highest occupied MO in the whole system is from the H^- ion. The vacant orbital closest to this in energy is the LUMO

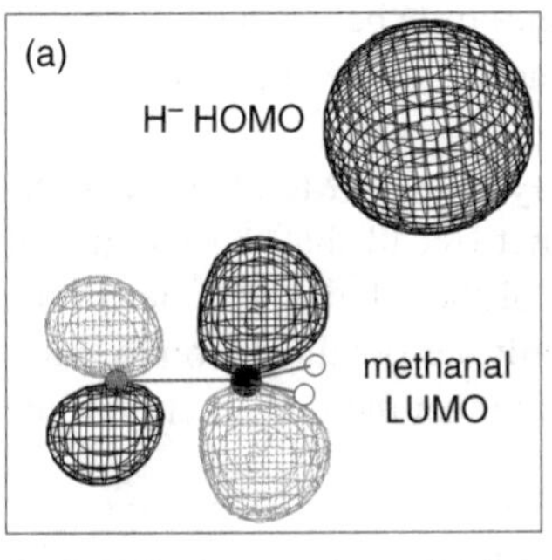

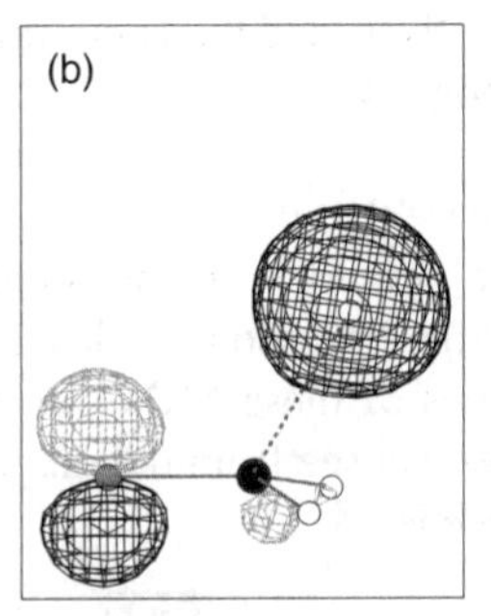

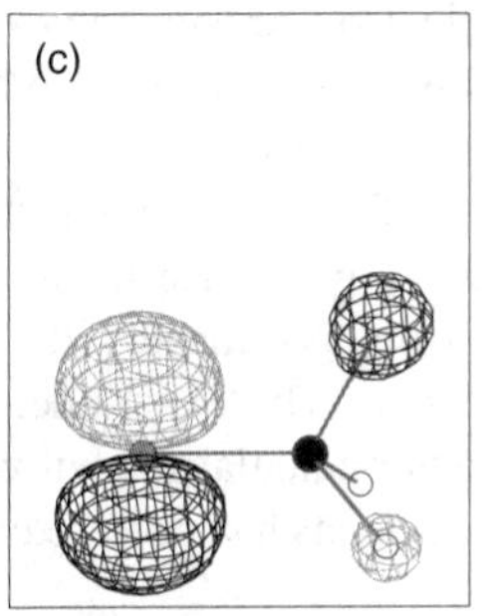

Fig. 7.11 The in-phase interaction between the methanal $\pi^\star$ and the hydride HOMO. In diagram (a) two orbitals are shown: the occupied hydride MO and the unoccupied methanal $\pi^\star$. Diagram (b) shows the orbital arising from the in-phase overlap of the HOMO and LUMO at an intermediate stage in the reaction. Electron density is building up on the oxygen at the expense of the hydride. The corresponding MO in the product is shown in diagram (c). This is essentially a *p* orbital on the oxygen, i.e. a lone pair; there is little electron density on the hydrogen when compared to (a).

of the methanal, the $\pi^\star$. Figure 7.11 (a) shows the initial interaction of these two orbitals, and (b) shows the orbitals at an intermediate stage in the reaction. The MO formed from the in-phase interaction between the hydride HOMO and methanal LUMO shows that electron density is being transferred from the hydride onto the oxygen. This is even more apparent in the product MOs shown in Fig. 7.11 (c) where there is much more electron density now on the oxygen and relatively little on the hydrogen.

The hydride HOMO and the methanal LUMO also combine out-of-phase to give an anti-bonding orbital higher in energy than the methanal LUMO; this anti-bonding orbital is not occupied.

The hydride HOMO is closest in energy to the CO $\pi^\star$ MO and so has the strongest interaction with this orbital. However, it also has an interaction with the CO π MO, and it turns out that it is this interaction which explains how the CO π bond is broken. Fig. 7.12 shows the in-phase interaction between these two orbitals.

At an intermediate stage, shown in (b), we see the interaction between the carbon and oxygen begins to weaken. In the product, shown in (c), there is

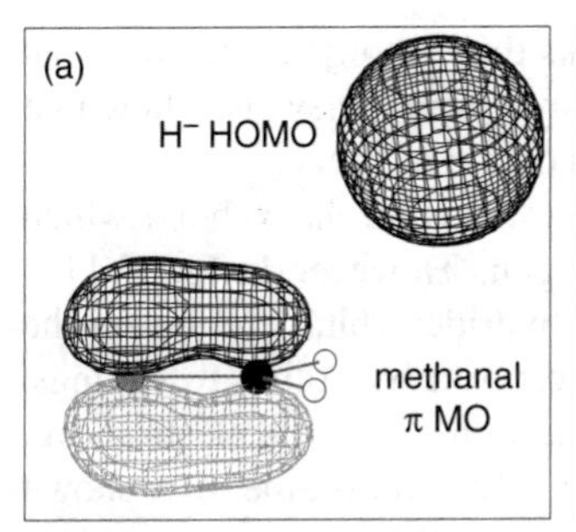

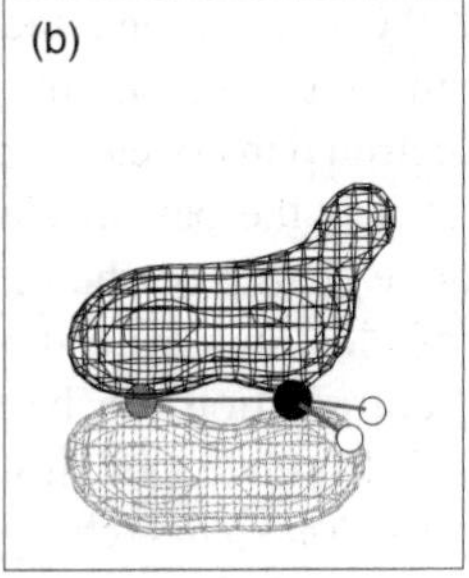

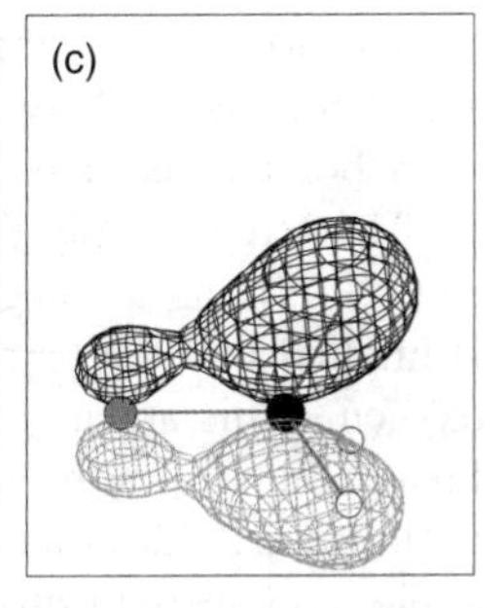

Fig. 7.12 The interaction between the methanal π MO and the hydride HOMO. Diagram (a) shows the two orbitals as the two reactants approach each other. Diagram (b) shows the orbital arising from the in-phase overlap of these two orbitals at an intermediate stage in the reaction. The bonding between the C and O is beginning to weaken while that between the C and the approaching H is strengthening. In the corresponding MO in the product, shown in (c), there is effectively no bonding between the C and O, only between the C and H.

electron density between the C and the H but rather little between the C and O. Just as we expected, a bond is formed between the C and the H and the C–O π bond is broken.

The interaction of these three orbitals – the hydride HOMO, the π and the $\pi^\star$ – gives rise to three new MOs in the product (recall that the number of orbitals is conserved); Fig. 7.13 shows a sketch of the relative energies of the MOs. The exact energies of these MOs can only be obtained from a computer calculation, but we can see from the diagram that the total energy of the electrons has been reduced.

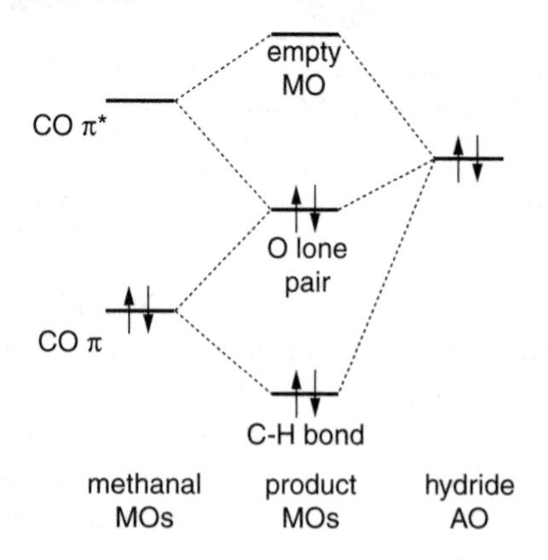

Fig. 7.13 Key orbital interactions between the hydride HOMO and carbonyl π and $\pi^\star$ MOs.

Curly arrows

Figure 7.14 uses curly arrows to show the important features of the changes that take place during the reaction between hydride and methanal.

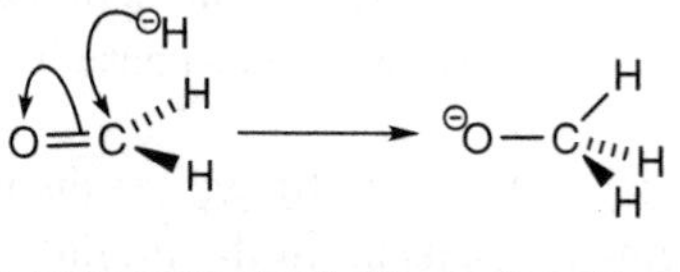

Fig. 7.14 The curly arrow mechanism for the reaction of methanal with hydride.

The arrows show that the hydride ion attacks the carbonyl carbon and that a new bond forms between the carbon and the hydrogen. They also show that the π bond breaks and electron density ends up on the oxygen.

The arrows imply that it is the pair of electrons from the π bond which end up forming the negative charge on the oxygen. However, in Fig. 7.11 it is implied that the pair of electrons from the hydride orbital end up on the oxygen. This apparent contradiction can be resolved by noting two things. Firstly, the arrows in Fig. 7.14 relate to a structure in which the electrons are localized in 2c-2e bonds or as lone pairs (charges). In contrast, the MOs shown in Fig. 7.11 are delocalized over all three atoms, and so there is not a one-to-one correspondence between the two pictures. Secondly, it is not really possible to say from which reactant MO a pair of electrons in a product comes from. For example, looking at Fig. 7.13, how can we decide which pair of electrons ends up in the middle energy MO when *all* of the reactant orbitals contribute to this?

So there really is no contradiction between the MO picture and the curly arrows. They are just different ways of representing the same thing – the curly arrows relate to a strictly localized view of bonding whereas the MOs relate to a delocalized approach. The important point is that the curly arrows give the same result as the MO approach: in this case a new bond is formed between the carbon and the attacking hydride nucleophile, the C–O π bond is broken, and a negative charge ends up on the oxygen.

The geometry of attack

For the hydride to interact with the π and $\pi^\star$ orbitals, it must approach from the right direction. If it were to approach the carbonyl in the same plane as the molecule, there will be no net interaction since the constructive and destructive contributions will cancel one another out. This is shown schematically in Fig. 7.15 (a).

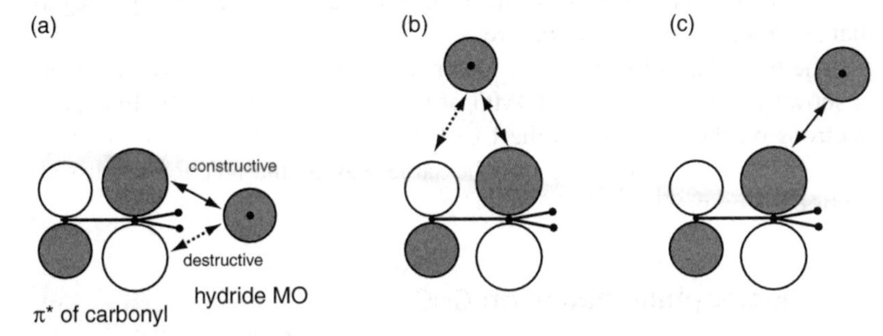

Fig. 7.15 The angle which the attacking nucleophile approaches from is important. In (a) the hydride approaches in the same plane as the carbonyl – there are equal amounts of constructive and destructive overlap leading to no net interaction. Similarly there is no net overlap in (b) where the hydride approaches from directly over the carbonyl group. The best angle of attack is shown in (c).

Similarly, if the hydride were to approach from above and directly in between the carbon and oxygen, as shown in (b), there would be no net overlap. Detailed calculations and experimental evidence suggests the best angle of approach is approximately 107° from the carbonyl, as shown in Figs. 7.15 (c) and 7.16. This maximizes the degree of constructive overlap and also minimizes the repulsion between the filled π orbital and the incoming nucleophile.

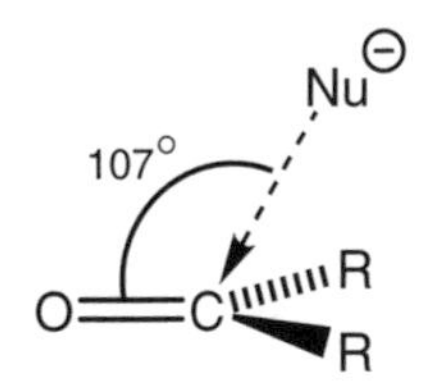

Fig. 7.16 The optimum angle of attack by a nucleophile on a carbonyl group.

Borohydride reduction of a ketone

We will finish this section by describing a reaction which can be carried out in the laboratory and which effectively results in the addition of H^- to a carbonyl group. Rather than using H^- directly, we use the reagent sodium borohydride, $NaBH_4$, which reacts with aldehydes and ketones to give the corresponding alcohol. The reaction is, of course, a reduction as it involves the addition of hydrogen.

The MOs for $NaBH_4$ are very similar to those of $LiBH_4$ which are depicted in Fig. 7.4 on p. 99. The highest energy occupied MO (MO-P6) is one which is involved in the bonding between the boron and the hydrogens. When borohydride acts as a nucleophile, it will be the electrons in this HOMO which are primarily involved.

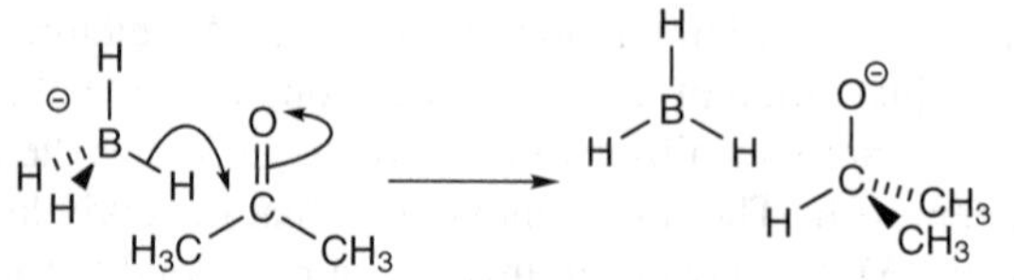

Fig. 7.17 The reaction between borohydride, BH_4^-, and propanone.

Having identified the HOMO of the system, we need to identify the LUMO of the other reagent. For a ketone such as propanone (acetone) the LUMO is once again the $\pi^\star$. As before, the interaction with the HOMO results in the π bond being broken and the electron pair moving to the oxygen; a curly arrow mechanism for this is shown in Fig. 7.17.

Figure 7.17 shows the electrons from the B–H bond attacking the carbon of the carbonyl. This results in the B–H bond breaking and a new C–H bond being formed. At the same time, the C–O π bond is broken, leaving a negative charge on the oxygen just as before.

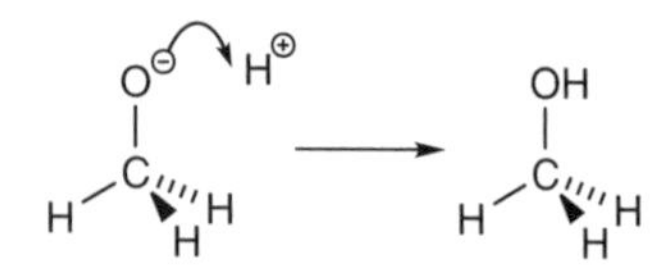

Fig. 7.18 A curly arrow mechanism for the protonation of the anion formed from the reduction of methanal.

The final step is to add acid which reacts with the $-O^-$ to give an alcohol, as shown in Fig. 7.18. The HOMO of the anion is certainly the lone pair of electrons on the oxygen and the LUMO is the $1s$ on H^+. These interact to form a new bond to H^+ in much the same way as the reaction discussed in Section 7.1 on p. 97.

7.4 Nucleophilic attack on C=C

A carbon–carbon double bond has the same π orbitals as a carbon-oxygen double bond so we might expect nucleophiles to attack C=C in the same way they attack C=O. However, this is not the case – nucleophiles do *not* usually attack alkenes. In fact, alkenes often act as nucleophiles themselves, preferring to donate electrons rather than accept them.

We can understand these differences by considering the nature of the C=C bond and the orbitals involved. Firstly, unlike in the carbonyl, the C=C bond does not have a dipole moment so there is no electrostatic attraction to bring the alkene and nucleophile together. The next main point to consider is the energies of the orbitals involved; these are shown in Fig. 7.19.

The $\pi^\star$ MO of the C=C bond is higher in energy than the $\pi^\star$ of the C=O

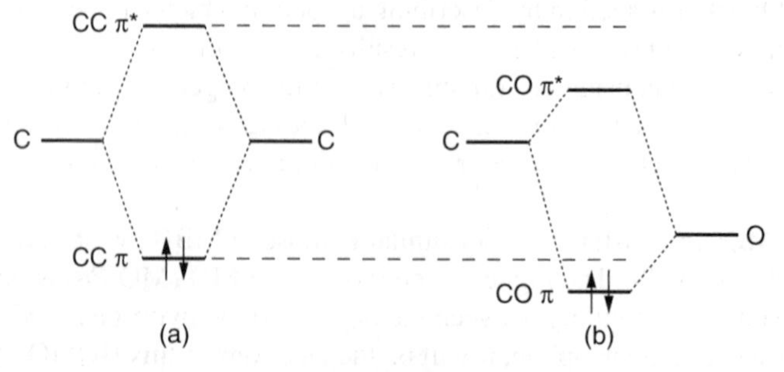

Fig. 7.19 Due to the energy differences between the atomic orbitals of carbon and oxygen, the C=C π and $\pi^\star$ MOs are higher in energy than the corresponding orbitals of C=O.

bond. This means that there is a poorer energy match between the C=C $\pi^\star$ and the HOMO of any attacking nucleophile. As we have seen, the poorer the energy match, the weaker the interaction will be.

The final point is that if a nucleophile did attack a C=C, a new bond will form between the carbon and the nucleophile and the carbon–carbon double bond will break leaving the charge on a carbon, as shown in Fig. 7.20. This is not as favourable as the attack on the carbonyl where the negative charge ends up on the oxygen atom, the reason being that the oxygen orbitals are lower in energy – put another way the charge is best located on an electronegative atom. Taking these factors together provides a rationale for why nucleophilic attack on C=C does not occur whereas attack on C=O occurs readily.

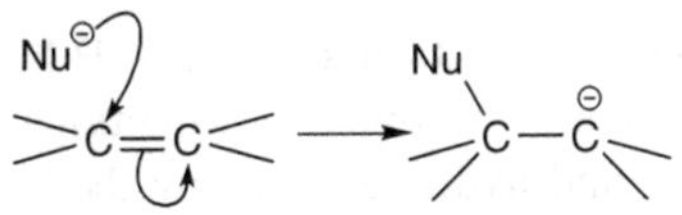

Fig. 7.20 If a nucleophile did attack a C=C π bond, a negative charge would form on the other carbon atom. Unlike having a negative charge on oxygen, this is not favourable.

7.5 Nucleophilic substitution

We are now going to look at a reaction in which the nucleophile donates electrons not into an empty AO or a $\pi^\star$ MO but into a $\sigma^\star$ MO. This will lead to the breaking of the σ bond, so an atom or group will break away from the molecule. The reaction is therefore called nucleophilic *substitution* as opposed to the addition reaction we have seen so far.

A typical example is where an incoming nucleophile, Nu^-, bonds to the carbon of chloromethane, displacing Cl^- in the process:

$$Nu^- + CH_3Cl \longrightarrow CH_3Nu + Cl^-.$$

In chloromethane the only unoccupied MOs will be $\sigma^\star$ MOs, of which there are two different types: the C–H and the C–Cl $\sigma^\star$. The nucleophile HOMO will interact most effectively with whichever of these is lower in energy; as we will see in the next paragraph, it turns out that this lowest energy MO is the C–Cl $\sigma^\star$.

The energy of the $\sigma^\star$ MO will depend on the strength of the interaction between the two AOs involved, and this depends primarily on the energy separation of these orbitals. As the chlorine AOs are lower in energy than those of the hydrogen, there will be a better match between the AOs involved in the C–H interaction than in the C–Cl interaction. As a result the $\sigma^\star$ MO for the C–Cl bond will be lower in energy than the corresponding orbital for C–H; this is illustrated in Fig. 7.21.

In larger halides such as chloroethane, there are also C–C interactions. However, these C–C interactions also involve well-matched AOs, so again we expect that the resulting $\sigma^\star$ MO will be higher in energy than for C–Cl. In fact, of the $\sigma^\star$ MOs in such molecules, it will *always* be the case that the MO associated with a bond to an electronegative element (such as Cl, N or O) will be

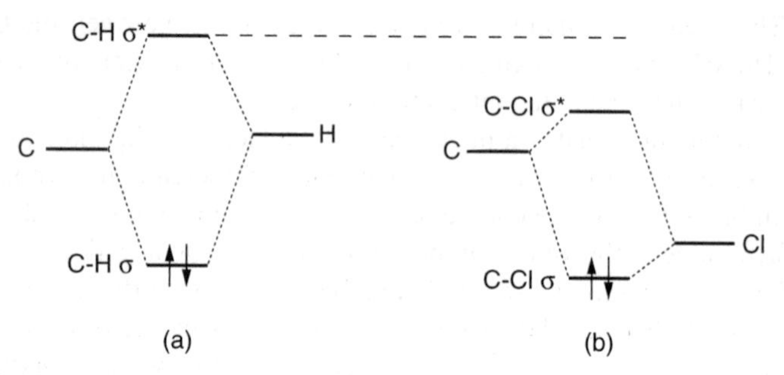

Fig. 7.21 Due to the larger difference in energy between the carbon and chlorine AOs than between the carbon and hydrogen AOs, the C–Cl $\sigma^\star$ will be lower in energy than the C–H $\sigma^\star$.

lower in energy than the $\sigma^\star$ MOs from C–C or C–H bonds. This is why much organic chemistry often centres around the functional groups which contain atoms other than carbon and hydrogen, leaving the basic C–C / C–H framework intact. The anti-bonding MOs involving the heteroatoms are inevitably lower than those in the basic framework and hence more readily involved in reactions.

A contour plot and surface plot of the C–Cl $\sigma^\star$ MO (the LUMO) from chloromethane is shown in Fig. 7.22 together with a schematic representation.

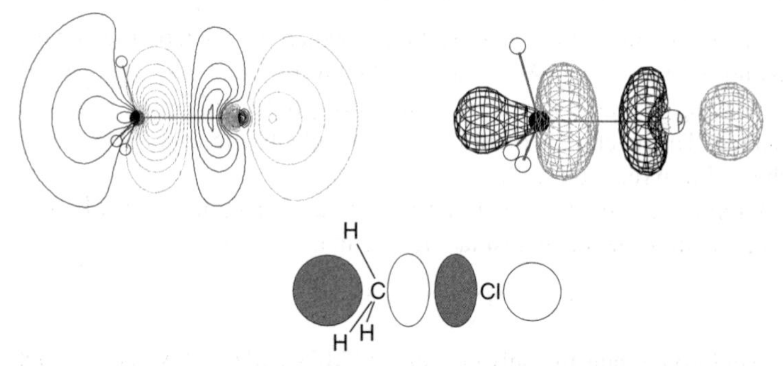

Fig. 7.22 Three representations of the LUMO (the C–Cl $\sigma^\star$ MO) in chloromethane: a contour plot, a surface plot and a sketch. Note the node between the C and Cl, and the lobe on C which points away from the bond.

The reaction between hydroxide ion and bromomethane

We shall look at the reaction between a halogenoalkane, in this case bromomethane and hydroxide ion. This is a simple substitution reaction where the hydroxide takes the place of the bromide ion:

$$HO^- + CH_3Br \longrightarrow CH_3OH + Br^-.$$

Having identified the LUMO as the C–Br $\sigma^\star$ MO, we need to identify the HOMO that this will interact with. Not surprisingly, this is the orbital occupied by one of the lone pairs on the oxygen of the hydroxide ion. The overall negative charge raises the energy of this orbital further. For simplicity we will assume that the lone pair occupies an sp^3 orbital on the oxygen – the details are not really important for this discussion.

The angle of approach

Bromomethane has a dipole moment with the carbon atom slightly positively charged and the bromine atom slightly negatively charged (Fig. 7.23). The negatively charged hydroxide ion is therefore attracted towards the carbon. The best interaction of the HOMO of the hydroxide and the LUMO of the bromomethane is achieved if the nucleophile attacks the carbon from the opposite side to the bromine. This attack is shown schematically in Fig. 7.24.

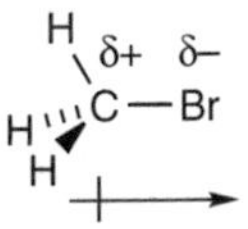

Fig. 7.23 Due to the differences in the electronegativities of carbon and bromine, bromomethane has a dipole moment with the carbon atom being slightly positive and the bromine atom slightly negative.

HOMO of hydroxide + LUMO of bromomethane ⟶ new σ MO between C and O + HOMO of bromide

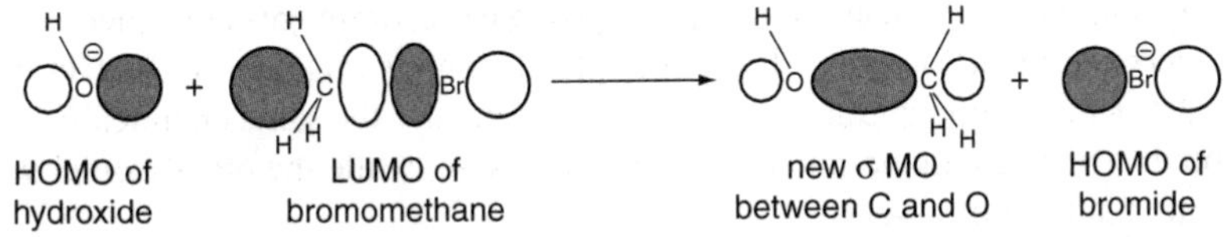

Fig. 7.24 The hydroxide ion attacks the bromomethane from directly behind the C–Br bond since this maximizes the interaction with the LUMO (the C–Br $\sigma^\star$) of the bromomethane.

If the nucleophile were to attack the carbon atom from the side, the orbital overlap would be poor due to the symmetry mismatch (Fig. 7.25).

During the reaction, the HOMO of the hydroxide interacts with the LUMO of the bromomethane, the C–Br $\sigma^\star$. A new σ bond forms between the carbon and oxygen at the same time as the old C–Br σ bond breaks. Eventually the bromide ion leaves, carrying with it the negative charge. The curly arrow mechanism for the reaction is shown in Fig. 7.26.

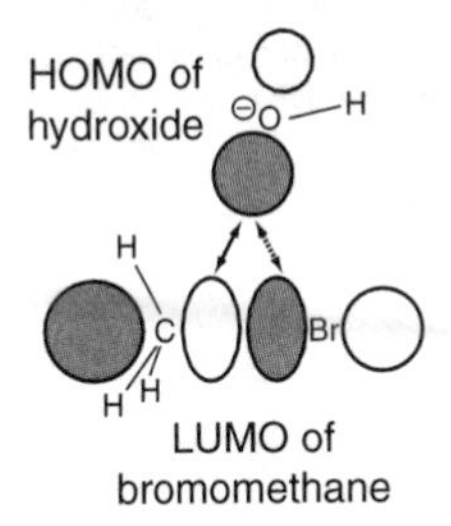

Fig. 7.25 If the hydroxide ion attacked the carbon more from the side, the orbital overlap is much worse due to the constructive and destructive interactions with the $\sigma^\star$ orbital.

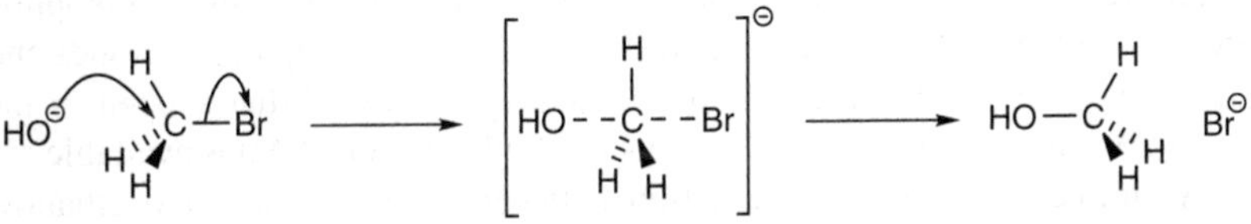

Fig. 7.26 A curly arrow mechanism for the attack of hydroxide ion on bromomethane. The species in the square brackets is a *transition state* (see Section 9.2 on p.133) which is not an isolable compound but just an arrangement of atoms through which the system passes during the reaction. Note that the carbon centre has been inverted during the reaction.

The inversion of the carbon centre

During the course of the reaction, the carbon centre undergoes an inversion – it is basically turned inside-out with the molecule passing through an arrangement in which all three hydrogen atoms lie in a plane perpendicular to the C–Br bond. This is similar to the inversion of ammonia described in Fig. 6.13 on p. 87.

For the reaction of bromomethane, there is no way that we could tell that the carbon atom has undergone an inversion. If, however, instead of three hydrogen atoms, the carbon has three different groups around it, then the inversion can be detected. This was demonstrated very neatly in a variation on the reaction involving radioactive iodide attacking 2-iodooctane as shown in Fig. 7.27.

The non-radioactive starting material 2-iodooctane and the product are not the same even when the radioactive atom is ignored. As can be seen in Fig. 7.28, they are in fact non-superimposable mirror images of each other.

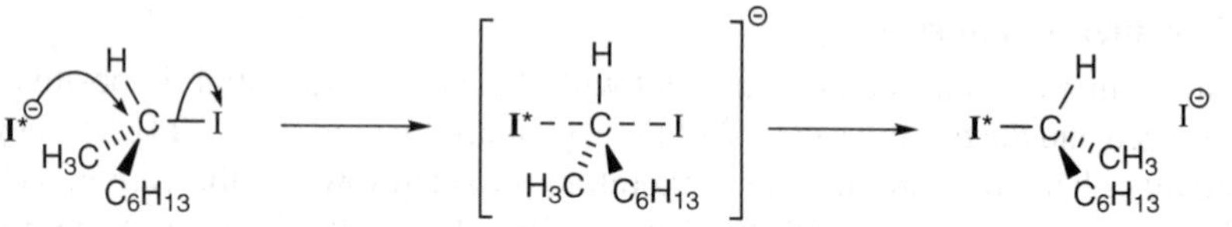

Fig. 7.27 The inversion at the carbon centre can be detected in this experiment involving the substitution of the iodine atom in 2-iodooctane by a radioactive iodide ion, denoted with an asterisk.

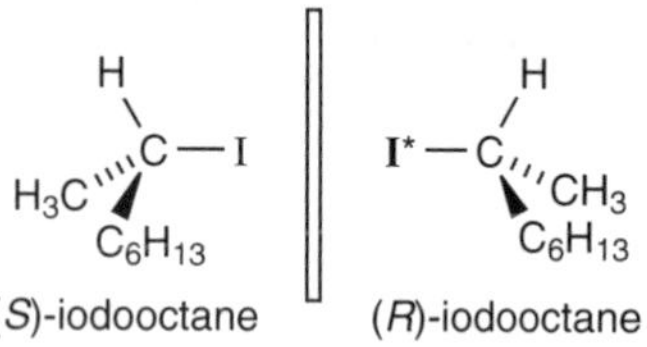

(*S*)-iodooctane (*R*)-iodooctane

Fig. 7.28 The starting material and product are non-superimposable mirror images of each other (ignoring the different isotopes of iodine). These isomers are differentiated with the labels *S* and *R*.

Compounds which are non-superimposable mirror images are called *enantiomers* or *optical isomers*. The term optical isomers arises from the fact that whilst (*R*)-2-iodooctane and (*S*)-2-iodooctane behave the same chemically, one isomer will rotate plane-polarized light clockwise, while the other isomer will rotate it anti-clockwise.

In the experiment, the rate of substitution in the reaction was followed by monitoring the incorporation of radioactive iodine into the 2-iodooctane; the rate of inversion was followed by measuring the degree to which plane-polarized light is rotated. The results clearly showed that each time a substitution occurred, the molecule underwent an inversion, which is what we would predict for this mechanism.

7.6 Nucleophilic attack on acyl chlorides

So far we have described reactions in which the HOMO of the nucleophile interacts with a LUMO which is either a $\pi^\star$ MO (for example, in ketones and aldehydes) or a $\sigma^\star$ MO (for example, in an alkyl halide). What happens if the molecule which the nucleophile attacks has both $\pi^\star$ and $\sigma^\star$ MOs available?

A simple example is the reaction between hydroxide ion and ethanoyl (acetyl) chloride, whose structure is shown in Fig. 7.29. The reactive carbon centre has both a double bond to oxygen and a bond to chlorine. With both these electronegative elements withdrawing electrons, the carbon has a partial positive charge but which orbital will a nucleophile such as hydroxide attack into? The hydroxide could either attack into the $\pi^\star$ of the C=O giving a tetrahedral species as shown in Fig. 7.30 (a) or it could attack into the C–Cl $\sigma^\star$ to form ethanoic (acetic) acid as shown in Fig. 7.30 (b).

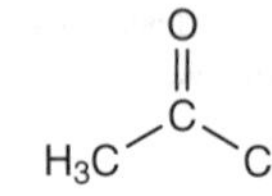

Fig. 7.29 Ethanoyl chloride.

The final product from the reaction of ethanoyl chloride with aqueous hydroxide is in fact the carboxylic acid so it is tempting to conclude that mechanism (b) is correct and that the hydroxide attacks into the C–Cl $\sigma^\star$ MO. It is tempting but sadly incorrect! In order to understand what does happen, we need to consider the orbitals involved.

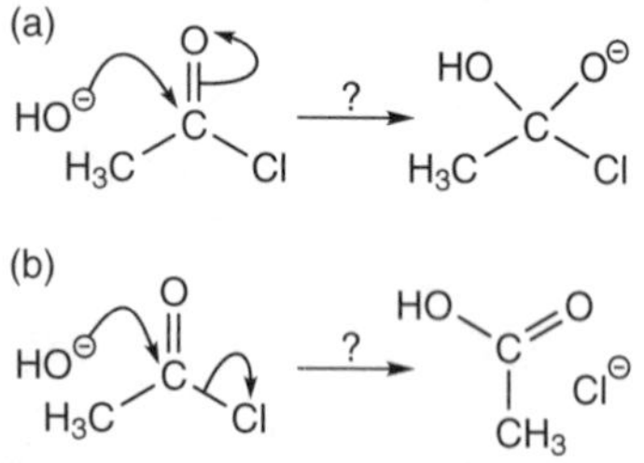

Fig. 7.30 Two possible routes of attack on ethanoyl chloride. In (a) the hydroxide attacks into the C–O $\pi^\star$ MO and in (b) it attacks into the C–Cl $\sigma^\star$ MO.

A computer calculation reveals that the LUMO for ethanoyl chloride is the $\pi^\star$ MO with the C–Cl $\sigma^\star$ MO lying somewhat higher in energy. These orbitals are shown in Fig. 7.31. The initial attack by the hydroxide nucleophile is into the C–O $\pi^\star$ MO as shown in Fig. 7.30 (a). There are two reasons why the nucleophile prefers to attack this MO rather than the C–Cl $\sigma^\star$.

Firstly, there is a better energy match between the HOMO of the hydroxide and the C–O $\pi^\star$ MO. This is because the C–Cl $\sigma^\star$ is higher in energy than the C–O $\pi^\star$.

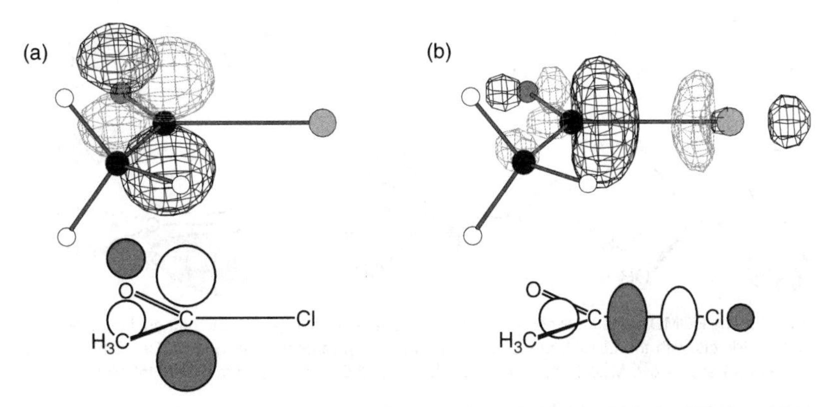

Fig. 7.31 Surface plots and cartoons of two MOs from ethanoyl chloride: (a) is the LUMO and (b) is the next highest energy MO. The LUMO is essentially just the C–O $\pi^\star$ MO and the orbital shown in (b) is essentially the C–Cl $\sigma^\star$ MO with some minor contributions from the other atoms.

Secondly, it is easier for the nucleophile to approach the $\pi^\star$ MO in the correct orientation. As we saw in Fig. 7.16 on p. 105, the best direction of approach for the nucleophile to attack the $\pi^\star$ MO is from above the plane of the C=O bond. This approach is unhindered in the case of hydroxide attacking ethanoyl chloride.

In contrast, when a nucleophile attacks a $\sigma^\star$ MO it must approach directly from behind, as shown in Fig. 7.24 on p. 109. In the case of hydroxide attacking ethanoyl chloride this approach is hindered by the methyl group and the oxygen, both of which are in the same plane as the $\sigma^\star$ MO.

Overall, then, attack on the $\pi^\star$ MO is the preferred route. However, as shown in Fig. 7.30 (a), this leads to what is called a tetrahedral intermediate. What we now need to work out is how this intermediate reacts further to give what we know to be the product, ethanoic acid, as shown in Fig. 7.32.

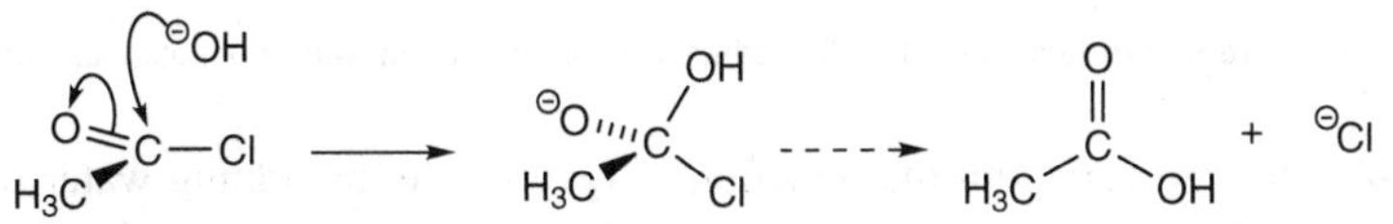

Fig. 7.32 The initial reaction between the hydroxide and ethanoyl chloride produces a tetrahedral intermediate. This then goes on to form the final products, ethanoic acid and a chloride ion.

The tetrahedral intermediate has no π bond and therefore no $\pi^\star$ MO either, so the C–Cl $\sigma^\star$ MO (which has remained almost unchanged after the attack of the hydroxide) is now the LUMO of the system. The lone pair of electrons on the negative oxygen is the HOMO in this molecule and, as we commented before, the negative charge raises the energy of this orbital. The interacting HOMO and LUMO are in this case in the same molecule; these orbitals interact to form some new orbitals in the product as shown in Fig. 7.33. The overall process is rather more complex than this figure suggests, as more orbitals are involved. A curly arrow mechanism for the whole reaction is shown in Fig. 7.34.

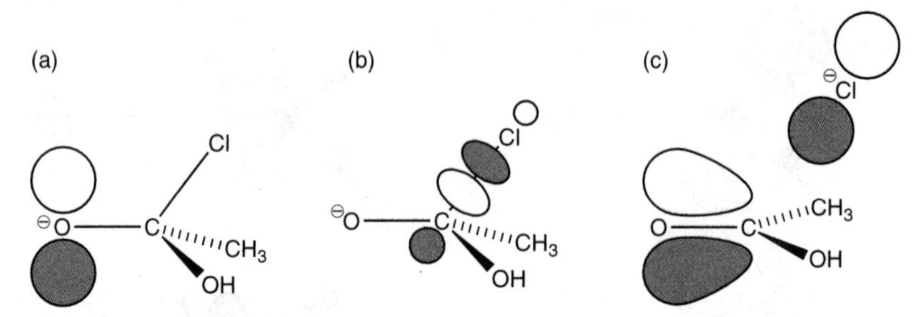

Fig. 7.33 The HOMO of the intermediate shown in (a) is essentially just a lone pair on the oxygen. This orbital interacts with the LUMO, the C–Cl σ^* shown in (b), to contribute to the MOs of the products. Diagram (c) shows two MOs in the products: the new C–O π bond and the *p* orbital on the chloride ion.

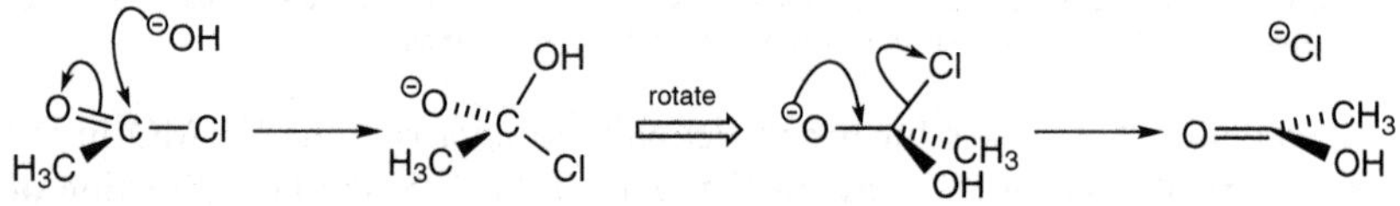

Fig. 7.34 A curly arrow mechanism for the reaction between hydroxide and ethanoyl chloride. The intermediate has been rotated after the initial attack to show more clearly how the final products are formed.

This attack on the C–O π^* by a nucleophile to give a tetrahedral intermediate and then the subsequent reformation of the C–O π bond is a common reaction which you will meet many other examples of in later chapters. One example is the reaction between a carboxylic acid, such as ethanoic acid, and an alcohol, such as methanol, to give an ester as shown in Fig. 7.35.

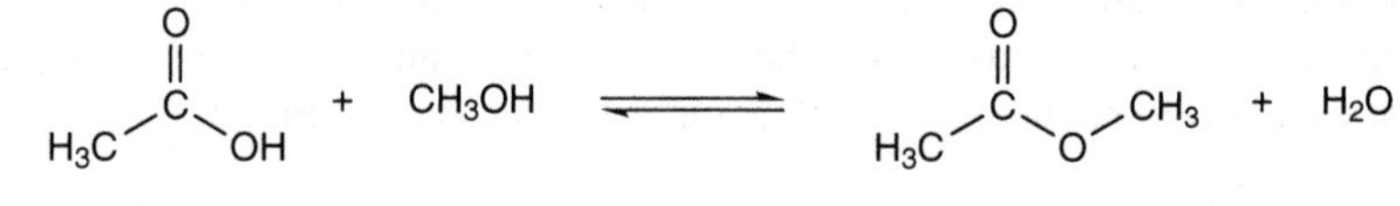

$$CH_3COOH + CH_3OH \rightleftharpoons CH_3COOCH_3 + H_2O$$

Fig. 7.35 The reversible formation of methyl ethanoate (an ester) and water from ethanoic acid and methanol.

We can also carry out this reaction in reverse, i.e. by adding water to the ester and allowing the hydrolysis reaction to produce the carboxylic acid and alcohol. In other words, this reaction is *reversible*.

Exactly why some reactions seem to be able to go in both directions and how chemists can try to alter the conditions to favour either the products or the reactants is the subject of the next chapter.

8 Equilibrium

In Chapter 2 we saw that whether or not a process will 'go' is controlled by the Second Law of Thermodynamics. What this law says is that a process will only be spontaneous – that is, take place on its own without continuous intervention from us – if it is associated with an *increase* in the entropy of the Universe (Section 2.2 on p. 7). We also saw that another and entirely equivalent way of applying the Second Law is to express it in terms of the Gibbs energy (Section 2.7 on p. 16); using this, a spontaneous process is one in which the Gibbs energy *decreases*.

Using either of these criteria we were able to explain why it is that water will freeze at -5 °C but not at $+5$ °C, since it is only at the lower temperature that water $\longrightarrow$ ice is accompanied by an increase in the entropy of the Universe or, equivalently, a decrease in the Gibbs energy.

Applying the same ideas to chemical reactions is rather more subtle, as we need to explain what determines the *position of equilibrium*, an idea introduced on p. 5 and illustrated in Fig. 2.1. Some reactions go almost entirely to products so the position of equilibrium lies very much towards the products; others hardly go at all and so the position of equilibrium lies towards the reactants. Other reactions come to equilibrium with significant amounts of reactants and products present. Our task in this chapter will be to work out what determines this position of equilibrium.

8.1 The approach to equilibrium

On p. 6 we discussed the equilibrium between NO_2 and N_2O_4:

$$2NO_2(g) \rightleftharpoons N_2O_4(g)$$

and pointed out that the reaction could come to the equilibrium position (at which significant amounts of both reactants and products are present) either starting from pure reactants (NO_2) or from pure products (N_2O_4). What this implies is that in going from either *pure products* or *pure reactants* to the equilibrium position the Gibbs energy must *decrease*. In other words, the equilibrium mixture of reactants and products has lower Gibbs energy than *either* pure products *or* pure reactants.

Shown in Fig. 8.1 is a plot of the Gibbs energy of the mixture of reactants and products (here NO_2 and N_2O_4) as a function of the composition. Pure reactants correspond to no reaction having taken place and so appear on the left; pure products correspond to complete reaction and so appear on the right. From the graph we see that the Gibbs energy goes to a minimum at some point intermediate between pure reactants and pure products; this minimum corresponds to the equilibrium position.

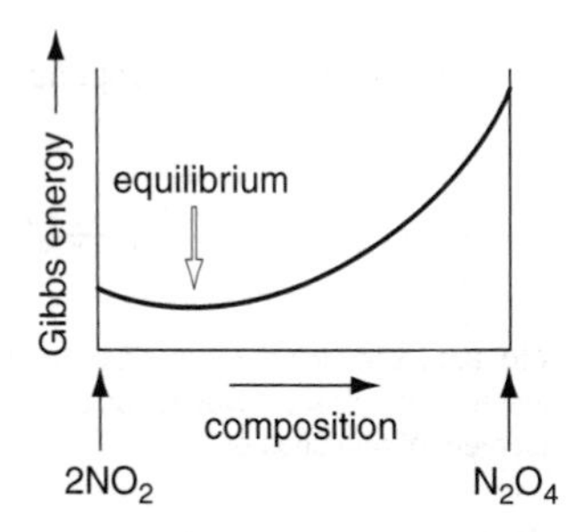

Fig. 8.1 Plot of the Gibbs energy of the reaction mixture as a function of its composition for the dimerization of NO_2. The left-hand side of the graph corresponds to pure reactants and the right to pure products. The minimum in the Gibbs energy corresponds to the position of equilibrium which can be approached either from pure reactants or pure products.

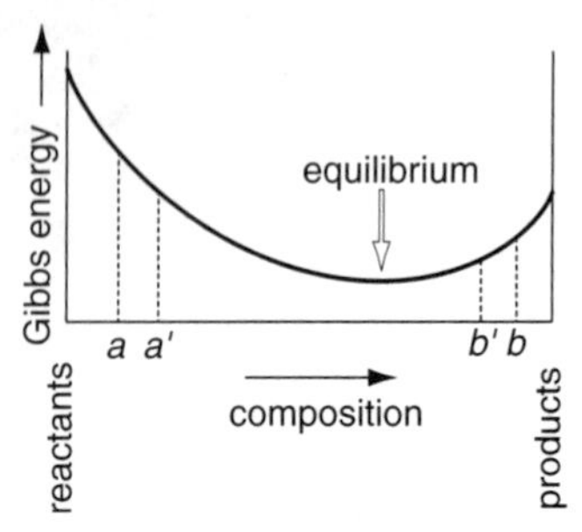

Fig. 8.2 Starting from point *a* the Gibbs energy will fall as we move to point *a'*; such a change will be spontaneous and will take us toward the position of equilibrium by increasing the amount of products. In contrast, from point *b* the direction which involves a reduction in Gibbs energy is towards point *b'*; for this spontaneous change the amount of products is reduced. Starting from either *a* or *b* the direction of spontaneous change is towards the equilibrium position, at which the Gibbs energy is a minimum.

No matter whether we start with pure products, or pure reactants, or some mixture of the two, the Gibbs energy falls as we approach equilibrium. For example, in Fig. 8.2 if we start at composition a, moving to a' is accompanied by a decrease in the Gibbs energy. The change from a to a' is therefore a spontaneous process which moves us towards the position of equilibrium by increasing the amount of products. Starting at b and moving to b' is also accompanied by a decrease in the Gibbs energy and so will be spontaneous; as for the change from a to a', the change from b to b' also moves us towards the equilibrium position but this time by *decreasing* the amount of products.

The shape of the graph therefore funnels the composition towards the value with the minimum Gibbs energy, which is the equilibrium position. From this point any change would involve an *increase* in the Gibbs energy, which is not permitted. So, once at its equilibrium value the composition cannot change. Our task is to try to understand why the plot of Gibbs energy as a function of composition has the form shown in Fig. 8.2 and also to identify the factors which influence the position of the minimum.

8.2 The equilibrium between two species

The simplest kind of equilibrium we can consider is when there is just one reactant (A) and one product (B):

$$\mathrm{A} \rightleftharpoons \mathrm{B}.$$

The equilibrium between isomers (illustrated in Fig. 8.3) is an example of this kind of reaction.

cyclopropane ⇌ propene

Fig. 8.3 Examples of equilibria involving just two chemical species which are isomers of one another.

Let us imagine that A and B are both gases and that we start out with one mole of pure A sealed in a container such that the pressure is 1 bar (1 bar is 10^5 N m^{-2}, very close to 1 atmosphere pressure). The reaction will come to equilibrium by some of the A converting to B, so we can specify the extent to which this has happened simply by quoting the percentage of B in the mixture. As the reaction involves no change in the number of moles, the pressure does not change as A and B interconvert.

Figure 8.4 shows how the Gibbs energy varies with percentage of B. On the left we have pure A, and so, as there is one mole of A present, the Gibbs energy is the *molar* Gibbs energy of *pure* A, $G_{\mathrm{m}}(\mathrm{A})$. On the right we have pure B, and so the Gibbs energy is just the *molar* Gibbs energy of *pure* B, $G_{\mathrm{m}}(\mathrm{B})$. To draw the graph we have arbitrarily chosen $G_{\mathrm{m}}(\mathrm{B})$ to be less than $G_{\mathrm{m}}(\mathrm{A})$. The task now is to try to explain why the curve has this shape, and in particular why it shows a minimum.

Suppose we start with pure A and then allow a very small amount of A to convert to B; as a consequence there will be a small change in the enthalpy and in the entropy. We will concentrate on the entropy change which we can think of as being due to two contributions.

The first contribution comes from the fact that we have converted some A to B. Given that the two substances are likely to have different molar entropies, changing some A into B is surely going to result in a change in entropy.

The second contribution comes from the *mixing* of this small amount of B into the bulk of A. So far we have not discussed this entropy of mixing, but you

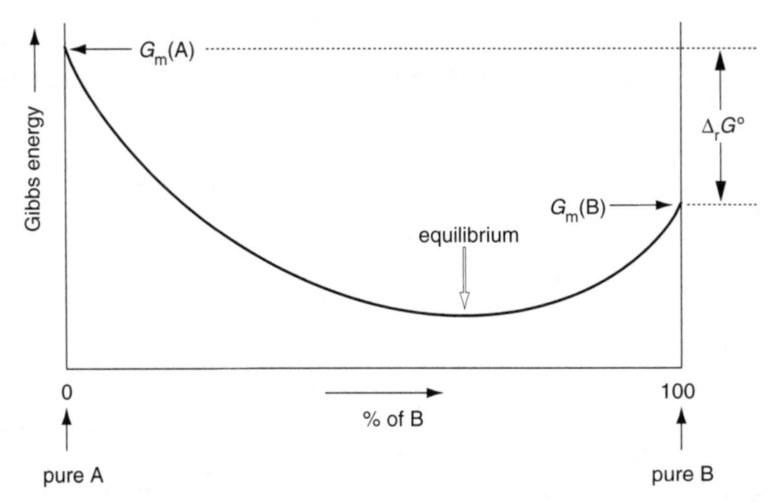

Fig. 8.4 Plot showing an example of how the Gibbs energy for the equilibrium between A and B varies with the composition, which in this case can be specified by the percentage of B. Arbitrarily, we have made the molar Gibbs energy of B lower than that of A.

can see that going from pure A to having a little B mixed in with A certainly gives rise to an increase in the entropy on the grounds that the mixture is more 'random' than pure A. We therefore expect this second contribution always to result in an increase in the entropy.

The first contribution might result in an increase or decrease in the entropy, but we argue that the entropy increase due to the second contribution – the one due to mixing – will *always* be dominant as there is a large increase in randomness (that is, in entropy) on going from pure A to a mixture with a small amount of B in it.

Given that ΔG is defined as $(\Delta H - T\Delta S)$ (p. 16), it follows that this increase in entropy will lead to a decrease in the Gibbs energy. Therefore, as we start from the left-hand side of the graph in Fig. 8.4 we expect the Gibbs energy to fall. Similarly, when starting from the right-hand side of the plot we can employ the same reasoning to argue that having a small amount of A mixed in with the bulk of B leads to an increase in the entropy and hence a decrease in the Gibbs energy.

Having now convinced ourselves why the Gibbs energy falls when we start from either side of this plot, we now need to work out what happens in between. A simple argument is to note that as the Gibbs energy falls when we start from the left or right, for the curve to join up there must be a minimum somewhere between these two points.

To determine the exact form of the curve we need some more detailed thermodynamics, which we do not have time to go into here – so for now you will simply have to take the form of these graphs on trust.

Locating the position of equilibrium

The remarkable thing about the plot in Fig. 8.4 is that it turns out that the location of the minimum – that is the percentage of B present at equilibrium – depends only on the *difference* between the molar Gibbs energies of A and B.

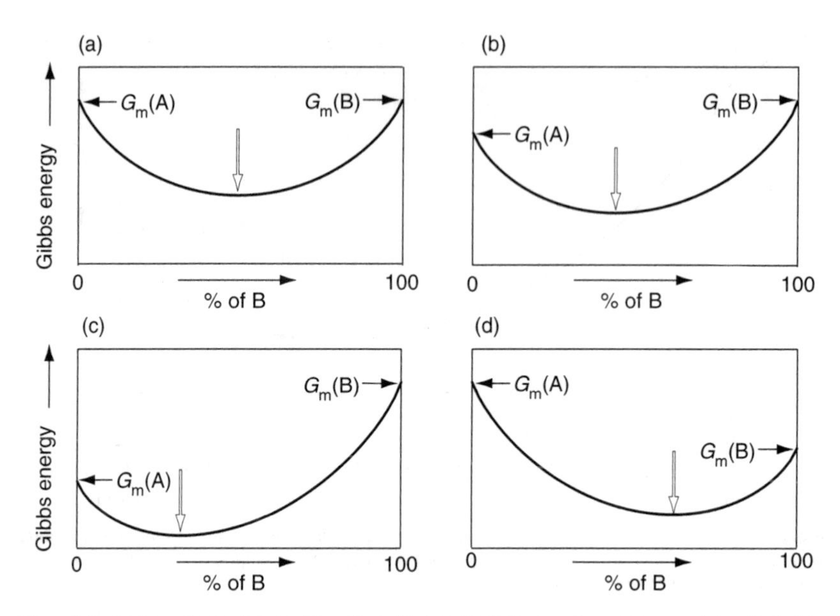

Fig. 8.5 Illustration of how the position of equilibrium (indicated by the open arrow) is affected by changing the relative molar Gibbs energies of A and B. In (a) G_m(A) and G_m(B) are equal, making the curve symmetrical and so the position of equilibrium lies at 50% B. In (b) G_m(A) is less than G_m(B) and so the position of equilibrium lies towards A; in (c) G_m(A) has been reduced further and the equilibrium therefore lies even further towards A. Finally, in (d), G_m(B) is less than G_m(A) and so the position of equilibrium lies towards B.

To prove that this is so needs a deeper study of the principles of thermodynamics than we have time to go into here, so we will simply have to take this result on trust.

Figure 8.5 illustrates how the position of the minimum shifts for different values of G_m(A) and G_m(B). In (a) the molar Gibbs energies of A and B are equal; the minimum in the Gibbs energy of the mixture is at 50% B. In (b) the molar Gibbs energy of A is lower than that of B, and so the position of equilibrium moves to the left, corresponding to the presence of more A than B; in (c) the molar Gibbs energy of A is lower still, and this shifts the position of equilibrium even further towards A. Finally, in (d) the molar Gibbs energy of B is lower than that of A so now the equilibrium lies towards B. We see that the position of equilibrium lies towards the species with the lowest molar Gibbs energy.

If you take a piece of string or chain and hold one end in your left hand and the other in your right, the string falls in a curve rather similar to those in Fig. 8.5. The height of your left hand represents the molar Gibbs energy of A, and the height of your right hand represents the molar Gibbs energy of B. If you hold your two hands level, your will see that the string hangs so as to make a minimum at a position midway between your hands: this is like plot (a). If you lower your left hand, the minimum moves towards the left, just as in plots (b) and (c). Conversely, if you lower your right hand, the minimum moves towards the right, as in plot (d).

As you know, when a reaction has come to equilibrium, the concentrations of products and reactants are related by the value of the *equilibrium constant*.

In the case of the equilibrium between A and B, the equilibrium constant, K, is given by

$$K = \frac{[\mathrm{B}]_{\mathrm{eq}}}{[\mathrm{A}]_{\mathrm{eq}}}$$

where $[\mathrm{A}]_{\mathrm{eq}}$ and $[\mathrm{B}]_{\mathrm{eq}}$ are the *equilibrium* concentrations of A and B respectively. If at equilibrium more B is present than A, K is greater than 1 and the equilibrium lies towards the products. Conversely, if the equilibrium lies towards the reactants there will be more A present than B and the equilibrium constant will be less than 1.

The percentage of B at which the minimum Gibbs energy occurs determines the value of the equilibrium constant. We have already described how the location of the minimum is determined solely by the difference in the molar Gibbs energies of A and B. It therefore follows that the value of the equilibrium constant must be related to this difference, and some more detailed thermodynamics shows us that:

$$G_{\mathrm{m}}(\mathrm{B}) - G_{\mathrm{m}}(\mathrm{A}) = -RT \ln K \tag{8.1}$$

where R is the gas constant (8.3145 J K^{-1} mol^{-1}) and T is the absolute temperature (in K).

This equation has the correct form, as if $G_{\mathrm{m}}(\mathrm{B})$ is less than $G_{\mathrm{m}}(\mathrm{A})$, the left-hand side will be negative and so $\ln K$ will be positive. This means that K is greater than 1 so that at equilibrium the product B is favoured, as depicted in Fig. 8.5 (d).

In contrast, if $G_{\mathrm{m}}(\mathrm{A})$ is less than $G_{\mathrm{m}}(\mathrm{B})$, the left-hand side of Eq. 8.1 will be positive and so $\ln K$ will be negative. This means that K will be a fraction between 0 and 1, favouring the reactant A at equilibrium, as depicted in Fig. 8.5 (b) and (c).

The surprising thing about Eq. 8.1 is that it tells us that the equilibrium constant depends only on the difference between the molar Gibbs energies of *pure* A and *pure* B. This really is quite a remarkable result and worth discussing a little more before we move on.

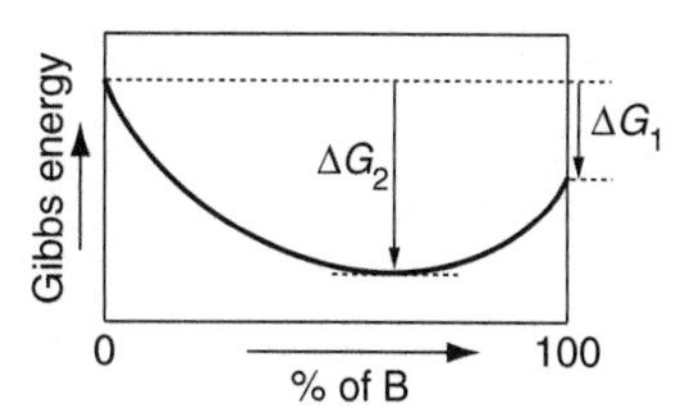

Fig. 8.6 The molar Gibbs energy of the product B is lower than that of the reactant A, so going from pure A to pure B involves a decrease in the Gibbs energy (ΔG_1). However, for certain ratios of A to B the Gibbs energy of the mixture is *lower* than that of pure B; the lowest value of the Gibbs energy corresponds to the equilibrium point and moving to this point from pure A involves the largest decrease in the Gibbs energy (ΔG_2).

Interpretation

Suppose that $G_{\mathrm{m}}(\mathrm{B})$ is less than $G_{\mathrm{m}}(\mathrm{A})$. This means that if we start with pure A and convert *all* of it to B the process would be accompanied by a decrease in Gibbs energy and therefore be spontaneous. This change in Gibbs energy is shown as ΔG_1 in Fig. 8.6.

However, what we have seen is that there are some ratios of A to B for which the mixture has lower Gibbs energy than *either* pure A *or* pure B; one of these mixtures has the minimum Gibbs energy and this is the one which corresponds to equilibrium. As is shown in Fig. 8.6, going from pure A to the equilibrium mixture involves a larger decrease in the Gibbs energy, ΔG_2, than going to pure B.

8.3 General chemical equilibrium

These ideas about the simple A $\rightleftharpoons$ B equilibrium are just a special case of a more general result which is that for any reaction the equilibrium constant, K, is given by

$$\Delta_r G^\circ = -RT \ln K. \tag{8.2}$$

$\Delta_r G^\circ$ is the *standard Gibbs energy change* which can be computed from

$$\Delta_r G^\circ = \Delta_r H^\circ - T\Delta_r S^\circ \tag{8.3}$$

where $\Delta_r H^\circ$ is the *standard enthalpy change* and $\Delta_r S^\circ$ is the *standard entropy change* for the reaction. The first thing we need to do is describe what we mean by a 'standard change'.

Standard states and standard changes

We first need to introduce the concept of a *standard state*; this is a particular state of the substance in question, defined according to the following agreed convention:

substance	standard state
gases	the pure gas at a pressure of 1 bar (10^5 N m^{-2})
solids	the pure solid
liquids	the pure liquid
solutions	the solution at unit concentration

The standard state given in the table for a solution only applies to an *ideal solution* in which there are no solute–solvent interactions; for real solutions in which there are such interactions the definition is more complex and we shall not go into this here.

The standard state is denoted by a superscript $^\circ$ or sometimes a Plimsoll line $^\ominus$. It is sometimes thought, erroneously, that the standard state implies a certain temperature: this is *not* the case. In fact, when quoting values for standard enthalpy, entropy or Gibbs energy changes we must *always* state the temperature at which this value applies as these quantities are temperature dependent.

Having introduced the standard state we can now define *standard changes* for reactions. Such changes refer to a particular *balanced chemical equation*. For example, $\Delta_r G^\circ$ for the reaction

$$H_2(g) + \tfrac{1}{2}O_2(g) \longrightarrow H_2O(l) \tag{8.4}$$

is the *change* in Gibbs energy when one mole of $H_2(g)$ reacts with half a mole of $O_2(g)$ to give one mole of $H_2O(l)$, *all of the species being present in their standard states and at the stated temperature*.

It is very important to understand that the value of $\Delta_r G^\circ$ refers to complete reaction, i.e. the hydrogen and oxygen must be converted completely to water. Note too how the values of the stoichiometric coefficients come into the definition of $\Delta_r G^\circ$.

So, if we doubled the stoichiometric coefficients in Eq. 8.4 and wrote it as

$$2H_2(g) + O_2(g) \longrightarrow 2H_2O(l)$$

$\Delta_r G^\circ$ would be the change in Gibbs energy when two moles of $H_2(g)$ react with one mole of $O_2(g)$ to give two moles of $H_2O(l)$. The numerical value of $\Delta_r G^\circ$ for this reaction would be twice the value of that for Eq. 8.4.

A second example is the formation of ammonia:

$$\tfrac{1}{2}N_2(g) + \tfrac{3}{2}H_2(g) \longrightarrow NH_3(g)$$

for which $\Delta_r G^\circ$ is the *change* in Gibbs energy when half a mole of $N_2(g)$ reacts with $\tfrac{3}{2}$ moles of $H_2(g)$ to give one mole of $NH_3(g)$, all in their standard states and at the stated temperature.

In many ways $\Delta_r G^\circ$ is a hypothetical quantity as it refers to the change in Gibbs energy when the reactants, in their standard states, are converted *completely* to products, also in their standard states. If we actually mixed the reactants together in their standard states there is no guarantee that the reaction would go completely to products in their standard states. All that we can be sure will actually happen is that the reaction will go to its equilibrium position, which often will not involve complete conversion to products.

For example, at standard pressure and at room temperature not much ammonia will be formed by mixing N_2 and H_2. However, this does not stop us from imagining what the change in Gibbs energy would be *if* the reaction went completely from N_2 and H_2 to NH_3.

The standard enthalpy change is defined in the same way as $\Delta_r G^\circ$: for the stated reaction it is the change in enthalpy when the reaction proceeds completely from reactants to products, all species being in their standard states. Similarly, the standard entropy change is the change in entropy for this process.

Returning to the simple equilibrium between A and B which was discussed in Section 8.2 on p. 114, we can see that because we took the pressure to be 1 bar, the difference in the molar Gibbs energies of A and B is in fact the same thing as $\Delta_r G^\circ$ for the reaction; this is shown in Fig. 8.4 on p. 115. So, Eq. 8.1 is just a special case of Eq. 8.2.

Determining standard changes

The usual way of finding $\Delta_r H^\circ$ is to use tabulated values of *standard enthalpies of formation*, $\Delta_f H^\circ$. These are defined in such a way that for any balanced chemical equation we can find $\Delta_r H^\circ$ by adding together the $\Delta_f H^\circ$ values of the products and subtracting those of the reactants, taking into account the stoichiometric coefficients. For example, to find $\Delta_f H^\circ$ for the reaction

$$SO_2(g) + \tfrac{1}{2}O_2(g) \longrightarrow SO_3(g)$$

we can imagine a Hess' Law cycle in which the reactants are broken down into their elements (in their standard states) and then these are re-formed to products:

$$\begin{array}{ccccc} SO_2(g) & + & \tfrac{1}{2}O_2(g) & \xrightarrow{\Delta_r H^\circ} & SO_3(g) \\ \downarrow -\Delta_f H^\circ(SO_2) & & \downarrow -\tfrac{1}{2}\Delta_f H^\circ(O_2) & & \uparrow \Delta_f H^\circ(SO_3) \end{array}$$

elements in their standard states

$\Delta_r H^\circ$ is thus given by

$$\Delta_r H^\circ = \Delta_f H^\circ(SO_3(g)) - \Delta_f H^\circ(SO_2(g)) - \tfrac{1}{2}\Delta_f H^\circ(O_2(g)).$$

As you will recall, the standard enthalpy of formation of an element is zero, so in the above equation $\Delta_f H^\circ(O_2(g)) = 0$.

Similarly, $\Delta_r S^\circ$ values are found by adding together the standard (absolute) entropies (S°) of the products and subtracting those of the reactants, taking into account the stoichiometric coefficients. So, for the above reaction:

$$\Delta_r S^\circ = S^\circ(SO_3(g)) - S^\circ(SO_2(g)) - \tfrac{1}{2}S^\circ(O_2(g)).$$

As with enthalpies of formation, extensive tabulations of standard entropies are available. Note that the standard absolute entropies of the elements are *not* zero.

Having found $\Delta_r H^\circ$ and $\Delta_r S^\circ$ we then find $\Delta_r G^\circ$ using Eq. 8.3

$$\Delta_r G^\circ = \Delta_r H^\circ - T\Delta_r S^\circ.$$

An example

Let us use this approach to find $\Delta_r G^\circ$ for

$$SO_2(g) + \tfrac{1}{2}O_2(g) \longrightarrow SO_3(g).$$

From tables we find (all at 298 K and for the gaseous state) $\Delta_f H^\circ(SO_3) = -396$ kJ mol^{-1} and $\Delta_f H^\circ(SO_2) = -297$ kJ mol^{-1}; so, remembering that $\Delta_f H^\circ$ for O_2 is zero,

$$\begin{aligned} \Delta_r H^\circ &= \Delta_f H^\circ(SO_3(g)) - \Delta_f H^\circ(SO_2(g)) - \tfrac{1}{2}\Delta_f H^\circ(O_2(g)) \\ &= -396 - (-297) - \tfrac{1}{2} \times 0 \\ &= -99 \text{ kJ mol}^{-1}. \end{aligned}$$

The same tables give us the following standard entropies: $S^\circ(SO_3) = 257$ J K^{-1} mol^{-1}, $S^\circ(SO_2) = 248$ J K^{-1} mol^{-1} and $S^\circ(O_2) = 205$ J K^{-1} mol^{-1}; so

$$\begin{aligned} \Delta_r S^\circ &= S^\circ(SO_3(g)) - S^\circ(SO_2(g)) - \tfrac{1}{2}S^\circ(O_2(g)) \\ &= 257 - 248 - \tfrac{1}{2} \times 205 \\ &= -93.5 \text{ J K}^{-1} \text{ mol}^{-1}. \end{aligned}$$

These values allow us to find $\Delta_r G^\circ$ at 298 K:

$$\begin{aligned} \Delta_r G^\circ &= \Delta_r H^\circ - T\Delta_r S^\circ \\ &= -99 \times 10^3 - 298 \times (-93.5) \\ &= -71 \times 10^3 \text{ J mol}^{-1}. \end{aligned}$$

Next we will see what this implies for the value of the equilibrium constant.

Equilibrium constants

The equilibrium constant and the standard Gibbs energy change for a reaction are related according to Eq. 8.2

$$\Delta_r G^\circ = -RT \ln K$$

which can be rewritten to give K as

$$K = \exp\left(\frac{-\Delta_r G^\circ}{RT}\right).$$

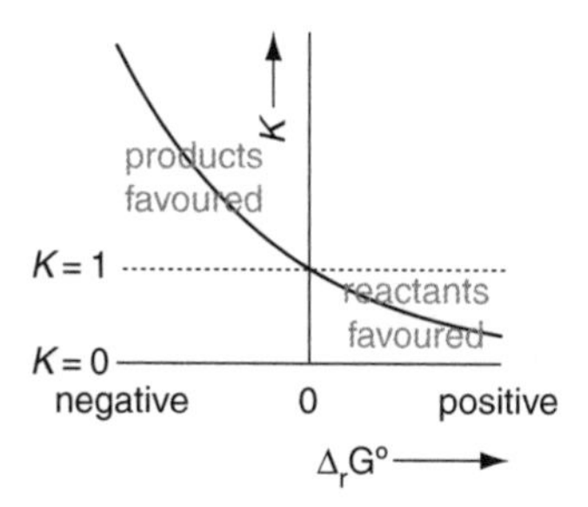

Fig. 8.7 Graph showing how the sign of $\Delta_r G^\circ$ affects the position of equilibrium. If $\Delta_r G^\circ$ is positive the equilibrium constant is less then 1 (but still positive) which means that the reactants are favoured. If $\Delta_r G^\circ$ is negative the equilibrium constant is greater than 1 and the products are favoured.

This relationship tells us that if $\Delta_r G^\circ$ is negative, the exponent (the expression in the brackets) will be positive and so the equilibrium constant will be greater than 1, i.e. the products are favoured. On the other hand, if $\Delta_r G^\circ$ is positive, the exponent will be negative, giving an equilibrium constant of less than 1, which means that the reactants are favoured. These points are illustrated in Fig. 8.7.

In the case of the formation of SO_3 from SO_2 and O_2 we found that at 298 K, $\Delta_r G^\circ = -71 \times 10^3$ J mol^{-1} and so

$$\begin{aligned} K &= \exp\left(\frac{-\Delta_r G^\circ}{RT}\right) \\ &= \exp\left(\frac{-(-71 \times 10^3)}{8.3145 \times 298}\right) \\ &= 2.29 \times 10^{12}. \end{aligned}$$

The equilibrium constant is very large, implying that the equilibrium will be totally in favour of the products in this reaction.

Before we move on there is one point of possible confusion we need to clear up. In Chapter 2 we showed that a spontaneous process must be accompanied by a decrease in the Gibbs energy, and we have used this criterion to discuss chemical equilibrium. Consider the situation shown in Fig. 8.8, in which $\Delta_r G^\circ$ is positive, i.e. the pure product, B, is higher in Gibbs energy than the pure reactant, A. It therefore follows that the conversion of pure A to pure B (at standard pressure) will *not* take place as this would be accompanied by an increase in the Gibbs energy.

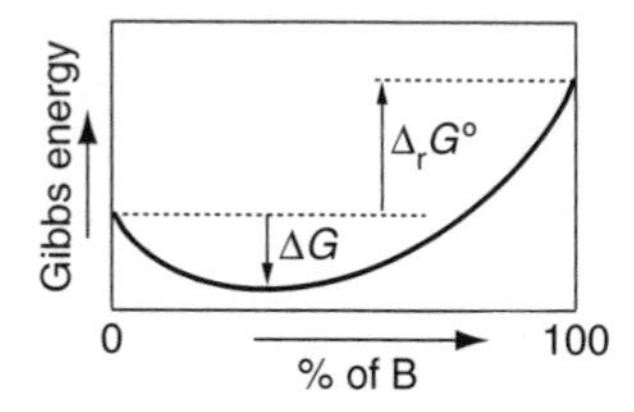

Fig. 8.8 Illustration of how a reaction which has a positive $\Delta_r G^\circ$ still takes place to a certain extent, i.e. comes to a position of equilibrium in which some product is present. Going from pure A to pure B would involve an increase in the Gibbs energy and so is not allowed. Nevertheless, going from pure A to the equilibrium mixture involves a decrease in the Gibbs energy and so is allowed. A reaction with a positive $\Delta_r G^\circ$ is not 'forbidden' but simply comes to a position of equilibrium which favours the reactants.

However, this does not mean that the conversion of *some* A to B is forbidden – far from it. We see this clearly in Fig. 8.8 where the conversion of pure A to the equilibrium mixture is accompanied by a decrease in the Gibbs energy, even though $\Delta_r G^\circ$ is positive.

So, if a reaction has a positive $\Delta_r G^\circ$ this does *not* mean that the reaction will not take place; rather, it means that the reaction will come to an equilibrium position which favours the reactants. Similarly, a reaction which has a negative $\Delta_r G^\circ$ will not go entirely to products, but it will go to an equilibrium position which favours the products.

In fact, because of the exponential relationship between $\Delta_r G^\circ$ and the equilibrium constant, once $\Delta_r G^\circ$ becomes more positive than a certain value the equilibrium constant becomes so small that the equilibrium lies entirely to the

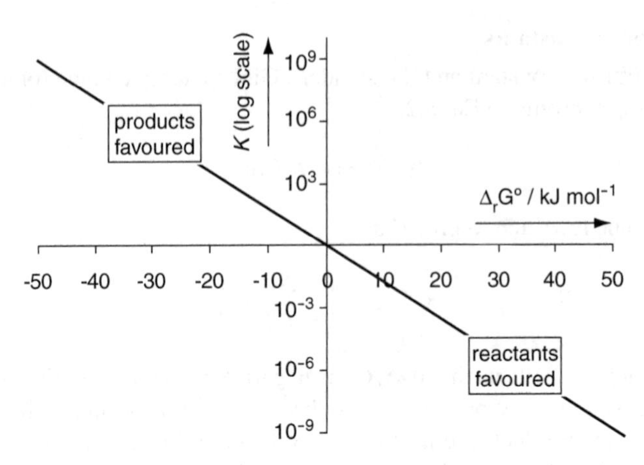

Fig. 8.9 Plot showing how the equilibrium constant, K, varies with the standard Gibbs energy change of the reaction ($\Delta_r G^\circ$) at a temperature of 298 K; note that the vertical scale is logarithmic. If $\Delta_r G^\circ$ is more negative than about -40 kJ mol^{-1} the equilibrium constant is so large that essentially the conversion to products is complete. Similarly, if $\Delta_r G^\circ$ is more positive that about +40 kJ mol^{-1} the equilibrium constant is so small that essentially no products are formed.

reactants, i.e. the reaction does not proceed to a significant extent. Similarly, once $\Delta_r G^\circ$ is more negative than a certain value, the equilibrium constant becomes so large that to all intents and purposes the equilibrium lies entirely in favour of the products.

Figure 8.9 shows how K varies with $\Delta_r G^\circ$; note that the vertical scale is logarithmic. We see from this plot that once $\Delta_r G^\circ$ exceeds around +40 kJ mol^{-1} the equilibrium constant is so small that essentially none of the reactants have become products. On the other end of the scale, if $\Delta_r G^\circ$ is more negative than about -40 kJ mol^{-1} the equilibrium constant is so large that to all practical intents and purposes the reaction has gone entirely to products. Finally, note that if $\Delta_r G^\circ = 0$ the equilibrium constant is one, implying an equal balance between products and reactants.

8.4 Influencing the position of equilibrium

For a given reaction the value of the equilibrium constant is fixed once we specify the temperature and the states of the reactants (solid, liquid, gas, etc.). This section describes a number of ways in which the actual amount of products can be increased – in other words, how to increase the yield of the reaction, something we often want to do for practical reasons.

Temperature

Both the equation relating $\Delta_r G^\circ$ to the equilibrium constant

$$\Delta_r G^\circ = -RT \ln K \tag{8.2}$$

and the definition of $\Delta_r G^\circ$

$$\Delta_r G^\circ = \Delta_r H^\circ - T\Delta_r S^\circ \tag{8.3}$$

involve temperature explicitly, so we can expect the temperature to have an influence on the value of the equilibrium constant. It turns out that the values of $\Delta_r S^\circ$ and $\Delta_r H^\circ$ also depend on temperature, although not very strongly. Over a modest temperature range it is reasonable to assume that they are constant, which is what we will do from now on.

As both Eq. 8.2 and Eq. 8.3 are expressions for $\Delta_r G^\circ$ we can equate their right-hand sides to give

$$-RT \ln K = \Delta_r H^\circ - T\Delta_r S^\circ.$$

Dividing both sides by $-RT$ gives

$$\ln K = -\frac{\Delta_r H^\circ}{R}\left(\frac{1}{T}\right) + \frac{\Delta_r S^\circ}{R}. \quad (8.5)$$

This equation tells us that the way in which the equilibrium constant varies with temperature depends on $\Delta_r H^\circ$.

If the reaction is *endothermic* ($\Delta_r H^\circ$ is positive) the term $(-\Delta_r H^\circ/RT)$ is negative and so increasing the temperature makes it *less negative*. This means that as T increases both $\ln K$ and K increase, as is illustrated in Fig. 8.10. In other words, for an endothermic reaction increasing the temperature shifts the equilibrium towards the products.

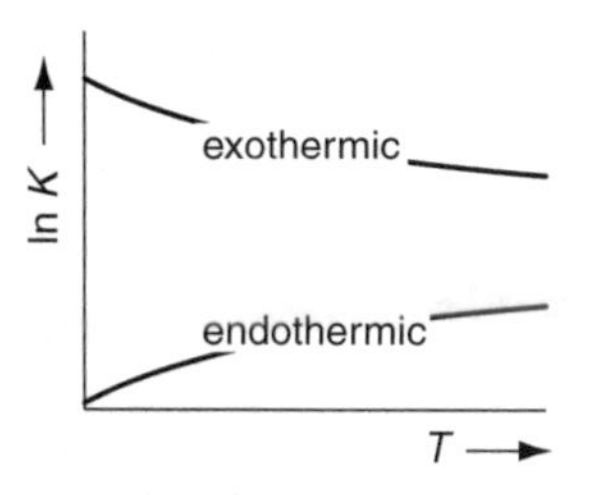

Fig. 8.10 Graph showing the different ways in which ln K varies with temperature (as predicted by Eq. 8.5) for an exothermic and an endothermic reaction. For an endothermic reaction the equilibrium constant increases with temperature; for an exothermic reaction, the opposite is the case. Thus, Eq. 8.5 is an expression of Le Chatelier's Principle.

If the reaction is *exothermic* ($\Delta_r H^\circ$ is negative) the term $(-\Delta_r H^\circ/RT)$ is positive and so increasing the temperature makes it *less positive*. As a result increasing the temperature makes both $\ln K$ and K smaller (see Fig. 8.10). In words, for an exothermic reaction, increasing the temperature shifts the equilibrium towards the reactants.

You are probably familiar with these conclusions as they are often described as being a consequence of *Le Chatelier's Principle*, one statement of which is that 'the equilibrium shifts in order to oppose the change'. So, for an endothermic reaction, an increase in the temperature is opposed by absorbing heat which means that the reaction must move further to products, i.e. the equilibrium constant must increase. Equation 8.5 puts this application of Le Chatelier's Principle on a quantitative footing.

We will look at two examples of the effect of temperature on the position of equilibrium.

Dimerization of NO_2

For our first example, let us look again at the dimerization of $NO_2(g)$:

$$2NO_2(g) \rightleftharpoons N_2O_4(g).$$

From tables, we find for this reaction that at 298 K, $\Delta_r H^\circ = -57$ kJ mol^{-1} and $\Delta_r S^\circ = -176$ J K^{-1} mol^{-1}. These values make sense as a bond is being made, so we expect the release of energy, and the reduction in entropy is associated with two moles of gas going to one.

At 298 K we can compute $\Delta_r G^\circ$ and then K:

$$\begin{aligned}\Delta_r G^\circ &= \Delta_r H^\circ - T\Delta_r S^\circ \\ &= -57 \times 10^3 - 298 \times (-176) \\ &= -4.6 \times 10^3 \text{ J mol}^{-1}.\end{aligned}$$

$$K = \exp\left(\frac{-\Delta_r G^\circ}{RT}\right)$$
$$= \exp\left(\frac{-(-4.6 \times 10^3)}{8.3145 \times 298}\right)$$
$$= 6.3$$

We see that $\Delta_r G^\circ$ is negative and so the equilibrium constant is greater than 1, showing that at equilibrium the products are preferred. However, $\Delta_r G^\circ$ is not that negative and so the equilibrium constant is modest (6.3) which tells us that although the equilibrium lies on the side of the products, there are still plenty of reactants present.

As the reaction is exothermic, lowering the temperature will increase the proportion of products. Let us repeat the same calculation at 273 K:

$$\Delta_r G^\circ = -57 \times 10^3 - 273 \times (-176) = -9.0 \times 10^3 \text{ J mol}^{-1}$$

$$K = \exp\left(\frac{-(-9.0 \times 10^3)}{8.3145 \times 273}\right) = 52.$$

The equilibrium constant is now much larger, meaning that at this lower temperature the proportion of dimer (N_2O_4) present is larger than it was at 298 K. Repeating the calculation at 373 K gives $K = 0.062$, showing that at this higher temperature the fraction of dimer present is very small. These calculations illustrate very nicely how we can shift the equilibrium one way or the other simply by altering the temperature.

Extraction of metals

Our second example is rather different, and concerns the commercially very important process of extracting metals from their oxides. Typically, this is done by heating the oxide with carbon which acts as a reducing agent, thus releasing the free metal and forming CO or CO_2. The chemical processes involved are often fairly complex, but for a simple divalent metal oxide (general formula MO) we can get a flavour of what is going on by just considering the reaction

$$\text{MO}(s) + \text{C}(s) \rightleftharpoons \text{CO}(g) + \text{M}(s). \quad (8.6)$$

The following table gives thermochemical data (at 298 K) for C, CO, the metals copper, lead and zinc and their oxides.

	C	CO	Cu	CuO	Pb	PbO	Zn	ZnO
$\Delta_f H^\circ$ / kJ mol^{-1}		−111		−157		−219		−348
S°/ J K^{-1} mol^{-1}	5.74	198	33.2	42.6	64.8	66.5	25.4	40.3

Using these we can compute $\Delta_r G^\circ$ for the reaction of Eq. 8.6 at any temperature (provided we assume that $\Delta_r H^\circ$ and $\Delta_r S^\circ$ are not temperature dependent). The results for reaction at 1000 °C and 1500 °C are shown in the following table (you can check to see if you agree with the numbers); similar data are also shown in Fig. 8.11.

oxide	$\Delta_r H^\circ$ / kJ mol^{-1}	$\Delta_r S^\circ$ / J K^{-1} mol^{-1}	at 1000 °C $\Delta_r G^\circ$ / kJ mol^{-1}	at 1500 °C $\Delta_r G^\circ$ / kJ mol^{-1}
CuO	46	182	−186	−277
PbO	108	190	−133	−229
ZnO	274	177	12.3	−76.2

As expected, for all the metals the reaction has a positive $\Delta_r S^\circ$ resulting from the generation of one mole of gas on the right-hand side of the equation. The reactions are all endothermic which means that, as $\Delta_r G^\circ = \Delta_r H^\circ - T\Delta_r S^\circ$, a negative value of $\Delta_r G^\circ$ can only be achieved by a positive $\Delta_r S^\circ$ (which is what we have here) and a sufficiently high temperature that the $-T\Delta_r S^\circ$ term is dominant.

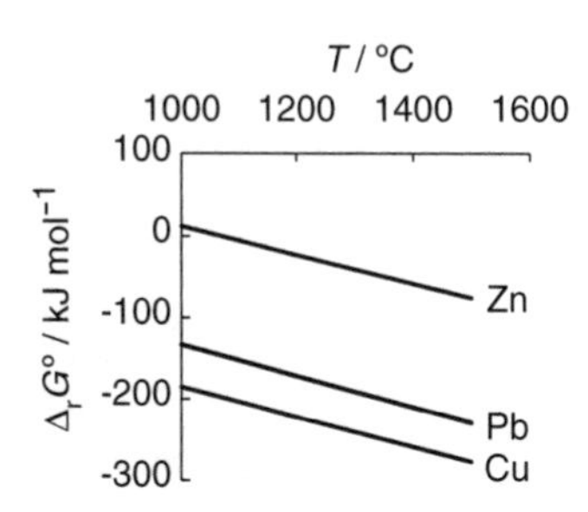

Fig. 8.11 Plot showing how the standard Gibbs energy change for the reaction of Eq. 8.6 varies with temperature for three different metals; the process will only be successful if $\Delta_r G^\circ$ is significantly negative. It is clear, therefore, that the extraction of Zn requires a higher temperature than for Cu or Pb. The values of $\Delta_r G^\circ$ have been computed assuming that $\Delta_r H^\circ$ and $\Delta_r S^\circ$ are independent of temperature, and so the plots are straight lines.

We need to be careful here not to fall into making an error about how temperature influences the *equilibrium constant*. Read casually, the previous paragraph makes it sound as if it is the sign of $\Delta_r S^\circ$ which determines how the equilibrium will change with temperature. This is not what the paragraph says: what it says is that the sign of $\Delta_r S^\circ$ determines whether $\Delta_r G^\circ$ rises or falls with temperature, but $\Delta_r G^\circ$ is *not* the equilibrium constant. To find the equilibrium constant from $\Delta_r G^\circ$ we need to use $\Delta_r G^\circ = -RT \ln K$, a relationship which introduces another temperature dependence; so knowing how $\Delta_r G^\circ$ varies with temperature is not the whole story.

In fact, as we have seen, the effect of temperature on the equilibrium constant is given by Eq. 8.5

$$\ln K = -\frac{\Delta_r H^\circ}{R}\left(\frac{1}{T}\right) + \frac{\Delta_r S^\circ}{R}.$$

This shows that it is the sign of $\Delta_r H^\circ$ which determines how the equilibrium constant depends on temperature as it is $\Delta_r H^\circ$ which is multiplying the $(1/T)$ term. So, although it is true that if $\Delta_r S^\circ$ is positive, increasing the temperature will cause $\Delta_r G^\circ$ to decrease, to understand the effect of temperature on the equilibrium constant we need to look at the sign of $\Delta_r H^\circ$. What the table tells us therefore is that *as the reactions are all endothermic*, increasing the temperature will move the equilibrium to the products.

This subtle point about the temperature dependence does not alter the basic idea that if $\Delta_r G^\circ$ is substantially negative the position of equilibrium will lie almost entirely toward the products. At 1000 °C, $\Delta_r G^\circ$ is substantially negative for copper and lead, showing that metal + carbon monoxide will be formed. However, for zinc $\Delta_r G^\circ$ is positive at this temperature, which tells us that the equilibrium will favour the reactants (ZnO + C) and so zinc metal will not be extracted.

In order for the reduction to produce zinc we have to raise the temperature higher, and we can see from the table that at 1500 °C, $\Delta_r G^\circ$ is now substantially negative for the ZnO reduction. We therefore conclude that to obtain copper or lead from their oxides, reduction with carbon at a temperature of 1000 °C will be sufficient, but to obtain zinc from its oxide a substantially higher temperature will be needed. Figure 8.11 presents the results in a slightly different way, and from this graph we can see that for Zn the switch over from a positive to a negative $\Delta_r G^\circ$ occurs at around 1100 °C. In practice we need to be at a higher temperature than this to ensure that $\Delta_r G^\circ$ is sufficiently negative for the equilibrium to favour products strongly.

Concentration

The value of the equilibrium constant determines the ratio of the concentrations of products and reactants, and this has a fixed value at a particular temperature. However, we can influence the *actual* concentration of a product by altering the concentrations of the other species. How this works is best illustrated by an example, such as the equilibrium shown in Fig. 8.12 which is the reaction used to form an ester (**E**) from an alcohol (**A**) and a carboxylic acid (**C**).

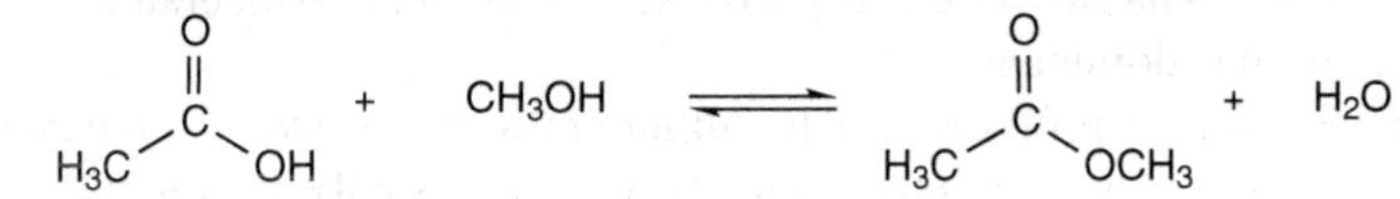

$$\underset{\mathbf{C}}{\mathrm{H_3C{-}C({=}O){-}OH}} + \underset{\mathbf{A}}{\mathrm{CH_3OH}} \rightleftharpoons \underset{\mathbf{E}}{\mathrm{H_3C{-}C({=}O){-}OCH_3}} + \mathrm{H_2O}$$

Fig. 8.12 An ester, **E**, is formed from the reaction between a carboxylic acid, **C**, and an alcohol, **A**; water is also produced in the reaction. Here the reaction is between ethanoic (acetic) acid and methanol to give methyl ethanoate (methyl acetate).

The equilibrium constant, K, for this reaction is given by:

$$K = \frac{[\mathbf{E}]_{eq}[\mathrm{H_2O}]_{eq}}{[\mathbf{C}]_{eq}[\mathbf{A}]_{eq}} \tag{8.7}$$

where $[\mathbf{E}]_{eq}$ means the equilibrium concentration of the ester, and so on. Remember that at a particular temperature the value of K is fixed, so the concentrations of the four species must adjust themselves so that the ratio on the right-hand side of Eq. 8.7 is equal to the value of K.

Suppose we let the reaction come to equilibrium, but then by some means we start to remove one of the products (say the ester); what will happen? The moment the concentration of the ester falls the ratio on the right-hand side of Eq. 8.7 will be too small; to restore it to the correct value we have to either *increase* the amounts of ester and water or *decrease* the amounts of carboxylic acid and alcohol. Both of these things are achieved if some of the acid and alcohol react to give more water and ester.

So, by removing the ester we disturb the equilibrium in such a way that the only way it can be restored is for more ester to be produced; we say that the reaction has been 'forced to the right' by removing the product. For the particular reaction shown in Fig. 8.12 it turns out that the ester is more volatile than any of the other species, so it can simply be distilled off constantly forcing the reaction to the right and so increasing the yield of the ester.

Another way of forcing the reaction to the right is to add more of one of the reactants. Once again, this reduces the value of the ratio on the right-hand side of Eq. 8.7 and so to restore the correct value some of the alcohol and carboxylic acid have to react to form more ester. If the alcohol is a simple one, such as methanol, we would probably use it as the solvent for the reaction, thus ensuring that its concentration was very high.

A similar example is in the formation of an imine from the reaction of a ketone with an amine, Fig. 8.13. In this reaction water is produced on the right-hand side of this equilibrium and so the yield of the imine can be increased by removing the water. This can be achieved by adding a solid material known as a *molecular sieve* which is a special kind of zeolite clay which has cavities into which water will bind tightly. This effectively removes the water, so forcing the reaction to produce more of the imine in an attempt to restore the equilibrium.

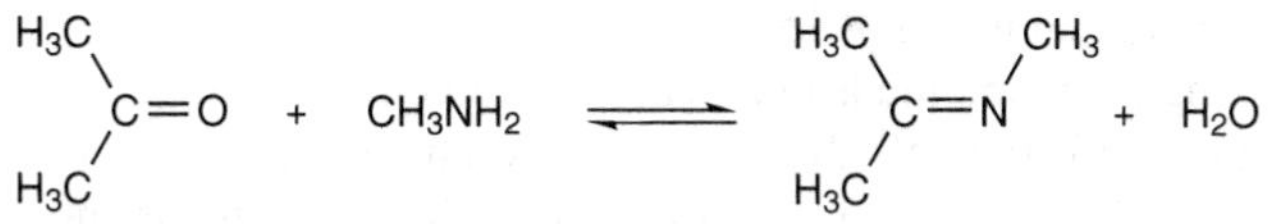

$$(H_3C)_2C{=}O + CH_3NH_2 \rightleftharpoons (H_3C)_2C{=}N{-}CH_3 + H_2O$$

ketone imine

Fig. 8.13 An imine is formed from the reaction of a ketone and an amine.

We can use the same strategy to shift the reaction the other way; suppose, for example, we want to hydrolyse an ester back to the alcohol and the carboxylic acid – this is the reverse reaction shown in Fig. 8.12. To force the reaction to the left, we need to increase the amount of the species on the right and this is easily done by making water the solvent. Now its concentration will be very large, and so the reaction will be shifted to the left, in favour of the hydrolysed products.

The formation of an acetal from the reaction of a ketone with an alcohol, shown in Fig. 8.14, is an equilibrium reaction which we sometimes want to force one way and sometimes the other.

$$(H_3C)_2C{=}O + CH_3OH \rightleftharpoons (H_3C)_2C(OCH_3)_2 + H_2O$$

ketone acetal

Fig. 8.14 The formation of an acetal from a ketone and an alcohol is a readily reversible reaction which we can 'force' one way or the other by altering the conditions.

If we wish to form the acetal we use the alcohol as the solvent and add a dehydrating agent – both choices force the equilibrium to the right. On the other hand, if we wish to regenerate the ketone from the acetal, we run the reaction in aqueous solution so that the excess of water will force the reaction to the left.

The Haber process

Our final example is the formation of ammonia from hydrogen and nitrogen (the *Haber process*):

$$N_2(g) + 3H_2(g) \rightleftharpoons 2NH_3(g)$$

which is the way in which millions of tonnes of ammonia are synthesized each year. In order to make the reaction go at a reasonable rate temperatures of around 400 °C are needed together with a solid iron oxide catalyst. Unfortunately, the reaction is exothermic ($\Delta_r H^\circ = -92$ kJ mol^{-1}) so raising the temperature shifts the equilibrium towards the reactants (Section 8.4 on p. 122); it is found that at this high temperature the equilibrium constant is about 10^{-2}.

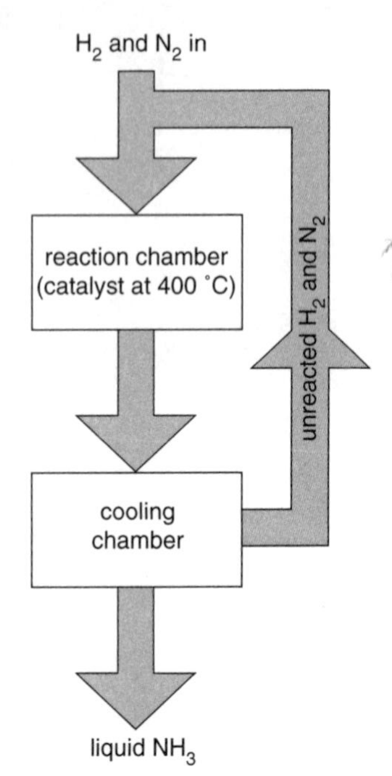

Fig. 8.15 Simplified picture of the arrangement for the Haber process for synthesizing ammonia. After coming to equilibrium over the catalyst, only a small fraction of the hydrogen and nitrogen has been converted to ammonia. The ammonia is separated from the other gases by liquefying it, and the unreacted nitrogen and hydrogen are returned to the reaction chamber.

The strategy for increasing the yield of ammonia is simply to remove the ammonia and then let the nitrogen and hydrogen come back to equilibrium, thus generating more ammonia. In practice this is done by passing the gases over the catalyst so that they come to equilibrium, then separating out the ammonia and finally recycling the unreacted gases back over the catalyst; the process is illustrated in Fig. 8.15.

Removing the ammonia is quite easy as ammonia will liquefy at a temperature (around −35 °C) well above the temperature at which either nitrogen or hydrogen will liquefy. So, simply by cooling the gases the ammonia can be separated as a liquid.

Coupling reactions together

Suppose that the reaction we are interested in has a positive $\Delta_r G^\circ$, which means that the equilibrium constant will be less than 1, and very little of the products will be formed. One way to force the formation of larger amounts of the products is to 'couple' the reaction to another reaction which has a negative $\Delta_r G^\circ$ so that, when taken together, the overall $\Delta_r G^\circ$ is negative. What we are doing is driving the reaction with the unfavourable $\Delta_r G^\circ$ using a reaction with a favourable $\Delta_r G^\circ$.

For the two reactions to be coupled in this way they must be able to influence one another; having the reactions take place in separate beakers on the bench or simply mixing all the reagents together will not achieve the required coupling. In Nature this coupling is achieved using enzymes which control and direct the chemistry, and indeed very many of the important chemical processes of life are driven by this kind of coupling. We do not have time here to go into the details of exactly how enzymes achieve this coupling, but will just illustrate the outcome for one very fundamental reaction in living systems – the formation of the peptide bond.

In living systems, proteins are synthesized by the polymerization of amino acids into long chains. The key reaction in forming the chain is for amino acids to condense together by forming *peptide bonds*, as shown in Fig. 8.16 for the formation of a dipeptide.

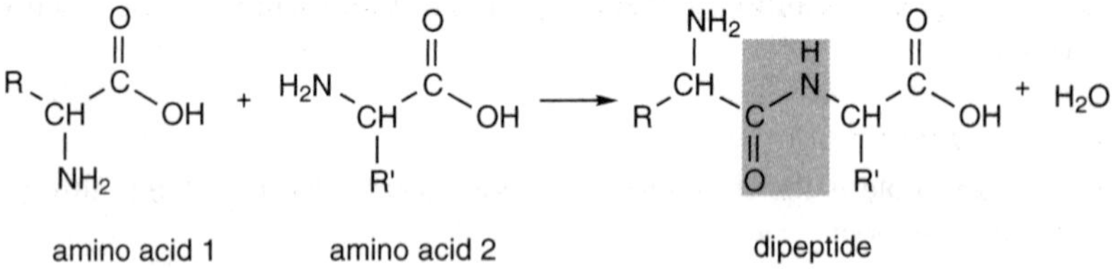

amino acid 1 + amino acid 2 ⟶ dipeptide + H_2O

Fig. 8.16 Two amino acids condense together to form a dipeptide, and in doing so form the –CONH– peptide bond, shown in the grey box. Different amino acids have different side groups, R and R′.

This reaction has a rather unfavourable $\Delta_r G^\circ$ of +17 kJ mol^{-1} and so at equilibrium very little of the dipeptide will be formed. To get round this, Nature couples this reaction with the hydrolysis of adenosine triphosphate (ATP) to adenosine diphosphate (ADP):

$$\mathrm{ATP} + \mathrm{H_2O} \longrightarrow \mathrm{ADP} + \text{inorganic phosphate}.$$

This reaction has a very favourable $\Delta_r G^\circ$ of -30 kJ mol^{-1}, sufficiently negative to outweigh the positive $\Delta_r G^\circ$ for the formation of the peptide bond. So, taken together the two reactions have an overall negative $\Delta_r G^\circ$ making the formation of the peptide bond thermodynamically feasible.

Nature frequently uses the hydrolysis of ATP to ADP to drive reactions which have unfavourable (positive) values of $\Delta_r G^\circ$, and so ATP is often regarded as the source of the energy which drives the chemistry of life. We can now see that by 'energy' we really mean Gibbs energy, as this is what drives reactions forward. So, when biochemists describe ATP as a high energy molecule, what they mean is a high *Gibbs* energy molecule.

For all of this to work, the cell has to be able to make ATP, for example from ADP by running the hydrolysis reaction backwards:

$$\text{ADP} + \text{inorganic phosphate} \longrightarrow \text{ATP} + \text{H}_2\text{O}$$

but this reaction has an unfavourable $\Delta_r G^\circ$ of +30 kJ mol^{-1} and so will not go on its own. As with peptide bond formation, Nature drives the formation of ATP by coupling it to another reaction, such as the oxidation of glucose.

This oxidation is often described as the fundamental source of 'energy' for many living systems, including ourselves. The overall reaction is

$$\text{C}_6\text{H}_{12}\text{O}_6 + 6\text{O}_2 \longrightarrow 6\text{CO}_2 + 6\text{H}_2\text{O}$$

and this has a very large negative $\Delta_r G^\circ$ of -2880 kJ mol^{-1}. In Nature the oxidation of one molecule of glucose is coupled by a very complex series of enzymatically controlled reactions to the formation of around 38 molecules of ATP from ADP, and in this way the high energy ATP molecules needed in many other processes in the cell are generated.

We obtain sugars, such as glucose, from plants which synthesize them from carbon dioxide and water, releasing oxygen in the process. The formation of glucose in this way is simply the reverse of the oxidation given above:

$$6\text{CO}_2 + 6\text{H}_2\text{O} \longrightarrow \text{C}_6\text{H}_{12}\text{O}_6 + 6\text{O}_2.$$

However, this reaction has a very large positive $\Delta_r G^\circ$ so how can it be made to go? The answer is that plants utilize the energy from light, in a process known as *photosynthesis*, to force this reaction. A very complex and subtle series of processes and reactions are used to harvest the energy from light and utilize it to form glucose – Nature has perfected this scheme over the millennia of evolution that have led to the green plants we know today.

The coupling of an unfavourable reaction to a favourable one is absolutely crucial in the chemistry of life. We can only marvel at the subtle and efficient way in which Nature is able to achieve this coupling.

8.5 Equilibrium and rates of reaction

Suppose that we have an equilibrium between reactants A and B and products C and D:

$$A + B \rightleftharpoons C + D.$$

Let us start out with just A and B mixed together; they will react at some rate to generate some of the products C and D. However, the moment these are formed they will start to react together and, via the reverse reaction, regenerate the reactants A and B.

The more of the products C and D that are formed, the faster the reverse rate becomes; in contrast the more of the reactants A and B that are used up the slower the forward rate becomes. We know that eventually we will reach the equilibrium point at which the concentrations are all constant – what must have happened therefore is that the rates of the forward and reverse reactions have become equal. When this is the case, as fast as A and B are removed by the forward reaction they are replenished by the reverse reaction in such a way that their concentrations do not change. Equilibrium is thus a dynamic situation – it is not that the reactions have stopped, it is just that the rate of the forward and back reactions are equal so that it appears that nothing is happening.

This brings us on to the whole topic of the rates of reactions – something which we have been carefully ignoring, or at least sidestepping, up to now. The problem is that even if a reaction is favourable in thermodynamic terms, i.e. has a large negative $\Delta_r G^\circ$, this does not guarantee that the reaction will actually take place at a measurable rate.

For example, the oxidation of glucose is accompanied by a very large negative $\Delta_r G^\circ$ of some -2880 kJ mol^{-1}, yet we can go out and buy glucose powder safe in the knowledge that it will not burst into flames spontaneously. The reaction is thermodynamically feasible, but kinetically very slow.

Other reactions are both thermodynamically feasible and fast. For example, the neutralization of acids and bases (essentially the aqueous phase reaction $H_3O^+ + OH^- \longrightarrow 2H_2O$) is both thermodynamically feasible and very fast. Even reactions in which the position of equilibrium lies close to the reactants can come to equilibrium quickly. For example, the dissociation of ethanoic (acetic) acid has an equilibrium constant of only 10^{-5} but nevertheless the equilibrium is established very quickly once the acid is dissolved in water.

Reactions involving ions – particularly if they are oppositely charged as in the case of the neutralization of acids and alkalis – tend to be quite fast, and similarly reactions involving simple proton transfer (for example from ethanoic acid to water) are often rapid. However, reactions involving the breaking and making of bonds (say to carbon) are likely to proceed much more slowly.

In the next chapter we will look at the factors which control the rates of reactions and then go on to see what a study of reaction rates can tell us about the mechanism of a reaction.

9 Rates of reaction

At the end of the previous chapter we commented on the fact that just because a reaction has a favourable $\Delta_r G^\circ$ (i.e. the position of equilibrium lies well to the products) this does not necessarily mean that it will proceed at a measurable rate. In this chapter we will look at the factors which determine the rates of reactions and how we can influence these. This is clearly a matter of great importance if we actually want to do some practical chemistry, as well as being a rather fundamental aspect of chemical reactions.

The second part of this chapter will move on to discuss how the study of reaction rates can give us information about what is going on in a reaction, particularly when the reaction involves more than one step. This is the experimental study of *reaction mechanisms*.

9.1 What determines how fast a reaction goes?

In this section we will focus on a particular reaction (one you have seen before on p. 108) which is the nucleophilic substitution of Br^- by OH^-, Fig. 9.1. Each factor which influences the rate will be considered in turn.

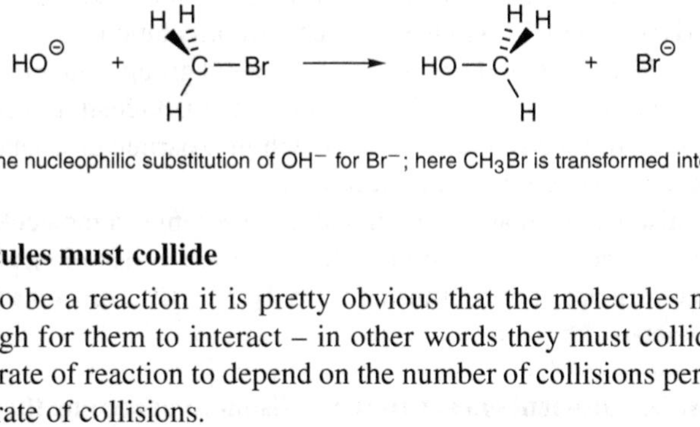

$$HO^{\ominus} + CH_3Br \longrightarrow HO{-}CH_3 + Br^{\ominus}$$

Fig. 9.1 The nucleophilic substitution of OH^- for Br^-; here CH_3Br is transformed into CH_3OH.

The molecules must collide

For there to be a reaction it is pretty obvious that the molecules must come close enough for them to interact – in other words they must collide. So we expect the rate of reaction to depend on the number of collisions per unit time, that is the rate of collisions.

In a gas or liquid the molecules are in a constant state of motion as a result of their thermal energy; they are rushing around in random directions, colliding with one another and hence changing their direction and speed. We therefore expect that the rate of collisions will depend on the *concentration* of the molecules, the *speed* at which they are moving and their *size*.

The rate of collisions between two different species, A and B, will clearly go up as the concentration of either of them increases, simply because as an A molecule rushes around the chance that it will encounter a B molecule goes up as the number of such molecules increases. It does not seem unreasonable that the rate of A–B collisions is proportional to both the concentration of A and that of B:

$$\text{rate of A–B collisions} \propto [A] \times [B].$$

The second thing which affects the rate of collisions is the speed at which the molecules are moving: we argue that the faster the molecules move the more likely they are to collide. As you may know, there is a distribution of speeds with which the molecules move, but the majority of molecules are moving at speeds which are not that far from the average. In a gas, this average speed increases as the temperature increases; this is hardly a surprise as increasing the temperature of a gas increases its energy which appears in the form of kinetic energy – the molecules therefore move faster.

The final factor which determines the collision rate is the size of the molecules: the larger they are the more chance there is of them colliding and so the higher the collision rate.

For simple molecules in the gas phase we can use gas kinetic theory to estimate the number of collisions as being around 10^{27} per second in a volume of 1 dm^3 at room temperature and pressure – a very large number by any standards. In liquids the situation is more complex because we also need to consider the rate at which the reactant molecules diffuse together; however, once in proximity the collision rate between two molecules is even greater than in a gas.

The molecules must collide with the correct orientation

As we discussed on p. 109, for the reaction shown in Fig. 9.1 to take place the nucleophile (OH^-) must approach the carbon from the *opposite* side to the Br; this requirement is dictated by the nature of the HOMO and LUMO which are involved in the reaction. Only a fraction of the total number of collisions between CH_3Br and OH^- will have this correct orientation.

We have also seen that the attack on a carbonyl group by a nucleophile has a strongly preferred geometry (Fig. 7.16 on p. 105). Indeed it is common for reactions to require the reactants to approach in a particular orientation – we call this the *steric requirement* of a reaction.

We can also expect that the larger and more complex a molecule becomes, the smaller the fraction of collisions with the correct orientation will become. For example, it would not be unusual for only 1 in 10^5 collisions to have the correct geometry.

There must be sufficient energy in the collision to overcome the energy barrier

For the vast majority of reactions there is an *energy barrier*, which is a minimum amount of energy that two colliding molecules need in order to react; simply colliding with the correct orientation is not sufficient – the molecules must bring with them sufficient energy to cross this barrier.

This energy comes mainly from the collision of the two molecules. Each has a certain kinetic energy as a result of its thermal motion – some molecules have more than the average energy and some less. So, some collisions will have higher energy than others, and it is the ones which have an energy which is high enough to overcome the barrier that can lead to reaction.

Typically, the energy barrier for a reaction is between 10 and 100 kJ mol^{-1}. At room temperature the *average* thermal energy of a molecule is around 4 kJ mol^{-1}, so you can see that only those molecules whose energy is very much greater than the average are likely to have enough energy to react when they collide with another molecule. The fraction of molecules with energies much greater than the average is very small, so that typically only 1 in 10^9 molecular collisions has sufficient energy to overcome the barrier.

Our picture of the reacting molecules is of there being many collisions between the reactants but of these only a minuscule fraction have the correct orientation and sufficient energy to lead to products. A reactive encounter between molecules is thus a very rare event.

9.2 Why is there an energy barrier?

The presence of an energy barrier means that even if the products have lower energy than the reactants, the energy of the reacting molecules must first go up as they start to rearrange themselves into product molecules. Once this rearrangement has proceeded to a certain extent the energy starts to fall, eventually reaching the energy of the product molecules when they are fully formed.

Figure 9.2 shows a mechanical analogy which is helpful in understanding this concept of an energy barrier. We start with a block stood on its short side and then try to push it over so that it lies on its long side. The end position has lower (gravitational) potential energy than the starting position, but to push the block over we first need to tilt it on its edge and in doing this the centre of gravity rises, thus increasing the potential energy. Only when the tilt is far enough will the block fall over and reach the low-energy position. Like a chemical reaction, this system has to first go to a state of higher energy before it can reach the state with the lowest energy.

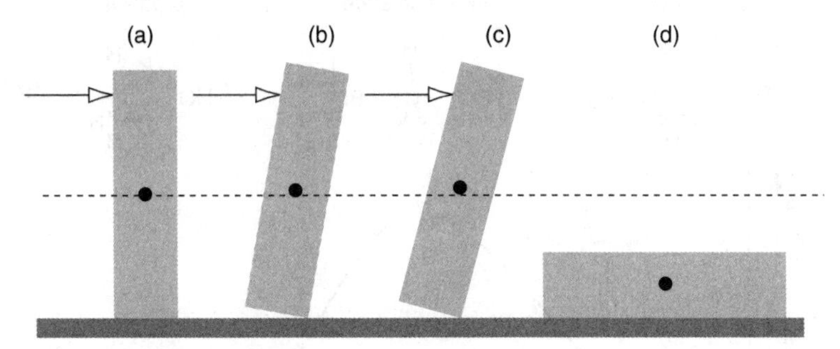

Fig. 9.2 Pushing a block over from position (a) to (d) results in a reduction in the (gravitational) potential energy as the centre of mass (indicated by the dot) is lowered. However, to get the block to topple over it first has to be tilted on its edge, as shown in (b) and (c). This raises the centre of mass and so the energy goes up; only when the block is past position (c) does the energy go down. This process is an analogy for a chemical reaction: the products are lower in energy than the reactants, but to get between the two the energy must first rise; in other words there is an energy barrier which has to be overcome.

A simple example of such an energy barrier occurs in inversion of ammonia, described on p. 87 and illustrated in Fig. 9.3. What happens here is that the molecule is going from a trigonal pyramidal geometry with the hydrogens on one side to an identical, mirror image, geometry with hydrogens on the other side; the whole process is rather like an umbrella being blown inside out. We can measure the progress along this route by giving the H–N–H bond angle;

this starts at 107°, goes to 120° at the planar geometry and then reduces back to its original value as the hydrogens move to the other side of the molecule and end up in mirror image positions.

We know that the equilibrium geometry of ammonia has an H–N–H bond angle of 107°, which means that this is the bond angle which gives the lowest energy; as the bond angle opens out the energy must therefore increase, as shown in Fig. 9.3. Our interpretation of this is that the overlap of the orbitals is less than optimum as we move away from a bond angle of 107°, and so the energy increases.

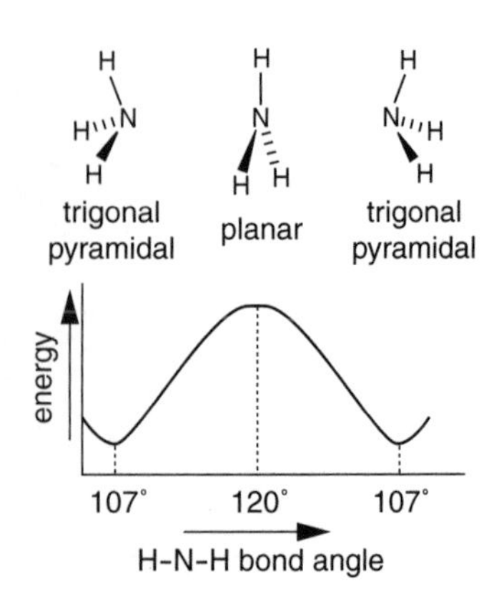

Fig. 9.3 In the inversion of ammonia the progress from the equilibrium trigonal pyramidal geometry to its mirror image is characterized by the H–N–H bond angle. A plot of the energy against this angle shows a maximum at 120°; this corresponds to the planar *transition state*.

The symmetry of the problem dictates that the highest energy point is when the bond angle is 120°, as this lies midway between the two trigonal pyramidal geometries. The species at this highest energy point is called the *transition state*; in this case it is an ammonia molecule with a planar geometry.

For the reaction of Fig. 9.1 we can trace out how the energy varies as the reaction proceeds from reactants to products just as we did for the inversion of ammonia. However, for this reaction the progress from reactants to products cannot be specified by a single angle since as the reaction proceeds several bond angles as well as internuclear distances are changing. So, we plot the energy against what is called the *reaction coordinate* which is a complex combination of distances and angles, the precise definition of which need not concern us; all we need to know is that this coordinate starts at the reactants and ends at the products. The plot of energy against reaction coordinate is known as an *energy profile* for the reaction; such a profile for the nucleophilic substitution reaction is shown in Fig. 9.4.

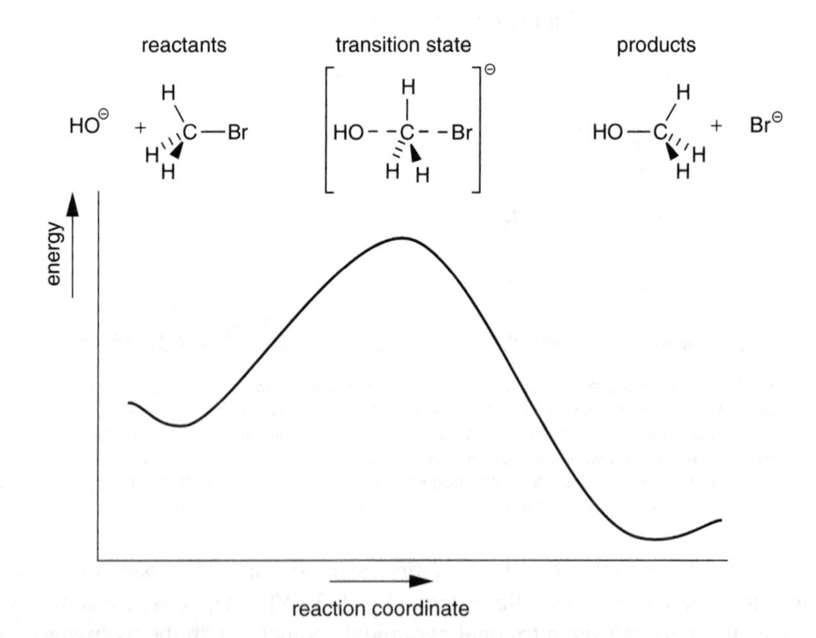

Fig. 9.4 Energy profile for the reaction shown in Fig. 9.1; the reactants appear on the left and the products on the right. At the energy maximum is the transition state, a suggested structure for which is shown; there are partial bonds (shown dashed) to the OH and Br. The products are shown as having lower energy than the reactants, and so the reaction is exothermic; nevertheless, there is still an energy barrier to be overcome in going from reactants to products.

At the top of the profile there is a suggestion as to what the transition state might look like. The carbon atom is five-fold coordinated with partial bonds to both the incoming OH^- and the departing Br^-. Over the course of the reaction, the hydrogen atoms have to move from one side to the other and so in the suggested structure we have shown them in the same plane as the carbon atom. It is important to realize that there is no way we can guess at the precise structure of the transition state – all we can do is to speculate about its general form.

It is clear from these pictures why the energy has to rise when the transition state is formed. The bonding is far from optimum – not least because there are five groups around the carbon; in addition, there are partially made bonds to the OH and the Br. This unusual geometry and the presence of partial bonds is typical for a transition state; it is not surprising, therefore, that its energy is higher than that of the reactants.

It is very important to realize that the transition state is not a real molecule: we cannot isolate it or study its physical or chemical properties. Rather, it is just a transient arrangement of the atoms through which the reaction passes on its way to products. The transition state has a fleeting existence, perhaps having a lifetime of 10^{-12} s or even less.

Reactions with no barrier

The usual situation is for reactions to have a barrier, but there are some that do not. A typical example is the reaction in the gas phase between oppositely charged ions, such as:

$$Li^+ + F^- \longrightarrow LiF.$$

In this case we can attribute the lack of a barrier to the fact that the oppositely charged reactants attract one another and also that in going to the products no bonds are broken.

A second example involves the gas-phase recombination of radicals:

$$^\bullet CH_3 + {}^\bullet CH_3 \longrightarrow CH_3{-}CH_3.$$

We can attribute the lack of a barrier here to the fact that a bond is being made and none broken. The C–H bonds only have to adjust themselves in a minor way to accommodate the new bond.

9.3 Rate laws

We described on p. 131 how the rate of collisions between two molecules A and B is expected to be directly proportional to the concentration of each. So, if our nucleophilic substitution reaction (Fig. 9.1) really does take place by the OH^- and the CH_3Br colliding in a reactive encounter, we would expect to find that the rate of reaction is proportional to the concentration of each of these species.

Experiment confirms this expectation; it is found that

$$\text{rate of reaction} = k[OH^-][CH_3Br]. \tag{9.1}$$

This is called the *rate law* for the reaction.

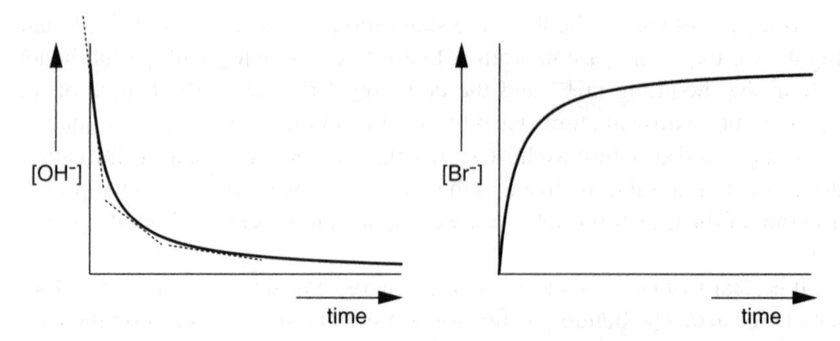

Fig. 9.5 Plots of the concentration of the reactant OH^- and the product Br^- as a function of time for the reaction of Fig. 9.1. Note how the concentration of the reactant falls towards zero and that of the product rises. Also shown as dotted lines on the left-hand plot are tangents to the curve; these give the slope of the curve which is the same thing as the rate. The way in which the absolute value of the rate falls as time increases can clearly be seen. In drawing these plots it has been assumed that at time zero the two reactants have the same concentration and that the concentrations of the products are zero.

In the rate law k is the *rate constant* for the reaction; its value depends on all of the factors described in Section 9.1 (p. 131 onwards) apart from the concentration, i.e. the size of the molecules, the speed they are moving at, and the fraction of collisions with the correct orientation and sufficient energy to overcome the barrier. The value of k has to be determined experimentally and it is usually found that it depends strongly on temperature, as will be discussed in the next section.

The rate of reaction is how the concentration varies with time:

$$\text{rate of reaction} = \frac{\text{change in concentration during time interval } \Delta t}{\Delta t};$$

the dimensions of the rate are thus concentration × time^{-1}. The concentration of any of the reactants or products can be used to specify the rate. However, if we use a reactant whose concentration *decreases* with time, the rate will be a negative quantity; in contrast, if we use a product whose concentration *increases* with time, the rate will be positive.

Looking at the rate law, Eq. 9.1, we can see that at the start of the reaction the rate will have its highest (absolute) value. As the reaction proceeds, the amount of the reactants OH^- and CH_3Br decreases, and so the rate decreases, eventually reaching zero at infinite times. Figure 9.5 shows plots of how the concentration of the reactants and products varies with time for this reaction.

The slope of a tangent to the curve gives us the rate at any time: this is because the slope is the change in concentration over the change in time, which is precisely the same thing as the rate. In Fig. 9.5 we see that for the plot of $[OH^-]$ the slopes are negative, as expected for a reactant, and that as time proceeds the slopes are becoming less negative, i.e. the rate of the reaction is falling.

Order

Another way of writing the rate law of Eq. 9.1 is

$$\text{rate of reaction} = k[OH^-]^1[CH_3Br]^1$$

where we have raised the concentrations to the power of 1. Of course this is rather superfluous as $[OH^-]^1$ means exactly the same thing as $[OH^-]$. However, the notional presence of this power in the rate law allows us to define something called the *order*.

The *order* is the power to which the concentration is raised in the rate law. So, in the above rate law the order with respect to OH^- is 1, and similarly the order with respect to CH_3Br is also 1. We can also define an overall order, which is the sum of the orders of each species present in the rate law. So for this reaction the overall order is 2.

An order of 1 is often called *first order* and an order of 2 is called *second order*. Thus the rate law for the reaction between hydroxide and bromomethane (methyl bromide) can be described as being first order with respect to hydroxide and bromomethane and second order overall.

9.4 The Arrhenius equation

The rate constant generally depends quite strongly on temperature, and experimental studies have shown that this temperature dependence often obeys the *Arrhenius equation*:

$$k = A \exp\left(\frac{-E_a}{RT}\right). \tag{9.2}$$

In this equation A is the *pre-exponential* or *A factor*, E_a is the *activation energy* and, as usual, T is the absolute temperature and R is the gas constant. This strong temperature dependence of the rate constant means that when quoting its value it is essential to state the temperature at which the measurements were made.

The interpretation of this equation is that the activation energy is the height of the energy barrier over which the reactants must pass – to be specific, it is the energy difference between the reactants and the top of the barrier (the transition state), as shown in Fig. 9.6.

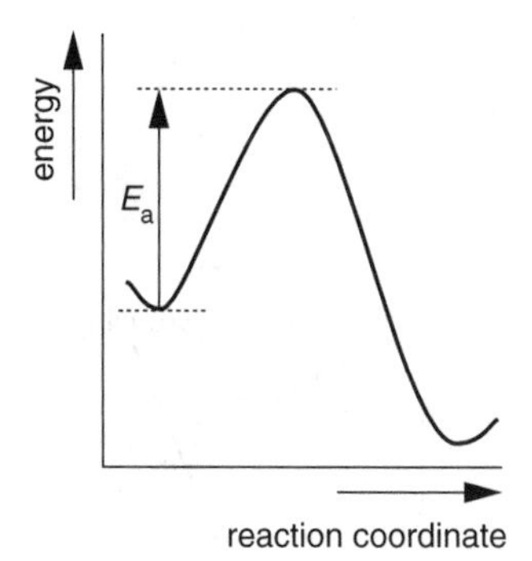

Fig. 9.6 The activation energy, E_a, is the difference in energy between the reactants and the top of the barrier, i.e. the transition state. For reaction to take place, the molecules must collide with at least this much energy and from gas kinetic theory we can show that the fraction of collisions satisfying this requirement is $\exp(-E_a / RT)$.

From gas kinetic theory it can be shown that the fraction of collisions with energy greater than or equal to E_a is given by $\exp(-E_a/RT)$ and, as we discussed on p. 132, only these collisions are energetic enough to give rise to products. We therefore interpret the exponential term in the Arrhenius equation as arising from the fraction of collisions with sufficient energy to pass over the barrier.

If there is no energy barrier all of the collisions have sufficient energy to become products; in mathematical terms if $E_a = 0$ the exponential term, which is the fraction of collisions with sufficient energy, becomes 1. Under these circumstances the Arrhenius equation (Eq. 9.2) tells us that the rate constant is equal to the pre-exponential factor, A. So, we can interpret the value of A as giving the rate constant in the absence of an energy barrier to reaction.

From the discussion in Section 9.1 (starting on p. 131) we can see that the factors which contribute to the value of A are: the fraction of collisions which have the correct geometry, the size of the molecules and the speed with which they are moving. As we noted, this last factor depends on temperature but it turns out that this dependence is much weaker than that due to the

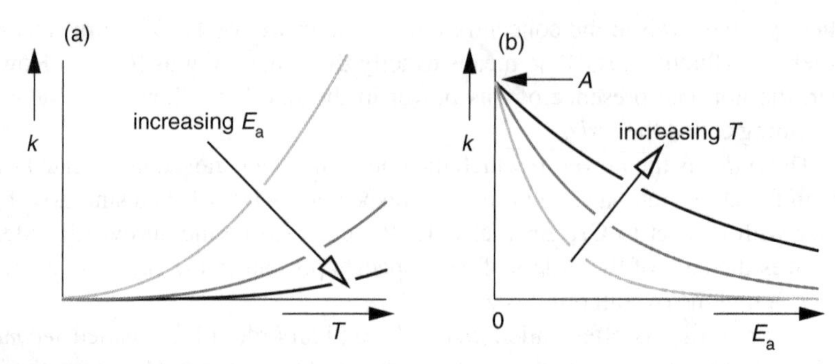

Fig. 9.7 Plot (a) shows the Arrhenius equation (Eq. 9.2) prediction for the variation of the rate constant, k, with temperature, T. The three lines are computed assuming the same value of the A factor but different values of the activation energy; the darker the line the larger the activation energy (the temperature scale does not start at zero). At high enough temperatures, the rate constant will reach the value of the pre-exponential factor, A; for the curves shown here the temperature is well below this limit. Plot (b) shows the variation of the rate constant, k, with activation energy, E_a; the three lines are computed assuming the same value of the A factor, but different temperatures; the darker the line the higher the temperature. Note how the rate constant responds rapidly to changes in the temperature and the activation energy.

$\exp(-E_a/RT)$ factor in the Arrhenius equation. So, to a reasonable approximation we can assume that the A factor is temperature independent.

Figures 9.7 (a) and (b) show the predictions from the Arrhenius equation (Eq. 9.2) for how the rate constant varies with temperature and activation energy, respectively. The plots show how the exponential term makes the dependence of the rate constant on the values of these two parameters rather strong.

We can turn the Arrhenius equation into a straight line plot by taking (natural) logarithms of both sides:

$$k = A \exp\left(\frac{-E_a}{RT}\right)$$

$$\text{hence} \qquad \ln k = \ln A + \left(\frac{-E_a}{R}\right)\left(\frac{1}{T}\right).$$

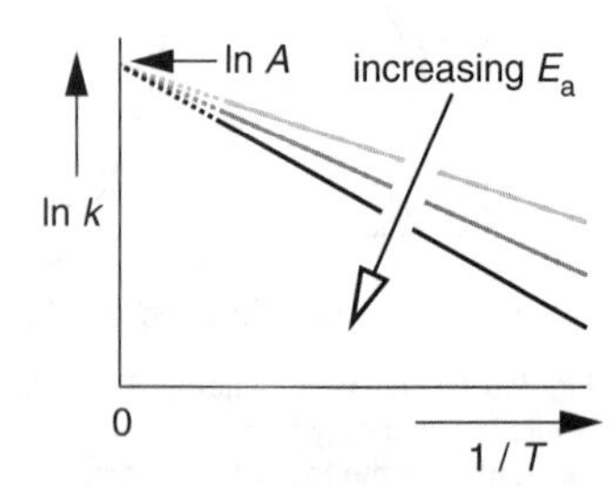

Fig. 9.8 An Arrhenius plot of ln k against ($1/T$) which, as explained in the text, is expected to give a straight line. The three lines are computed assuming the same value of the A factor but different values of the activation energy; the darker the line the larger the activation energy. As the slope of the line is $-E_a/R$ the greater the activation energy the more negative the slope. The intercept with the vertical axis (when $1/T$ goes to zero) gives ln A; this extrapolation is shown by dotted lines.

So, a plot of $\ln k$ against $1/T$ will be straight line with slope $(-E_a/R)$; the intercept with the vertical axis, when $1/T$ goes to zero, will be $\ln A$. Figure 9.8 shows examples of this straight line plot; note how increasing the activation energy results in a more negative slope.

We can interpret the intercept with the vertical axis in the following way: when the temperature becomes very high, i.e. when $1/T$ goes to zero, each collision is sufficiently energetic to overcome the barrier and so each leads to reaction. The rate constant then has the value A because, as we noted before, this is the value of the rate constant in the absence of a barrier. For most reactions this limit in which the rate constant becomes equal to the A factor occurs at such high temperatures that either the reactants will have dissociated into atoms or other reactions will have taken over. The limit is therefore of theoretical, rather than practical, interest.

Given some experimental data of rate constants at different temperatures we can determine E_a and A by plotting $\ln k$ against $1/T$, just as in Fig. 9.8. The slope is $(-E_a/R)$ and the intercept with the vertical axis is $\ln A$. In practice, the range of temperatures over which data is available is usually rather limited so a long extrapolation is needed to find $\ln A$, making this value rather inaccurate.

9.5 Elementary and complex reactions

There is plenty of experimental evidence which leads us to believe that the nucleophilic substitution reaction:

$$OH^- + CH_3Br \longrightarrow CH_3OH + Br^-$$

really does take place as it is written, i.e. the products are formed directly from a single reactive encounter between an OH^- ion and a CH_3Br molecule.

There are many other reactions for which the experimental evidence is that they take place as written. For example, the two gas-phase reactions:

$$Cl + O_3 \longrightarrow ClO + O_2$$
$$ClO + O \longrightarrow Cl + O_2$$

which are key reactions in the destruction of atmospheric ozone by halogens, are thought to proceed as written.

On the other hand, there are reactions which certainly do not take place as written. For example, the gas-phase reaction between H_2 and Br_2 has the stoichiometric equation

$$H_2 + Br_2 \longrightarrow 2HBr$$

but there is much experimental evidence that the reaction does *not* occur by a single reactive encounter between an H_2 and a Br_2 molecule.

Similarly, the oxidation of glucose in living systems has the overall stoichiometry

$$C_6H_{12}O_6 + 6O_2 \longrightarrow 6CO_2 + 6H_2O$$

but certainly does not take place by one molecule of glucose reacting directly with six of oxygen!

What we are describing here is the distinction between an *elementary* and a *complex* reaction. An elementary reaction is one which we believe to take place *in a single encounter* and *as the chemical equation is written*. For example, the gas phase reaction between a hydrogen atom and Br_2 written as

$$H + Br_2 \longrightarrow HBr + Br$$

is elementary and so takes place in a reactive encounter between an H and a Br_2 molecule.

Complex reactions take place by a set of elementary reactions, called a *mechanism*; typically such a mechanism will involve the generation of molecules, called *intermediates*, which do not appear in the stoichiometric equation.

For example, the decomposition of ozone into oxygen has the stoichiometry:

$$2O_3 \longrightarrow 3O_2$$

and it is thought that this overall process has a mechanism involving the following three elementary steps involving an oxygen atom as an intermediate:

$$\begin{aligned} O_3 &\longrightarrow O_2 + O \\ O_2 + O &\longrightarrow O_3 \\ O + O_3 &\longrightarrow 2O_2. \end{aligned}$$

A great deal of effort has been put into determining the mechanisms of chemical reactions, i.e. identifying the intermediates and the contributing elementary reactions. Determining a reaction mechanism is by no means a simple task and usually requires the application of many techniques, such as a study of how the rate of the reaction varies with the concentrations of the species involved and the detection of intermediates.

Rate laws for elementary and complex reactions

As an elementary reaction takes place as written we can write down the rate law directly from the chemical equation. For example:

$$\begin{aligned} &OH^- + CH_3Br \longrightarrow CH_3OH + Br^- &&\text{rate} = k_a[OH^-][CH_3Br] \\ &Cl + O_3 \longrightarrow ClO + O_2 &&\text{rate} = k_b[Cl][O_3] \\ &H + Br_2 \longrightarrow HBr + Br &&\text{rate} = k_c[H][Br_2]. \end{aligned}$$

In contrast, for a complex reaction there is no way that we can write down the rate law just by looking at the stoichiometric equation; in such cases the rate law has to be determined by experiment. A good example is the reaction between H_2 and Br_2 in the gas phase

$$H_2 + Br_2 \longrightarrow 2HBr$$

which has the experimental rate law:

$$\text{rate of formation of HBr} = \frac{k_d[H_2][Br_2]^{3/2}}{[Br_2] + k_e[HBr]}$$

which is, to say the least, far from obvious. Indeed, the observation of such a complex rate law is good evidence for the reaction not being elementary.

However, the opposite is not necessarily true: some reactions have simple experimental rate laws but turn out to have complex mechanisms! For example the gas-phase reaction between H_2 and I_2

$$H_2 + I_2 \longrightarrow 2HI$$

is found to have the rate law

$$\text{rate of loss of } H_2 = k_f[H_2][I_2].$$

Despite the simplicity of this rate law it is thought that the mechanism for the reaction may involve the intermediate IH_2 in the following steps (although there are certainly other possible mechanisms)

$$\begin{aligned} I_2 &\longrightarrow 2I \\ 2I &\longrightarrow I_2 \\ I + H_2 &\longrightarrow IH_2 \\ I + IH_2 &\longrightarrow 2HI. \end{aligned}$$

9.6 Intermediates and the rate-determining step

We mentioned on p. 139 that complex reactions often involve the formation of species, called intermediates, which are neither reactants nor products. The nature and properties of these intermediates often play a crucial role in the reaction, and as a result you will find that a lot of attention is focused on them.

Let us take as an example the reaction of the nucleophile OH^- with the acyl chloride CH_3COCl (discussed in Section 7.6 on p. 110). The reaction is shown in Fig. 9.9 and proceeds in two steps via the formation of a tetrahedral intermediate, **I**.

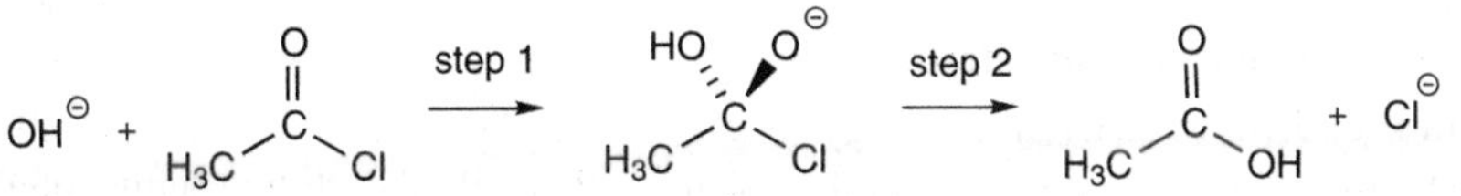

$OH^{\ominus}$ + $H_3C{-}C({=}O){-}Cl$ $\xrightarrow{\text{step 1}}$ $H_3C{-}C(OH)(O^{\ominus}){-}Cl$ (**I**) $\xrightarrow{\text{step 2}}$ $H_3C{-}C({=}O){-}OH$ + $Cl^{\ominus}$

Fig. 9.9 The reaction between OH^- and CH_3COCl proceeds in two steps. Step (1) involves nucleophilic attack on the carbonyl carbon to form a tetrahedral intermediate (**I**); in step (2) this intermediate collapses expelling Cl^- and re-forming the carbonyl group.

It is very important not to confuse intermediates with transition states (introduced on p. 134). A transition state only has a fleeting existence and is simply an arrangement of the atoms through which the reaction must pass in order to reach the products; a transition state is located at a maximum in the energy profile (see Fig. 9.4 on p. 134).

In contrast, an intermediate is as much a real molecule as any other – we can probe its structure using spectroscopic and physical techniques, and indeed we may even be able to isolate it. Admittedly, it is often the case that these intermediates are very reactive and rather short lived, so studying them can be difficult, but they do not have the fleeting existence of a transition state.

Intermediates – like product and reactant molecules – are found at *minima* in the energy profile of the reaction; this is in contrast to transition states which are found at *maxima*. These points are illustrated in Fig. 9.10 which shows the energy profile for the reaction between OH^- and CH_3COCl.

The first energy barrier in this reaction is between the reactants (OH^- + CH_3COCl) and the intermediate, **I**; at the top of the barrier we have the first transition state in which presumably the C–O π bond is partially broken and the bond between the OH^- and the carbonyl carbon is partially made. The activation energy for this first step is $E_{a,1}$. The second barrier in the reaction is between the intermediate and the products, and proceeds via the second transition state in which we imagine that the C–O π bond is partially re-formed and the C–Cl bond is partially broken.

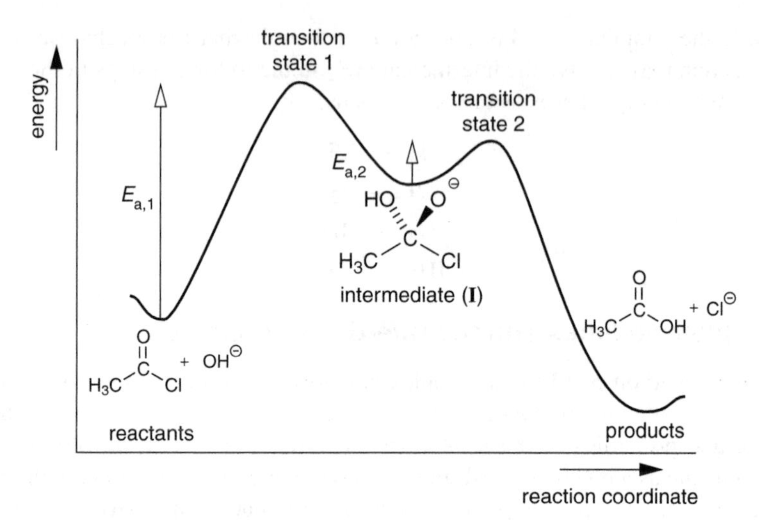

Fig. 9.10 Energy profile for the reaction shown in Fig. 9.9. The reactants, products and intermediate all occur at energy minima, whereas the two transition states occur at maxima. The activation energy for the first step, $E_{a,1}$, is shown as being much greater than that for the second, $E_{a,2}$; this means that the first step is the *rate-determining step.*

The rate-determining step

In Fig. 9.10 the activation energy for step (2) is shown as being considerably less than that for step (1). What this means is that, all other things being equal, the rate of step (2) is much greater than that of step (1). As a consequence, the moment a molecule of the intermediate is formed in step (1) it almost immediately reacts in step (2) to give the products. The rate at which the products are formed is thus determined *only* by the rate of step (1), which is therefore described as the *rate determining step.*

Assuming that step (1) is an elementary reaction we can write its rate as:

$$\text{rate of step (1)} = k_1[\text{CH}_3\text{COCl}][\text{OH}^-].$$

It follows that as step (1) is the rate-determining step, the rate of formation of the products is the same as the rate of step (1):

$$\text{rate of formation of the products} = k_1[\text{CH}_3\text{COCl}][\text{OH}^-].$$

We see that although a two-step mechanism is involved, the rate law is rather simple.

It is important to be careful not to fall into the mistake of describing this kind of reaction by saying that 'the first step is slower than the second, so the former is rate-determining'. What is wrong with this is that during the reaction, step (2) is in fact proceeding at the *same rate* as step (1).

If the concentrations of the reactants and the intermediate were equal it would undoubtedly be the case that the rate of step (2) would be greater than that of step (1). However, the concentrations are *not* equal – far from it. We expect to find that the concentration of the intermediate will be very low because it is whisked away to products as quickly as it is formed. No matter how low the activation energy for step (2) becomes, this reaction cannot take place until step (1) produces some of the intermediate; as a consequence, the *rate* of step (2) cannot be greater than that of step (1), and in fact they are equal.

9.7 Reversible reactions

So far we have ignored the possibility that a reaction can go backwards as well as forwards, i.e. that reactions are reversible. If a reaction has a position of equilibrium which favours the products very strongly then to a very good approximation we can ignore the reverse reaction, but if the position of equilibrium is more balanced between reactants and products we must consider the reverse reaction.

Like the forward reaction, the reverse (or back) reaction has an activation energy as is illustrated in the energy profiles shown in Fig. 9.11. The activation energy for the forward reaction, $E_{a,f}$, is the energy difference between the transition state and the reactants. For the reverse reaction, which takes us from products back to reactants, the activation energy, $E_{a,r}$, is the energy difference between the transition state and the products.

The profile of Fig. 9.11 (a) is for an exothermic reaction, i.e. one in which the products are lower in energy than the reactants; it follows that the activation energy of the reverse reaction must be greater than that of the forward reaction. In contrast, for an endothermic reaction, shown in Fig. 9.11 (b), it is the forward reaction which has the greater activation energy.

From the diagram it is clear that the energy difference between the products and reactants, ΔE, is related to the activation energies in the following way:

$$\Delta E = E_{a,f} - E_{a,r}. \tag{9.3}$$

The energy difference is not quite the same thing as $\Delta_r H^\circ$ for a reaction, but it is closely related to it.

Inspection of Fig. 9.11 (b) shows that for an endothermic reaction the activation energy (of the forward step) cannot be less than ΔE, i.e. the activation energy is greater than the endothermicity. For an exothermic reaction, Fig. 9.11 (a), there is no particular relationship between the activation energy of the forward reaction and ΔE.

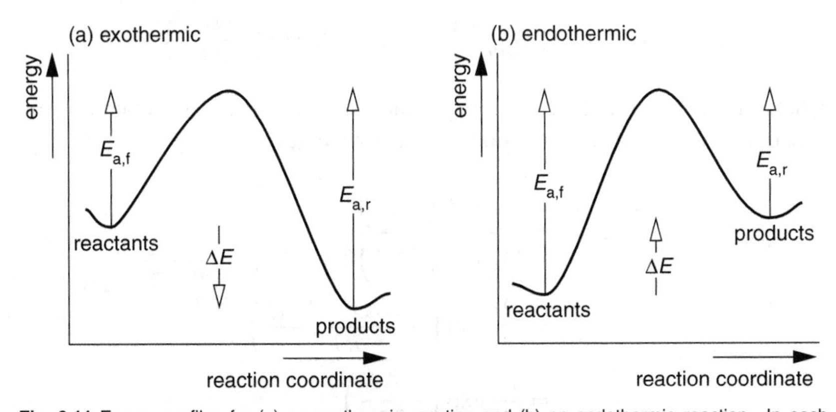

Fig. 9.11 Energy profiles for (a) an exothermic reaction and (b) an endothermic reaction. In each case the energy difference between the reactants and products, ΔE, is equal to the difference in the activation energies of the forward ($E_{a,f}$) and reverse ($E_{a,r}$) reactions.

Relation to the equilibrium constant

Let us suppose that we have a reversible reaction between A and B to give P and Q:

$$A + B \underset{k_r}{\overset{k_f}{\rightleftarrows}} P + Q.$$

We will also suppose that the forward and reverse reactions are elementary (with rate constants k_f and k_r, respectively) so that, as described on p. 140, we can write the rates of the two reactions as:

$$\text{rate of forward reaction} = k_f[A][B]$$
$$\text{rate of reverse reaction} = k_r[P][Q].$$

In Section 8.5 on p. 130 we argued that at equilibrium the rates of the forward and reverse reactions are equal. So, we can write:

$$k_f[A]_{eq}[B]_{eq} = k_r[P]_{eq}[Q]_{eq}$$

where we have added the subscript 'eq' to remind ourselves that this relationship is *only* true when the concentrations are at their equilibrium values. Some rearrangement of this equation gives:

$$\frac{k_f}{k_r} = \frac{[P]_{eq}[Q]_{eq}}{[A]_{eq}[B]_{eq}}.$$

We recognize the quantity on the right-hand side as the equilibrium constant, K_{eq}, and so see that the ratio of the rate constants for the forward and reverse reactions is equal to the equilibrium constant:

$$K_{eq} = \frac{k_f}{k_r}. \tag{9.4}$$

We have already seen in Section 9.4 on p. 137 that rate constants can be expressed in terms of an A factor and activation energy using the Arrhenius equation, Eq. 9.2. If we do this for the forward and reverse reactions we have:

$$k_f = A_f \exp\left(\frac{-E_{a,f}}{RT}\right) \qquad k_r = A_r \exp\left(\frac{-E_{a,r}}{RT}\right)$$

where A_f and A_r are the A factors for the forward and reverse reactions, respectively. Using these expressions in Eq. 9.4 we have:

$$\begin{aligned} K_{eq} &= \frac{A_f \exp\left(\frac{-E_{a,f}}{RT}\right)}{A_r \exp\left(\frac{-E_{a,r}}{RT}\right)} \\ &= \frac{A_f}{A_r} \exp\left(\frac{-(E_{a,f} - E_{a,r})}{RT}\right) \\ &= \frac{A_f}{A_r} \exp\left(\frac{-\Delta E}{RT}\right) \end{aligned}$$

where, to obtain the last line, we have used Eq. 9.3.

What the final line tells us is that the equilibrium constant varies with temperature in a way determined by the value of ΔE, the energy change for the reaction. If ΔE is positive (i.e. an endothermic reaction), an increase in the temperature results in an increase in the equilibrium constant. In contrast, for an exothermic reaction which has a negative ΔE, increasing the temperature reduces the equilibrium constant.

If we identify ΔE with $\Delta_r H^\circ$ these conclusions are exactly the same ones we came to in Section 8.4 on p. 122, albeit using a very different approach.

9.8 Pre-equilibrium

Look again at the energy profile for the $CH_3COCl + OH^-$ reaction (Fig. 9.10 on p. 142), and consider the 'choices' which are open to the intermediate, **I**. It can either go on to products, which is what we assumed would happen, or it could take the reverse of step (1) and go back to reactants. However, as the profile is drawn the activation energy for **I** going back to reactants is larger than for it going on to products; we therefore expect that the latter will be the dominant process as it will have a larger rate constant. With hindsight you can see that it was no accident that the energy profile was drawn in this way!

Figure 9.12 shows two different energy profiles for a two-step reaction. Profile (a) is just like that in Fig. 9.10; the dominant process for the intermediate is for it to go on to products and, as we have seen, this leads to step (1) being the rate-determining step. In profile (b) the activation energy for the intermediate returning to reactants is lower than for it going to products, and this leads to quite a different result.

In the case of profile (b), once an intermediate is formed its most likely fate is to return to reactants in the reverse of step (1), which we will call step (−1). The slower conversion from the intermediate to products via step (2) does take place but only a small fraction of the intermediates take this route.

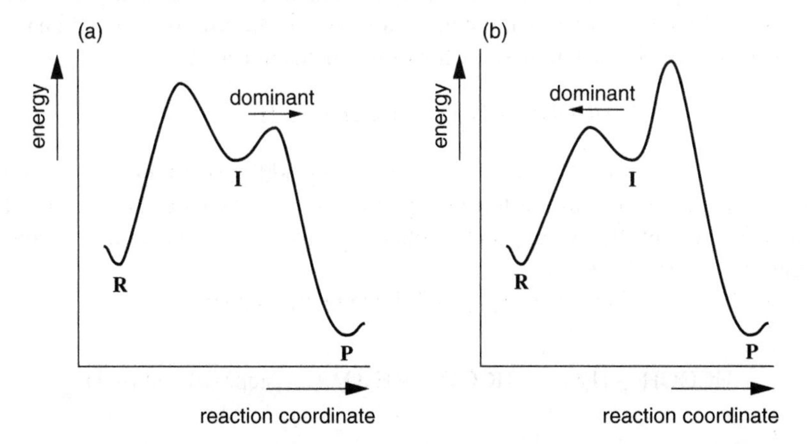

Fig. 9.12 Alternative energy profiles for a two-step reaction. In (a) the pathway with the lowest activation energy open to the intermediate **I** is to go to products, **P**. In (b), the lowest energy pathway is for the intermediate to return to reactants, **R**, by the reverse of the first step. Profile (a) leads to kinetics in which the first step is rate-determining; profile (b) leads to *pre-equilibrium* kinetics, as described in the text. The dominant fate of the intermediate, **I**, is indicated by the arrow.

The picture of what is happening is therefore this: the reactants and the intermediate are rapidly interconverting via steps (1) and (-1), with a small, slow, 'bleed off' of some of the intermediate to give products via step (2). Under these circumstances we are justified in assuming that the reactants and the intermediate are in equilibrium; even though *some* of the intermediate goes to form products, the equilibrium between reactants and intermediates is quickly re-established. Step (2) is the rate-limiting step as its rate constant controls the rate of formation of the product.

This idea that the intermediate (or intermediates) are in equilibrium with the reactants is called the *pre-equilibrium hypothesis*. The 'pre-' comes about because this equilibrium precedes the rate-determining step. This kind of situation is quite commonly encountered, particularly if the pre-equilibrium steps involve protonation and deprotonation; let us see an example of how this works out.

Example: oxidation of methanoic acid with bromine

In aqueous solution, methanoic (formic) acid is oxidized to CO_2 by Br_2 according to the stoichiometric equation:

$$HCOOH + Br_2 + 2H_2O \longrightarrow CO_2 + 2Br^- + 2H_3O^+.$$

This reaction has been investigated experimentally and it is found that the rate of the reaction is sensitive to the concentration of H_3O^+ in the solution (this can be varied by adding a strong acid, such as HCl). The rate law is found to be:

$$\text{rate of consumption of } Br_2 = \frac{k[Br_2][HCOOH]}{[H_3O^+]}. \quad (9.5)$$

This rate law tells us that the rate goes *down* as the concentration of H_3O^+ goes up, i.e. the reaction is inhibited by H_3O^+. This gives us a clue as to what might be happening in this reaction: suppose that the bromine actually oxidizes the anion $HCOO^-$ rather than the methanoic acid itself; this anion would have to be formed by the usual dissociation of the methanoic acid:

$$HCOOH + H_2O \rightleftharpoons HCOO^- + H_3O^+.$$

Increasing the amount of H_3O^+ (for example by adding a strong acid) would shift this equilibrium to the left (see p. 126) and so decrease the amount of $HCOO^-$ present; this would then explain why the rate decreases as the concentration of H_3O^+ increases.

With this in mind we propose the following mechanism:

$$HCOOH + H_2O \underset{k_{-3}}{\overset{k_3}{\rightleftharpoons}} HCOO^- + H_3O^+ \qquad \text{steps (3) and } (-3)$$

$$HCOO^- + Br_2 \xrightarrow{k_4} \text{products} \qquad \text{step (4).}$$

We have numbered the steps (3), (-3) and (4) to avoid confusion with steps (1) and (2) in the reaction in Fig. 9.9 on p. 141.

The first line is simply the acid dissociation equilibrium. We have written the rate constants of the forward and reverse reactions as k_3 and k_{-3}, respectively, but recall from the discussion on p. 144 that the equilibrium constant is the ratio of these two rate constants:

$$K_a = \frac{k_3}{k_{-3}}. \tag{9.6}$$

If we assume that pre-equilibrium applies, then we can find an expression for $[HCOO^-]$ using the definition of the equilibrium constant, K_a:

$$K_a = \frac{[H_3O^+][HCOO^-]}{[HCOOH]}.$$

This can be rearranged to give

$$[HCOO^-] = \frac{K_a[HCOOH]}{[H_3O^+]}.$$

The rate of consumption of bromine is just the rate of step (4), so

$$\begin{aligned}\text{rate of consumption of } Br_2 &= k_4[HCOO^-][Br_2] \\ &= \frac{k_4 K_a[HCOOH][Br_2]}{[H_3O^+]}.\end{aligned}$$

The rate law we have derived is therefore

$$\text{rate of consumption of } Br_2 = \frac{k_{\text{eff}}[HCOOH][Br_2]}{[H_3O^+]}$$

where k_{eff} is

$$\begin{aligned}k_{\text{eff}} &= k_4 K_a \\ &= k_4 \frac{k_3}{k_{-3}}\end{aligned}$$

and on the last line we have used Eq. 9.6.

The rate law we have predicted has the same form as the experimental rate law, Eq. 9.5, which at least means that our mechanism is consistent with the experimental observations and that the pre-equilibrium assumption is valid for this mechanism. Finally, it shows us that the experimental rate constant is in fact a composite of three rate constants, or – looked at another way – a composite of a rate constant and an equilibrium constant.

When can we apply the pre-equilibrium assumption?

For the pre-equilibrium assumption to be valid the rate constants have to be such that an intermediate, once formed, is much more likely to return to where it came from than to go on to react further. In other words, the further reactions of the intermediate must be slow compared to the intermediate returning to the species from which it was formed.

A common situation in which these conditions hold is when the intermediate is formed by protonating or deprotonating one of the reactants. Such proton transfers often have rather large rate constants – they are sometimes described as being 'facile' or 'easy' processes.

9.9 The apparent activation energy

For any reaction – be it an elementary reaction or one with a complex mechanism – we can always determine the activation energy by measuring how the rate constant varies with temperature and then fitting the data to the Arrhenius equation in the way described on p. 138 and illustrated in Fig. 9.8. If the reaction is a complex one proceeding via a number of elementary steps, the question is what is the relationship between this measured activation energy (we will call this the *apparent activation energy*) and the activation energies of the elementary steps? We will use the two reactions we have been discussing in this chapter to illustrate this relationship and this will lead us to a rather neat general conclusion about the size of the apparent activation energy.

The reaction between CH_3COCl and OH^- (Fig. 9.9 on p. 141) proceeds by a two-step mechanism:

$$CH_3COCl + OH^- \xrightarrow{k_1} \mathbf{I} \qquad \text{step (1)}$$

$$\mathbf{I} \xrightarrow{k_2} CH_3COOH + Cl^- \qquad \text{step (2).}$$

We argued that if the first step is rate-limiting the overall rate just depends on the rate constant for step (1):

$$\text{rate of formation of the products} = k_1[CH_3COCl][OH^-].$$

We can write the rate constant k_1 using the Arrhenius equation (Eq. 9.2 on p. 137) as:

$$k_1 = A_1 \exp\left(\frac{-E_{a,1}}{RT}\right).$$

Therefore the apparent activation energy for the overall reaction is exactly the same as the activation energy for step (1), the rate-limiting step.

Identifying the activation energy for the oxidation of methanoic acid is a little more complicated; let us remind ourselves of the reaction scheme:

$$HCOOH + H_2O \underset{k_{-3}}{\overset{k_3}{\rightleftarrows}} HCOO^- + H_3O^+ \qquad \text{steps (3) and (−3)}$$

$$HCOO^- + Br_2 \xrightarrow{k_4} \text{products} \qquad \text{step (4).}$$

We found that by applying the pre-equilibrium assumption the overall rate law was:

$$\text{rate of consumption of } Br_2 = \frac{k_{\text{eff}}[HCOOH][Br_2]}{[H_3O^+]}$$

where

$$k_{\text{eff}} = k_4 \frac{k_3}{k_{-3}}.$$

We now write each rate constant in the expression for k_{eff} using the Arrhenius law:

$$k_{\text{eff}} = \frac{A_4 \exp\left(\frac{-E_{a,4}}{RT}\right) A_3 \exp\left(\frac{-E_{a,3}}{RT}\right)}{A_{-3} \exp\left(\frac{-E_{a,-3}}{RT}\right)}$$
$$= \frac{A_4 A_3}{A_{-3}} \exp\left(\frac{-(E_{a,4} + E_{a,3} - E_{a,-3})}{RT}\right).$$

From this we can identify the apparent activation energy as the energy term in the exponent:

$$\begin{aligned} E_{a,\text{apparent}} &= E_{a,4} + E_{a,3} - E_{a,-3} \\ &= E_{a,4} + \Delta E_3 \end{aligned} \qquad (9.7)$$

where on the last line we have used the relationship between the energy change for a reversible reaction and the difference in activation energies between the forward and reverse reactions: $\Delta E = E_{a,3} - E_{a,-3}$ (Eq. 9.3 on p. 143).

We see in the case of this mechanism that the apparent activation energy depends on the activation energies of all three steps or, alternatively, on the energy change of the reversible reaction and the activation energy of the final step.

The general result

In fact, there is a simple result which describes both reaction schemes. It is that the apparent activation energy for a reaction is the energy difference between the reactants and the *highest energy transition state* on the reaction pathway: this is illustrated in Fig. 9.13.

Profile (a) is appropriate for the reaction between CH_3COCl and OH^- in which the first step is rate-determining and so the highest energy transition state

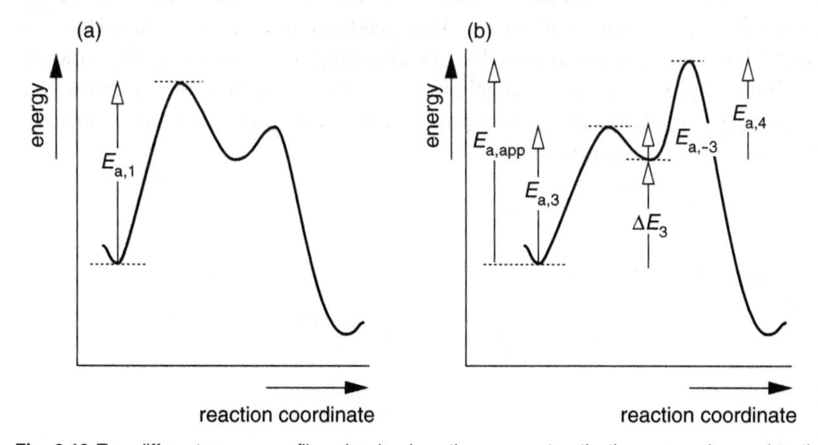

Fig. 9.13 Two different energy profiles showing how the apparent activation energy is equal to the energy separation between the reactants and the highest energy transition state along the reaction pathway. Profile (a) is for the reaction between CH_3COCl and OH^-; the first step is rate-determining and so has the highest activation energy. The apparent activation energy is therefore equal to the activation energy of the first step. Profile (b) is for the reaction between HCOOH and Br_2; the second transition state is the highest energy point and so the apparent activation energy depends on the activation energies of all of the reactions, as described in the text.

on the reaction profile is the first transition state we encounter. The apparent activation energy is simply the activation energy of the first step.

Profile (b) is for a reaction in which the first step is reversible, such as the reaction between HCOOH and Br_2 which we have been discussing. The second transition state, between the intermediate and the products, is now the highest energy point on the pathway and so this determines the apparent activation energy. As we have discussed, this activation energy can either be expressed in terms of the activation energies of all three reactions or the energy change of the first reaction and the activation energy of the second (Eq. 9.7).

9.10 Looking ahead

We mentioned at the start of this chapter that a knowledge of the rates of reactions can be helpful in teasing out the mechanisms of complex reactions. The oxidation of methanoic acid is a good example of this – we saw that the rather unusual rate law, in which H_3O^+ was seen to inhibit the reaction, suggested a mechanism which we went on to show was consistent with the experimental rate law.

In the coming chapters we will see a great deal more of how a study of kinetics helps us to unravel the mechanism. For example, consider the following two nucleophilic substitution reactions and their experimentally determined rate laws:

$$CH_3Br + Nu^- \longrightarrow CH_3Nu + Br^- \qquad \text{rate} = k[CH_3Br][Nu^-]$$

$$(CH_3)_3CBr + Nu^- \longrightarrow (CH_3)_3CNu + Br^- \qquad \text{rate} = k'[(CH_3)_3CBr].$$

Note that the reaction involving bromomethane is first order in both the nucleophile and the alkyl halide, and hence second order overall. In contrast, for the 2-bromo-2-methylpropane (t-butyl bromide) the concentration of the nucleophile does *not* appear in the rate law, making it first order overall. How can this be? We will see in Chapter 11 how changing the alkyl group from methyl to t-butyl causes a change in mechanism and hence in the rate law. However, before we can understand that we must look at a further electronic effect – delocalization and conjugation.

10 Bonding in extended systems – conjugation

In Chapter 6 we introduced hybrid atomic orbitals and used these as the basis for a localized description of bonding; such an approach is perfectly adequate for many molecules, especially those commonly encountered in organic chemistry. We then saw in Chapter 7 how we could use a knowledge of orbitals to help us to understand and rationalize reactions.

In the following chapters we want to move on to some reactions which involve molecules and intermediates whose bonding cannot be described adequately using only the localized approach of Chapter 6. The kind of molecules we are talking about typically have extended π systems and are often charged; we will see that a proper description of the bonding in such molecules needs the delocalized MO approach for parts, but not usually all, of the molecule.

For example, let us consider how to describe the bonding in propene and benzene (Fig. 10.1). For propene we choose sp^3 hybridization for C_1, and sp^2 hybridization for the other two carbons. The overlap of these hybrids with one another or with hydrogen 1*s* AOs gives us the C–C and C–H σ bonds, and the π bond is formed by the overlap of out-of-plane 2*p* orbitals on C_2 and C_3. For propene, this localized description is entirely adequate.

For benzene we choose all of the carbons to be sp^2 hybridized and once again form C–C and C–H σ bonds by overlapping these HAOs with one another and with the hydrogen 1*s* AOs. The six out-of-plane 2*p* orbitals then overlap in three pairs to form three π bonds, thus giving alternating single and double bonds round the ring.

However, the problem with this localized description is that there is much physical and chemical evidence that all the C–C bonds in benzene are the *same* (for example, they are all the same length); a structure in which there are alternating single and double bonds is simply not consistent with these observations. Benzene is therefore an example of where a delocalized description is required.

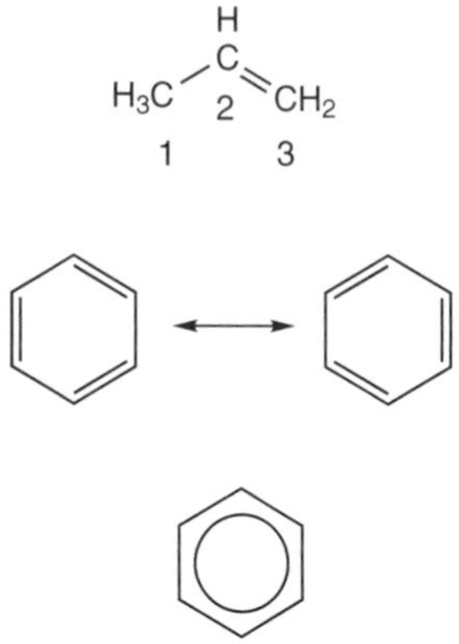

You may have come across the idea of using *resonance structures* to explain the delocalized nature of the bonding in benzene. The idea is illustrated in Fig. 10.1 where two different structures for benzene are shown which only differ in the placement of the double bonds. On their own, both structures are an inadequate representation of the bonding in benzene, but the two taken together convey the crucial point that there is π bonding between *all* the adjacent carbons. In a sense, the 'real' structure of benzene lies in between these two, and this is what the structure at the bottom of Fig. 10.1 tries to convey – the circle indicating that the π bonds are delocalized round the ring.

It is tempting to think that the molecule flips between these resonance structures, but this is a mistaken view. These structures are simply our attempts to

H
H_3C C 2 CH_2
1 3

Fig. 10.1 The bonding in propene can be described adequately using 2c-2e bonds, but benzene requires a delocalized approach for the π system. The two *resonance structures* for benzene, connected by a double headed arrow, are an attempt to describe this delocalized bonding using structures with localized bonds. The structure at the bottom, in which a circle is drawn in place of the separate π bonds, is another way of representing the delocalized bonding in benzene.

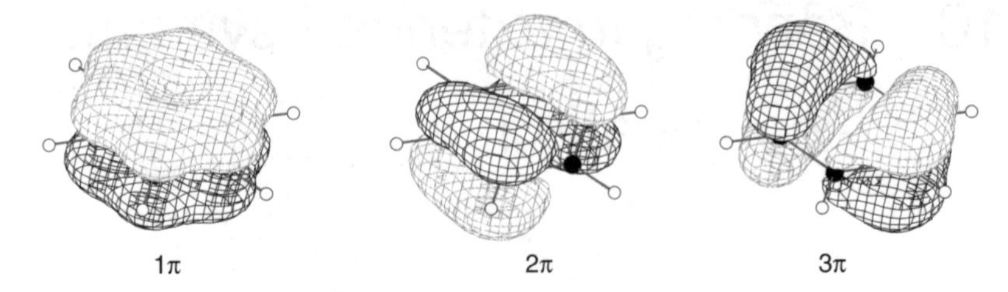

Fig. 10.2 Surface plots of the three highest energy occupied π MOs of benzene; positive and negative parts of the wavefunction are indicated by different shading of the grid lines. The skeleton of the benzene molecule is also indicated. These three MOs are clearly formed from the overlap of the out-of-plane $2p$ orbitals; the lowest energy MO, 1π, is reminiscent of the 'doughnut' picture often used to explain the delocalized π bonding in benzene.

make a localized description of the bonding in a molecule which in fact has delocalized bonding. No one of the structures is an adequate representation of the actual bonding, but taken together they convey something closer to the truth.

A useful analogy here is that of an *mule*, which we get by crossing a horse with a donkey. To describe an mule as a horse or a donkey would be inaccurate and it certainly does not interconvert rapidly between the two! Rather we need to recognize that although an mule has parts that are reminiscent of a horse and parts that are reminiscent of a donkey, it is something altogether different.

The MO picture of benzene naturally takes a delocalized view of the bonding, and the three highest energy occupied MOs shown in Fig. 10.2 illustrate this very well. It is clear from looking at these MOs that they are formed from the overlap of $2p$ orbitals which point out of the plane of the molecule, and indeed you may recognize the 1π MO as being like the 'doughnut' picture often given to explain the equivalence of the C–C bonds in benzene. However, this doughnut is just one of the six MOs which are formed from the overlap of the six out of plane $2p$ orbitals; only three of the MOs are occupied, and these are the ones shown in Fig. 10.2.

It is not really necessary to use the delocalized MO approach to describe *all* of the bonding in benzene. By using sp^2 hybrids on the carbons we can generate a σ *framework* of 2c-2e C–H and C–C bonds, just as we described above. The MO approach is only needed to treat the six out of plane $2p$ orbitals.

This way of describing benzene is typical of the approach we will adopt. The majority of the bonding in a molecule can usually be described using 2c-2e bonds to give the σ framework. This leaves a few orbitals, typically pointing out of the plane of the σ framework, whose interaction we need to treat using the MO approach.

What we will see in the next section is that it is quite easy to construct delocalized π MOs which are formed from $2p$ orbitals arranged in a row; such an arrangement is surprisingly common in molecules and so we will find immediate use for these MOs in describing aspects of structure and reactivity. We will start with four p orbitals in a row to introduce the idea and then reduce the number to three; this might seem a bit illogical, but it turns out that it is easier to see what is going on if we start with four orbitals.

10.1 Four *p* orbitals in a row

A good example of what we mean by a 'row of *p* orbitals' is the case of butadiene, shown in Fig. 10.3. We can describe the bonding in this molecule by making all of the carbons sp^2 hybridized, and then using these orbitals to form 2c-2e bonds with the hydrogens, and carbon–carbon σ bonds between the adjacent carbons. There are 22 valence electrons in butadiene, four from each carbon and one from each hydrogen. The σ framework uses up 18 of these electrons, leaving four to go into the out of plane π system which is formed from four $2p$ orbitals.

We know that these four $2p$ orbitals will overlap to form four MOs (remember that in Section 5.4 on p. 68 we described how the number of MOs is equal to the number of AOs which are being combined). These four MOs will each be a linear combination of the four AOs:

$$\text{MO} = c_1\text{AO}_1 + c_2\text{AO}_2 + c_3\text{AO}_3 + c_4\text{AO}_4$$

where AO_1 is the AO on atom 1, and c_1 is the coefficient for this AO and so on. These coefficients are a measure of the contribution that each AO makes to the MO. It is relatively easy to compute the energies of the MOs and the corresponding coefficients; the results are presented in cartoon form in Fig. 10.4.

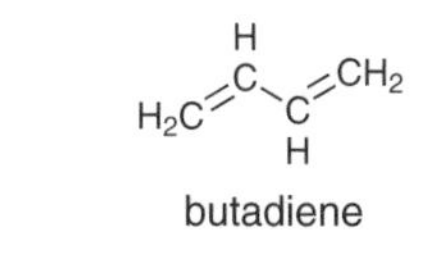

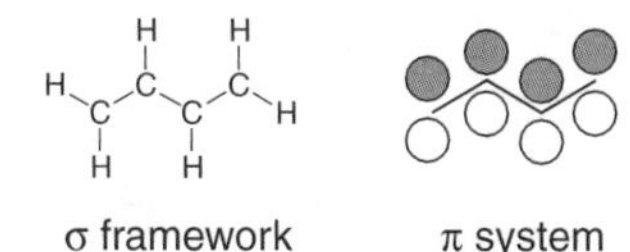

Fig. 10.3 At the top is shown the conventional representation of butadiene, with two C–C double bonds separated by a C–C single bond. If the carbons are sp^2 hybridized the bonding can be described as a σ framework formed from the overlap of these hybrids with one another and with the hydrogen $1s$ AOs, together with a delocalized π system formed from the overlap of the out-of-plane $2p$ AOs. These orbitals are an example of four *p* orbitals in a row.

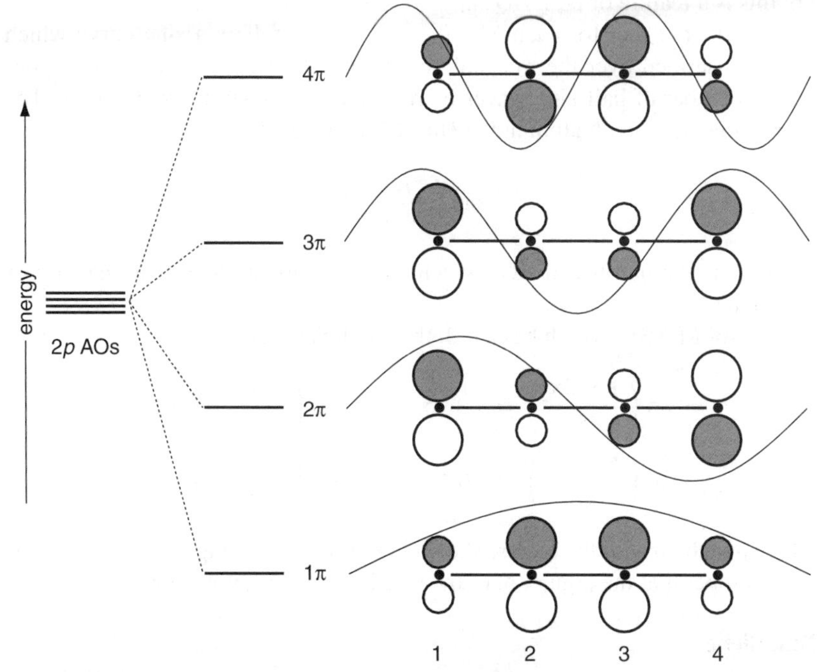

Fig. 10.4 Cartoons of the four π MOs resulting from four $2p$ AOs overlapping 'in a line'. The coefficient (or contribution) of each AO is indicated by the size of the AO; note that some of the coefficients are negative so the sign of the AO has been inverted. MO 1π is the lowest in energy and has no nodes between the atoms; 2π is next lowest and has one node between atoms 2 and 3. As the energy rises to MOs 3π and 4π the number of nodes increases to 2 and 3, respectively. The orbital coefficients are given by a sine wave which can be drawn over the molecule, as shown on the right; for 1π there is one half sine wave, for 2π there are two and so on.

The σ framework lies in a nodal plane for each MO; this is why the MOs are designated π. First look at the energies of the MOs: there are two MOs lower in energy than the AOs and two higher. The first two are therefore bonding MOs and the latter two are anti-bonding.

The coefficients of the AOs fall into a simple pattern as the MOs go up in energy. In the lowest energy MO, 1π, all of the coefficients are positive; there is therefore constructive overlap between all of the adjacent orbitals and it is therefore not surprising that this MO is the most strongly bonding.

In 2π there is a nodal plane between atoms 2 and 3; there is thus constructive interference between the AOs on atoms 1 and 2, and between those on atoms 3 and 4, but destructive interference between atoms 2 and 3. Since there are two constructive interactions but only one destructive, the orbital is overall bonding, but not as strongly as 1π.

MO 3π has two nodal planes (between atoms 1 and 2, and 3 and 4), so there is destructive overlap between the AOs on atoms 1 and 2, and 3 and 4. There is constructive overlap between the AOs on atoms 2 and 3, but this time (in contrast to 2π) the orbital coefficients are such that the orbital is net anti-bonding.

Finally, 4π has a nodal plane between each adjacent atom, and so there is only destructive interference; it is therefore the most anti-bonding MO. The general pattern is that the number of nodal planes increases as the energy goes up; this is a feature of both AOs and MOs which we have seen before.

It turns out that for each MO the coefficients follow a sine curve which can be inscribed over the molecule in a particular way. As the energy goes up the number of half sine waves increases, as is shown in the diagram. The coefficients for the jth atom in the kth MO are given by

$$\sin\left(\frac{jk\pi}{N+1}\right)$$

where N is the number of AOs which are overlapping; $N = 4$ in the case of butadiene.

So, for MO 3π, which has $k = 3$, the coefficients are:

$$c_1 = \sin\left(\frac{1\times 3\pi}{5}\right) = +0.95 \quad c_2 = \sin\left(\frac{2\times 3\pi}{5}\right) = -0.59$$

$$c_3 = \sin\left(\frac{3\times 3\pi}{5}\right) = -0.59 \quad c_4 = \sin\left(\frac{4\times 3\pi}{5}\right) = +0.95$$

which match up with the picture shown in Fig. 10.4 (to compute these you need to remember that the argument of the sine is in radians, not degrees).

Butadiene

In butadiene we have four electrons to place in the π system; we therefore put two (spin paired) in 1π and two in 2π. All of the electrons are therefore accommodated in bonding MOs.

MO 1π gives rise to π bonding across the *whole* molecule so, in contrast to the simple picture shown in Fig. 10.3, there is π bonding between C_2 and C_3. In fact, because in 1π the coefficients on C_2 and C_3 are the greatest (see

Fig. 10.4), this MO contributes more to the bonding between C_2 and C_3 than it does to the bonding between C_1 and C_2 or C_3 and C_4.

MO 2π has a node between C_2 and C_3 and so gives rise to an anti-bonding interaction between these two atoms. However, because of the smaller orbital coefficients on these atoms, the anti-bonding interaction does not outweigh the bonding interaction due to 1π. So, overall, the occupation of MOs 1π and 2π leads to some π bonding between C_2 and C_3.

MO 2π also gives rise to a bonding interaction between C_1 and C_2, and between C_3 and C_4. However, the contributions from MOs 1π and 2π do not add up to the same amount of π bonding as is present in the simple 2c-2e π bond in ethene. So in butadiene the degree of π bonding between C_1 and C_2 and between C_3 and C_4 is less than in ethene.

The overall picture for butadiene is of π bonding across all four carbons, but with a larger interaction between atoms 1 and 2, and 3 and 4. The bond length data give in Fig. 10.5 confirm this description. The C_2–C_3 bond length in butadiene is shorter than the C–C single bond in propene, but not as short as the C–C double bond. Similarly, the nominal double bond in butadiene between C_1 and C_2 is a little longer than the double bond in propene. These data point to partial π bonding between C_2 and C_3, and a weakening of the π bonding between C_1 and C_2, just as we predicted from the MOs.

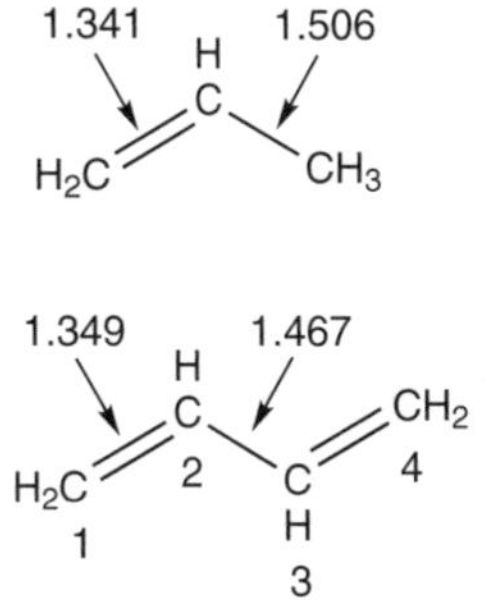

Fig. 10.5 Experimental bond length data, in Å, for the C–C bonds in propene and butadiene. The C_1–C_2 bond in butadiene is longer than the C–C double bond in propene, whereas the C_2–C_3 bond is shorter than the C–C single bond in propene. These data provide evidence for there being partial π bonding between C_2 and C_3.

The two π bonds which we drew in the simple structure of butadiene are described as being *conjugated*, which means that they form a pattern of alternating single and double bonds; the key feature of such systems is that they contain an *uninterrupted* chain of *p* orbitals lying out of the plane of the σ framework. In conjugated systems the π bonding is not localized between pairs of atoms, but rather is delocalized across the whole conjugated system. A proper description of the bonding in such a molecule requires the delocalized MO approach.

In contrast, a molecule such as CH_2=CH–CH_2–CH=CH_2 is *not* conjugated as the π bonds are separated by *two* σ bonds; a localized description of the bonding is perfectly adequate for such a molecule.

In summary, because the two π bonds in butadiene are conjugated the simple structure given in Fig. 10.3 does not really describe the bonding adequately. In fact, the four electrons assigned to the two π bonds in the simple structure are occupying π MOs which are delocalized over the whole molecule and so give rise to π bonding between *all* the adjacent carbon atoms.

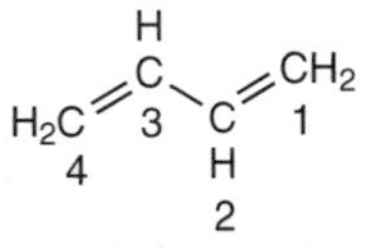

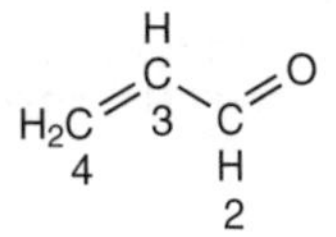

Propenal

Propenal (also called acrolein), whose structure is shown in Fig. 10.6, is closely related to butadiene; both have two conjugated double bonds with a total of four π electrons. Just as we did in butadiene, we can model the π system in propenal as four $2p$ orbitals in a row. The crucial difference, however, is that the four orbitals no longer have the same energy – the oxygen orbital is lower in energy than that on the carbon (see Fig. 4.34 on p. 57). The four $2p$ orbitals will still give rise to four MOs, but they will not be quite the same as those shown in Fig. 10.4; the presence of one oxygen AO will alter both the energies and the orbital coefficients of the MOs. It requires a computer calculation to work out the MOs which are shown, as surface plots, in Fig. 10.7.

Fig. 10.6 The structures of butadiene and propenal; to go from one to the other all we have to do is replace the CH_2 at position 1 in butadiene with an oxygen.

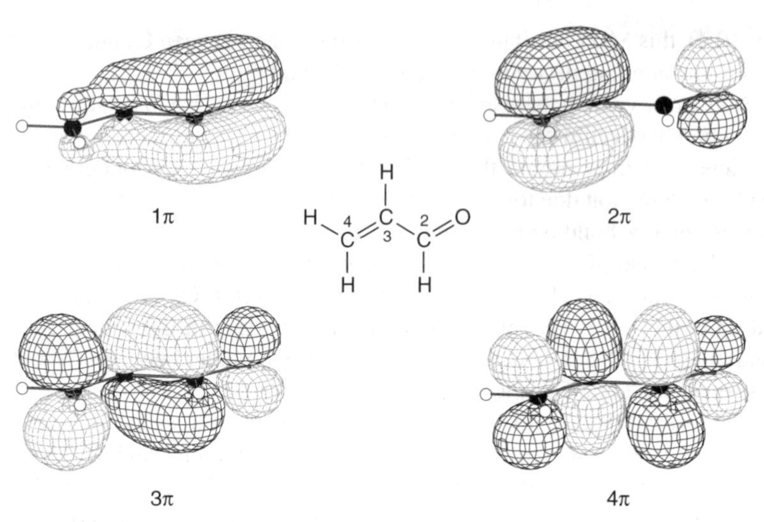

Fig. 10.7 Surface plots of the four π MOs in propenal; the numbering scheme matches that used in Fig. 10.8 and that of the corresponding MOs in butadiene shown in Fig. 10.4.

Broadly speaking the four MOs are comparable with those for butadiene shown in Fig. 10.4 on p. 153. MO 1π is the lowest in energy and has no nodal planes between the atoms. For propenal the coefficient of the AO from C_4 is so small that it hardly shows up on this plot.

MO 2π is the next highest in energy; as in butadiene it has constructive overlap between the AOs on C_3 and C_4, but in propenal the coefficient of the AO from C_2 is very small so there is little constructive overlap between this AO and the one from oxygen. Finally, just as in butadiene, MOs 3π and 4π have two and three nodal planes, respectively.

Figure 10.8 shows the computed ordering of the MO energies for propenal; compare these with the corresponding MOs in butadiene shown in Fig. 10.4 on p. 153. It always turns out that the lowest energy MO will be *lower* in energy than the lowest energy AO, and the highest energy MO is *higher* in energy than the highest energy AO; all of the MO diagrams you have seen conform to this pattern. The other two MOs lie somewhere in between, and it is not really possible to guess where they will fall. What we do know is that as the number of nodes in an orbital increases, so will the energy, and this is consistent with the ordering shown in Fig. 10.8.

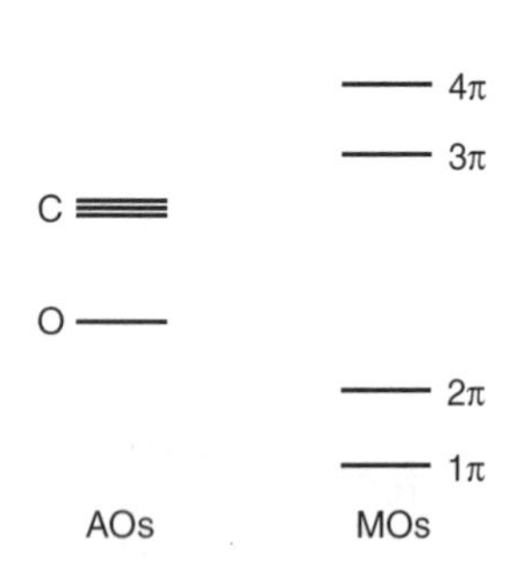

Fig. 10.8 MO diagram for the π system in propenal. As explained in the text, the lowest energy MO (1π) lies below the lowest energy AO and similarly the highest energy MO (4π) lies above the highest energy AO.

In Section 5.6 on p. 76 we described the MOs formed from the overlap of two AOs of unequal energies. What we found was that the two AOs do not contribute equally to a given MO, and that the AO which is closest in energy to a particular MO is the major contributor to that MO. This same idea carries forward to more complex situations, so we expect that the 1π MO in propenal will have a greater contribution from the oxygen AO than will 4π; the surface plots in Fig. 10.7 bear this out.

There are four electrons to accommodate in this π system, so the 1π and 2π MOs are occupied, making 2π the HOMO and 3π the LUMO. The 1π MO contributes to bonding between C_2 and the oxygen, with a smaller contribution between C_2 and C_3. The C_2–C_3 bond length of 1.484 Å, shorter than a typical C–C single bond, confirms the presence of partial π bonding between these two atoms. MO 2π contributes only to bonding between C_3 and C_4, making a π bond just as in the localized structure.

The way in which this partial double bond character between C_2 and C_3 can be represented using localized 2c-2e structures in shown in Fig. 10.9.

Fig. 10.9 Resonance structures for propenal which illustrate the partial π bonding between carbons 2 and 3; resonance structure **B** contributes less than structure **A**.

The fact that the C_2–C_3 bond length is only slightly shorter than a typical C–C single bond suggests that resonance structure **B** only makes a minor contribution. This structure also suggests that there is more electron density on the oxygen and less on C_4.

The removal of electron density from C_4 is supported by the observation that, whereas C=C bonds are not usually attacked by nucleophiles (Section 7.4 on p. 106), such attack is sometimes seen for conjugated C=C bonds. For example, CN^- can either attack the carbonyl carbon directly (like the reactions in Section 7.3 on p. 102), or it can attack at C_4, as shown in Fig. 10.10.

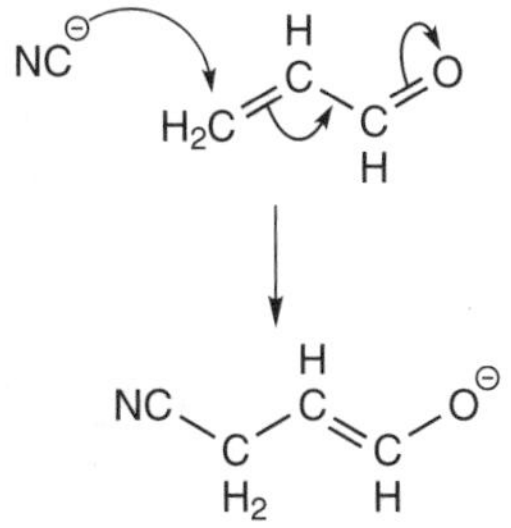

Fig. 10.10 In the reaction between CN^- and propenal the nucleophile can also attack C_4 as well as the carbonyl carbon.

10.2 Three *p* orbitals in a row

When three *p* AOs overlap in a line, three MOs are formed and these can be constructed in just the same way that we did for four overlapping *p* orbitals; Fig. 10.11 shows the MOs in cartoon form. As before, the energy of the MOs increases as the number of nodes increases, and the orbital coefficients are given by sine waves which can be inscribed over the molecule.

The one slight difference between the overlap of three as opposed to four orbitals is that in the former case one of the MOs (2π) is non-bonding (it has approximately the same energy as the AOs). That this is so can also be seen from the form of the MO; the orbital coefficient on the central atom is zero so there is neither constructive nor destructive overlap between AOs on adjacent atoms. The end carbons are too far apart for their AOs to interact significantly.

Understanding the π MOs which are formed from the overlap of three *p* orbitals turns out to be very useful for rationalizing all sorts of observations, such as why amide bonds are flat and why the two C–C bonds in the allyl anion are of equal length.

Allyl anion

The allyl anion can be thought of as arising from the removal of H^+ from propene, as shown in Fig. 10.12; a very strong base is needed to remove a proton from this molecule. The bonding in the allyl anion can be described by having sp^2 hybridization on each carbon atom; these hybrids overlap with one another or hydrogen $1s$ orbitals to give a σ framework. The remaining out-of-plane $2p$ orbitals overlap to form the three MOs pictured in Fig. 10.11.

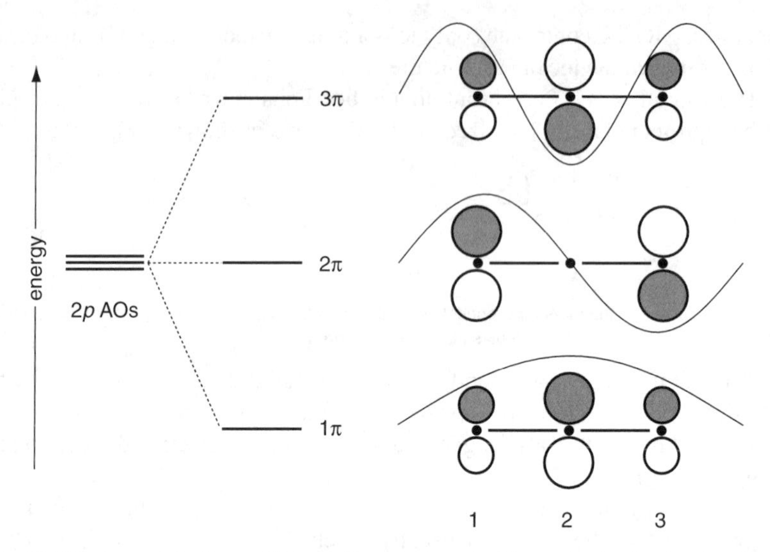

Fig. 10.11 Cartoons of the three π MOs resulting from three $2p$ AOs overlapping in a line; the diagram is drawn in the same way as Fig. 10.4. As before, the orbital coefficients are given by sine waves which can be drawn over the molecule; for the 1π MO there is one half sine wave, for 2π there are two and so on. As expected, the energies of the MOs increase as the number of nodes increases.

The (valence) electron count in the anion is four from each of the three carbons, plus one from each of the five hydrogens plus one more electron for the negative charge; this gives a total of 18 electrons. Of these, 14 are used in the σ framework and the remaining four go into the π system. We thus assign two electrons to MO 1π and two to 2π.

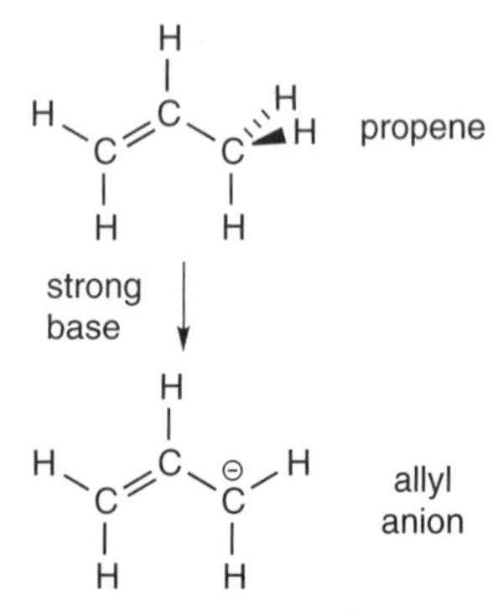

Fig. 10.12 The allyl anion can be formed by the action of a very strong base on propene.

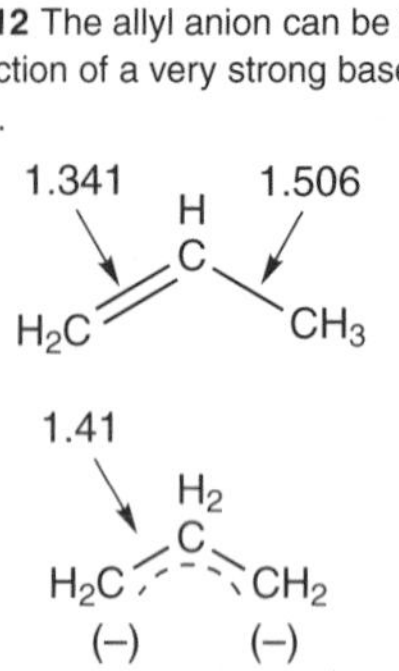

Fig. 10.13 Comparison of the C–C bond lengths (in Å) of propene and the allyl anion. The anion is symmetrical and has a bond length intermediate between the single and the double bond in propene.

Occupation of 1π contributes to the π bonding over all three atoms, whereas occupation of 2π contributes nothing to the π bonding. However, having 2π occupied does increase the electron density on the two end carbons preferentially over the centre carbon.

As 1π is symmetrical about the centre, we conclude that the two C–C bonds must be identical. Indeed, experiment bears this out and the bonds are found to be the same length. It is also interesting to compare this bond length with those in propene (Fig. 10.13); we see that the bond length in the anion is intermediate between that of the C–C single and double bonds in propene. Since there are only two electrons (those in 1π) which are contributing to the bonding across three atoms we expect the π bonding to be partial, which is just what the bond length data indicate.

The other interesting thing about the allyl anion is how the electron density differs between the carbons. MO 1π has the greatest orbital coefficient, and hence the greatest electron density, on the central carbon. However, MO 2π puts the electron density on the end carbons and none in the middle. It takes a more detailed calculation to work out which of these two wins – in fact it turns out that the end carbons have the greatest π electron density.

To explain these properties of the allyl anion using localized structures, we have to draw two resonance structures, as shown in Fig. 10.14. These structures convey the idea that the charge is delocalized and may be 'found' on either of the terminal carbons; similarly, the double bond may be located between C_1

and C_2 or between C_2 and C_3. This picture is consistent with the MO approach which, as we have explained, predicts a symmetrical structure with partial π bonding and maximum electron density on the end carbons. Sometimes the anion is represented using a dashed line across all three atoms to indicate the partial π bond; such a structure is also shown in Fig. 10.14.

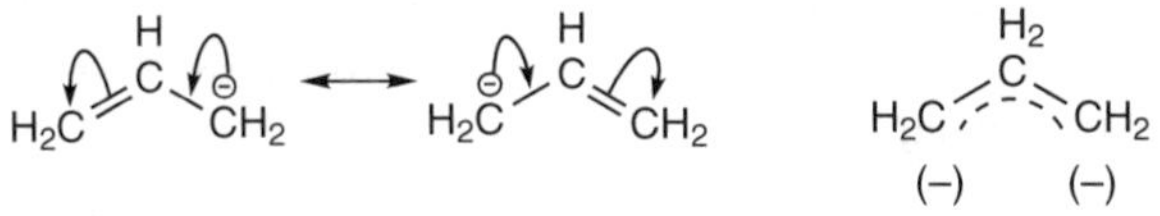

$H_2C{=}CH{-}CH_2^{\ominus} \longleftrightarrow {}^{\ominus}H_2C{-}CH{=}CH_2 \qquad H_2C{\cdots}CH_2 \, (-) \; (-)$

Fig. 10.14 On the left, connected by a double-headed arrow, are two resonance structures for the allyl anion. On the right is an alternative representation in which the dashed line shows the partial π bond across all three atoms. In this structure the minus signs in brackets (–) indicate the positions where negative charge can be found, as shown by the resonance structures on the left.

It is interesting to compare the two resonance structures with the form of the HOMO. Recall that this MO, the 2π, has the electron density confined to the end carbons. This is exactly where the negative charge is located in the resonance structures. There are some parallels, therefore, between the form of the HOMO and the charge distribution predicted by the resonance structures.

Carboxylate anion

Closely related to the allyl anion is the carboxylate anion, formed by removing H^+ from the OH group of a carboxylic acid, as shown in Fig. 10.15.

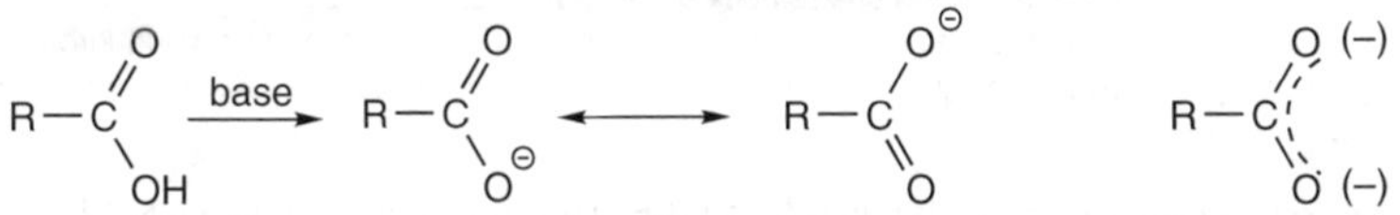

$R{-}C({=}O){-}OH \xrightarrow{\text{base}} R{-}C({=}O){-}O^{\ominus} \longleftrightarrow R{-}C(O^{\ominus}){=}O \qquad R{-}C(\cdots O\,(-))(\cdots O\,(-))$

Fig. 10.15 A carboxylate anion is formed when a base removes H^+ from a carboxylic acid. The π system in the anion is delocalized and can be described using two resonance structures, or represented using a dashed line to indicate a partial bond.

The resulting anion can be described in very similar terms to those used for the allyl anion, the only difference being that instead of the π system being formed from three carbon $2p$ orbitals it is formed from one on carbon and two on oxygen. Surface plots of the three π MOs are shown in Fig. 10.16 – they are very reminiscent of the cartoons shown in Fig. 10.11 on p. 158.

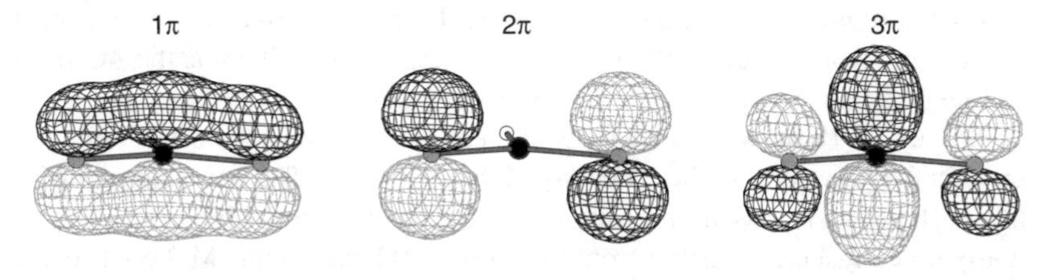

Fig. 10.16 Surface plots of the three π MOs of the methanoate ion, $HCOO^-$. The molecule is viewed sideways on with the two oxygens coming toward and the hydrogen going away from us. Note the similarity between these MOs and the cartoons shown in Fig. 10.11 on p. 158. MO 2π is the HOMO and 3π is the LUMO.

Of these MOs, only the 1π and 2π are occupied, so the conclusions about the carboxylate are much the same as for the allyl anion, i.e. there is partial π bonding across all three atoms, the negative charge is greatest on the two oxygens and the two C–O bonds are equivalent. Experimental data confirm that these two C–O bonds have the same bond length of 1.242 Å. This value is intermediate between a typical C=O bond length of 1.202 Å and a typical C–OH bond length of 1.343 Å, confirming that there is partial π bonding across all three atoms.

As is shown in Fig. 10.15, these features of the bonding in the carboxylate ion can be represented either using resonance structures, or a dashed line to represent a partial bond.

Amides

Studies of the structures of amides using experimental techniques such as X-ray diffraction show us that the bonds to the carbonyl carbon, the nitrogen and the two hydrogens attached to the nitrogen all lie in the same plane, as shown in Fig. 10.17.

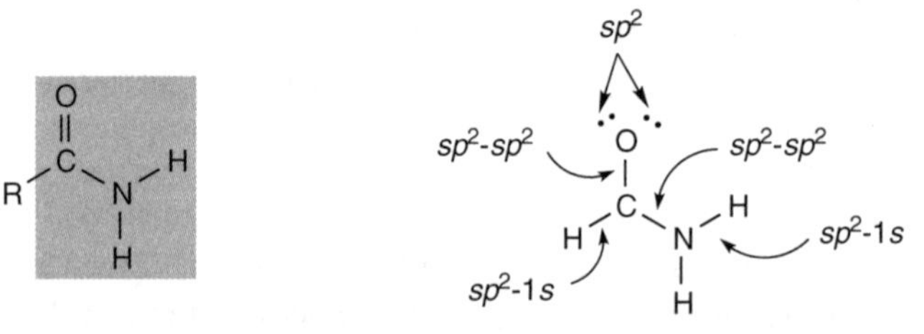

Fig. 10.17 *All* of the bonds within the shaded box of this amide structure are found to lie in the same plane. This being the case, the σ framework of methanamide is best described as being formed from the overlap of sp^2 hybrids on oxygen, carbon and nitrogen. Two lone pairs occupy the remaining two sp^2 hybrids on oxygen.

Given the planarity of the amide, to describe the bonding we should choose sp^2 hybrids for both the carbonyl carbon and the nitrogen. As we discussed on p. 92 it does not matter too much whether we choose the oxygen to be sp or sp^2 hybridized, so we will choose sp^2. With this hybridization scheme the σ framework of methanamide is straightforward to form and is shown Fig. 10.17.

The planarity of the amide group means that the remaining three p orbitals (one each on carbon, nitrogen and oxygen) are all perpendicular to the plane and so in the correct geometry to overlap with one another to form π MOs. The π system can be modelled as three p orbitals in a row but, unlike the allyl anion, the AOs do not have the same energy. From Fig. 4.34 on p. 57 we expect that the AO from oxygen will be lowest in energy, next will come the AO from nitrogen and the highest will be the AO from carbon.

Figure 10.18 shows surface plots of these three π MOs; there are four electrons in the π system, so these occupy MOs 1π and 2π. The lowest energy MO (1π) is bonding across all three atoms. At the level plotted, MO 2π shows virtually no contribution visible from the carbon AO and so this MO contributes little to the bonding. Finally, as in the allyl anion, 3π has two nodal planes between the atoms.

As there is experimental evidence that amides invariably adopt this planar geometry we conclude that this must have the lowest energy. The alternative arrangement, in which the nitrogen is pyramidal and sp^3 hybridized with the lone pair not involved in the π system must therefore be higher in energy. The

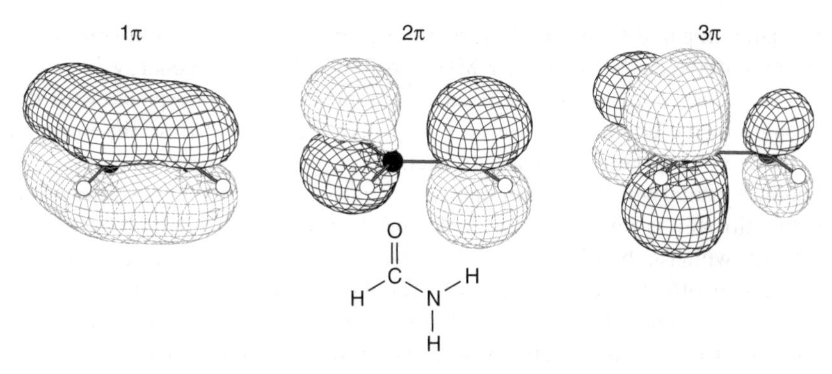

Fig. 10.18 Surface plots of the three π MOs from methanamide.

question is, therefore, why is it that participation of the nitrogen lone pair in the π system lowers the energy?

Figure 10.19 offers an explanation of this by considering the effect of overlapping the nitrogen lone pair with the carbonyl π MOs. In (a) we see the situation in a hypothetical molecule in which there is no interaction between these orbitals. The AOs from nitrogen are intermediate in energy between those of carbon and oxygen and so in the diagram we have placed the energy of the nitrogen AO between that of the two π orbitals.

If we 'switch on' the interaction the resulting MOs are as shown in (b). As noted on p. 156, the highest and lowest energy MOs must come, respectively, above and below the highest and lowest energy AOs. The 2π MO falls quite close in energy to the $2p$ on nitrogen, but we need a computer calculation to find whether it is above or below the energy of the AO; arbitrarily, we have shown it at slightly lower energy than the nitrogen $2p$.

The pair of electrons which start in the carbonyl π MO are lowered in energy as they end up in the 1π MO. Whether or not the *overall* energy is lowered on going from (a) to (b) also depends on how the energy of the lone pair changes as it moves into the 2π MO. Since we know that the planar geometry is favoured and that the feature of this geometry is the participation of the lone pair in the π system, we conclude that (b) must have the lower energy. It could

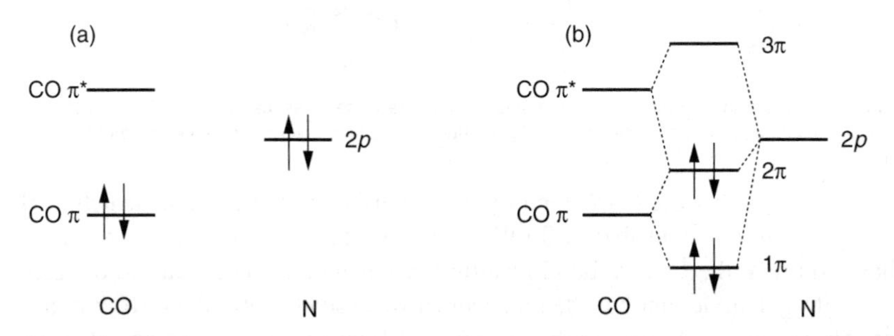

Fig. 10.19 MO diagrams illustrating the interaction between the nitrogen lone pair and the π MOs of the carbonyl group in an amide. In (a) is shown the situation before interaction; the nitrogen orbital appears at an energy intermediate between the π and π^* MOs on the grounds that the nitrogen AOs lie in energy between those of carbon and oxygen. MO diagram (b) shows the case after interaction between the orbitals; as drawn, the interaction results in a lowering of the overall energy of the electrons.

be that, compared to (a), the lowering in energy of the 1π MO outweighs any increase in the 2π or that the 2π MO is itself also lowered in energy.

The creation of this delocalized π system also has consequences for the strength of the bonding. If we do not involve a $2p$ orbital from the nitrogen in the π system there are just two electrons and these occupy the carbonyl π bonding orbital which is just bonding the carbon and the oxygen. However, if the nitrogen is involved, there are four electrons to consider. Two occupy MO 1π which is bonding across all *three* atoms, and two occupy 2π which contributes little to the bonding. The overall effect is that the C–O bond in the amide is *weakened* when compared to methanal as the two electrons in the π bond in methanal are effectively spread over three atoms in methanamide; in other words, the C–O π bond in an amide is partial.

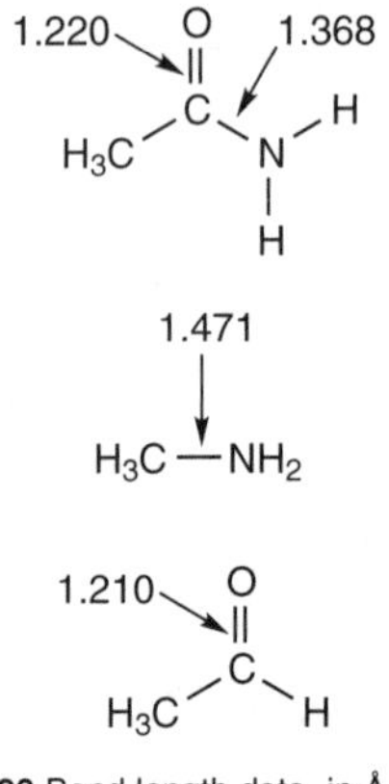

1.220 1.368 O C H_3C N H H

1.471 H_3C—NH_2

1.210 O C H_3C H

Fig. 10.20 Bond length data, in Å, for ethanamide, methylamine and ethanal. Most striking is the shortening of the C–N bond in the amide when compared to that in the amine – something we can attribute to the partial π bonding involving the nitrogen lone pair and the carbonyl group.

In contrast, the C–N bond is strengthened somewhat because occupation of the 1π MO leads to some π bonding between the carbon and nitrogen which, for example, is not present in molecules like methylamine. The partial π character of this bond also explains why it is more difficult to rotate about this bond than for a simple σ bond. The barrier to rotation is not as large as for a full π bond (see p. 90), such as that in ethene, but it is large enough to have consequences for the structures of molecules containing amide bonds. As we discussed on p. 128, amide bonds are formed when amino acids are joined together to form proteins, and it turns out that the restricted rotation about the C–N bond is a crucial factor in determining the three-dimensional structures of proteins.

The bond length data given in Fig. 10.20 confirms these predictions. The C–N bond length in the amide is considerably shorter than the straightforward C–N single bond in methylamine. Also, there is some lengthening of the C–O bond in ethanamide when compared to ethanal, although the effect is not as marked as for the C–N bond.

In terms of localized structures, we can represent the participation of the nitrogen lone pair in the π system by writing a resonance structure in which a π bond is formed between the carbon and the nitrogen, as shown in Fig. 10.21.

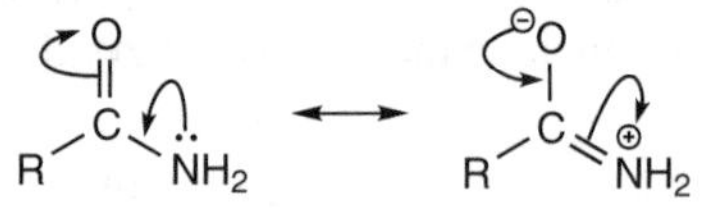

O C R NH_2 ⟷ $O^{\ominus}$ C R $\overset{\oplus}{N}H_2$

A **B**

Fig. 10.21 Resonance structures for an amide; in structure **B** the presence of a double bond between carbon and nitrogen explains the observed planarity of the amide unit and the shortening of the C–N bond.

We know that the C–N bond only has partial π character, so structure **B** must be a minor contributor. In this structure, to form the C–N π bond we have to break the C–O π bond resulting in a negative charge on the oxygen. The nitrogen no longer has the lone pair (it was used to form the C–N π bond) but shares these electrons with the carbon. Effectively it has lost one electron and so becomes formally positively charged.

10.3 Chemical consequences – reactions of carbonyl compounds

As we have seen in Section 7.3 on p. 102, the carbonyl group is susceptible to nucleophilic attack leading, in the first instance, to a tetrahedral intermediate as shown in Fig. 10.22. Changing the group X has a large effect on how readily this reaction occurs and it is the explanation of this effect which we will focus on in this section.

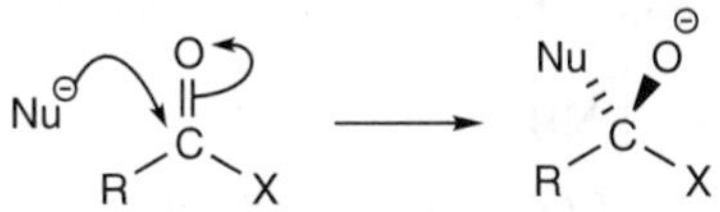

Fig. 10.22 Carbonyl compounds are attacked by nucleophiles (Nu^-) to give, in the first instance, a tetrahedral intermediate. The group X has a profound influence on how readily this reaction goes.

Amides and aldehydes

Amides and aldehydes show very different reactivity towards water as a nucleophile. As shown in Fig. 10.23, aldehydes are readily attacked by the water to form a tetrahedral intermediate which leads to what is known as a *hydrate*. In contrast, amides show no perceptible reaction with water.

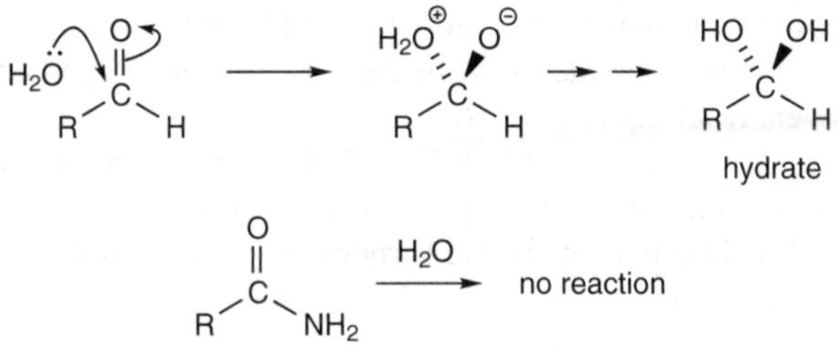

Fig. 10.23 Aldehydes react readily with water to give significant amounts of the hydrate; amides do not react with water.

The initial interaction in these reactions is between the HOMO of the nucleophile (in this case an oxygen lone pair) and the LUMO of the carbonyl compound. In the case of an aldehyde, we saw on p. 91 that the LUMO is the $\pi^\star$ MO formed from the overlap of $2p$ orbitals on the carbon and oxygen.

In the amide, the nitrogen lone pair is involved with the π system of the carbonyl, forming three π MOs. We discussed how this interaction leads to the 3π MO (the LUMO) being higher in energy than the carbonyl $\pi^\star$ MO, a point illustrated in Fig. 10.19 on p. 161.

Figure 10.24 illustrates the relative energies of the LUMO of the aldehyde, the LUMO of the amide and the HOMO on the nucleophile. Recall that as the HOMO is a lone pair on oxygen, we expect it to lie lower in energy than the $\pi^\star$ LUMO (see p. 95). We can clearly see from this diagram that the HOMO of the nucleophile has a much better energy match with the LUMO of the aldehyde than it does with the LUMO of the amide; we therefore expect the aldehyde to be the more reactive of the two species, which is precisely what is found.

In summary, the presence of a nitrogen substituent on the carbonyl leads to reduced reactivity towards nucleophiles – we can attribute this to the raising in energy of the LUMO which results from the participation of the nitrogen lone pair in the carbonyl π system. The effect is often called the *conjugative effect*.

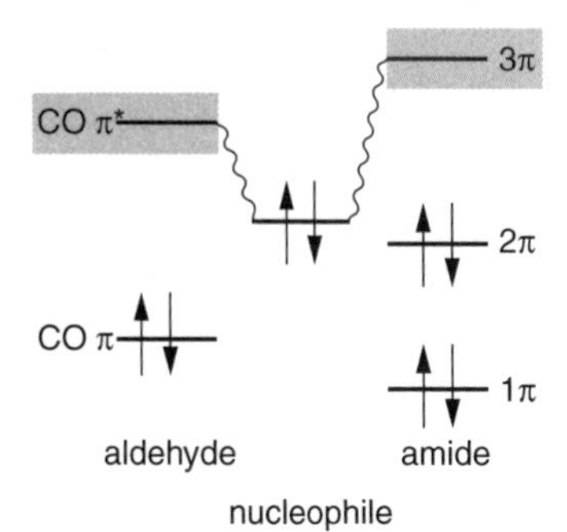

Fig. 10.24 On the left are shown the π MOs of an aldehyde, in the middle the HOMO of a nucleophile and on the right the π MOs of an amide. Possible HOMO/LUMO interactions are shown by the wavy lines. Due to the interaction of the carbonyl π system with the nitrogen lone pair, the LUMO of the amide (3π, shown in the shaded box) is higher in energy than the LUMO of the aldehyde (the CO $\pi^\star$ also shown shaded). The aldehyde is more reactive towards the nucleophile than is the amide because the LUMO of the aldehyde is closer in energy to the HOMO of the nucleophile.

Amides as bases

Before moving on to other carbonyl reactions, it is interesting to look at the behaviour of an amide when it acts as a base, since it turns out that this is also affected by the participation of the nitrogen lone pair in the π bonding system.

As you will recall, a base is a compound which can accept ('pick up') a proton from another molecule. It is usually an atom with a lone pair which is involved in bonding to the incoming H^+; for example, in the case of amines and alcohols the H^+ becomes bound to the nitrogen and oxygen, respectively:

$$\mathrm{RNH_2} + \mathrm{H^+} \longrightarrow \mathrm{RNH_3^+}$$

$$\mathrm{ROH} + \mathrm{H^+} \longrightarrow \mathrm{ROH_2^+}.$$

As we saw on p. 106, in these reactions the HOMO is a lone pair and the LUMO is the $1s$ orbital on the H^+. Amines turn out to be much more basic than alcohols, meaning that amines are more ready to pick up a proton; an explanation for this in terms of orbitals is as follows.

In these reactions, the HOMO is a lone pair on nitrogen or oxygen and these orbitals clearly lie lower in energy than the LUMO of the H^+. However, the oxygen lone pair is lower in energy than the nitrogen lone pair, and so the energy separation between the HOMO and LUMO is greatest for oxygen. There is thus a poorer interaction with the lone pair on oxygen compared to that on nitrogen, resulting in the alcohol being the weaker base.

Given that this is the case, it is at first sight surprising that when an amide acts as a base it is the carbonyl *oxygen* which is protonated first, not the nitrogen (Fig. 10.25). In other words, the carbonyl oxygen is more basic than the nitrogen.

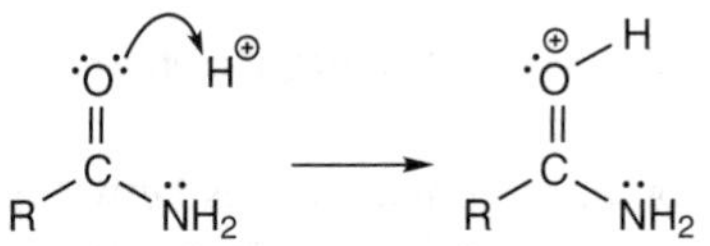

Fig. 10.25 In an amide the carbonyl oxygen is more basic than the nitrogen.

To understand what is going on here we first need to identify the HOMO of the amide. There are two candidates: either the lone pair on the nitrogen, or one of the lone pairs on the oxygen. As we discussed before, the carbonyl oxygen can be considered as sp^2 hybridized and two of these hybrids are occupied by lone pairs; these hybrids lie in the plane of the molecule and therefore are not involved in the π system.

From what we know about the energies of AOs (see Fig. 4.34 on p. 57) we would expect the oxygen lone pair to be lower in energy than that on nitrogen, so the latter would form the HOMO. However, as we have seen the interaction of the nitrogen lone pair with the carbonyl π system results in a lowering of the energy of the lone pair as it moves into the 2π MO (see Fig. 10.19 on p. 161). In the case of the amide, the nitrogen lone pair is so lowered in energy that it comes below the oxygen lone pairs thus making the latter the HOMO. So the best energy match to the LUMO of the proton is the lone pair orbital on the oxygen.

We can also formulate an explanation of this effect using resonance structures. In Fig. 10.21 on p. 162 it is illustrated how the lone pair can become involved in the formation of a C–N π bond. This decreases the 'availability' of the nitrogen lone pair, thus reducing the basicity of the nitrogen.

Amides and acyl chlorides

As we have seen in Section 7.6 on p. 110 acyl chlorides react very rapidly with nucleophiles, usually leading to subsequent elimination of Cl^-, whereas, as we have already noted, amides are rather unreactive. The reaction, or lack of it, with water (shown in Fig. 10.26) offers a clear demonstration of this difference.

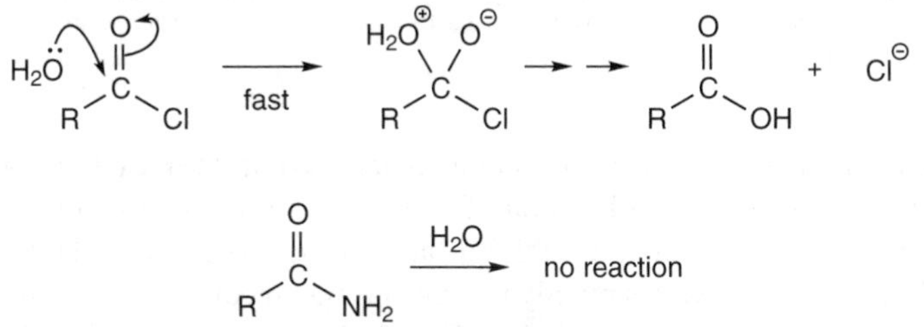

Fig. 10.26 Acyl chlorides react rapidly with water as a nucleophile, ultimately generating a carboxylic acid and chloride ions; amides are unreactive to water.

An explanation you will often find given for the high reactivity of the acyl chloride is that the chlorine is electron-withdrawing and so this increases the partial positive charge on the carbonyl carbon, making it more reactive towards nucleophiles. The problem with this argument is that nitrogen is *also* electron-withdrawing; in fact chlorine and nitrogen have about the same electronegativity, so if the presence of an electron-withdrawing chlorine substituent increases the reactivity of the carbonyl, a nitrogen substituent should do the same – which plainly it does not.

To sort out what is going on here we have to realize that the presence of the substituent X in RCOX has two effects. The first is that the electron-withdrawing nature of X increases the positive charge on the carbonyl carbon thus increasing its reactivity toward nucleophiles; this is called the *inductive effect*. The second is the conjugative effect described on p. 163; in this the delocalization of a lone pair from X into the carbonyl π system decreases the reactivity of the carbonyl towards nucleophiles by raising the energy of the $\pi^\star$ LUMO.

As nitrogen and chlorine have about the same electronegativity, and hence presumably the same inductive effect, we must look to the conjugative effect to explain the difference in reactivity between the acyl chloride and the amide. It must be that the conjugative effect (which decreases reactivity) is much greater for the amide than for the acyl chloride.

The explanation for this is that the orbital involved in forming the delocalized system is a $3p$ in chlorine and a $2p$ in nitrogen. The $3p$ is much larger than the $2p$ (see p. 49) and so the overlap with the $2p$ orbitals which form the carbonyl π system is poorer (see p. 67) for chlorine than it is for nitrogen.

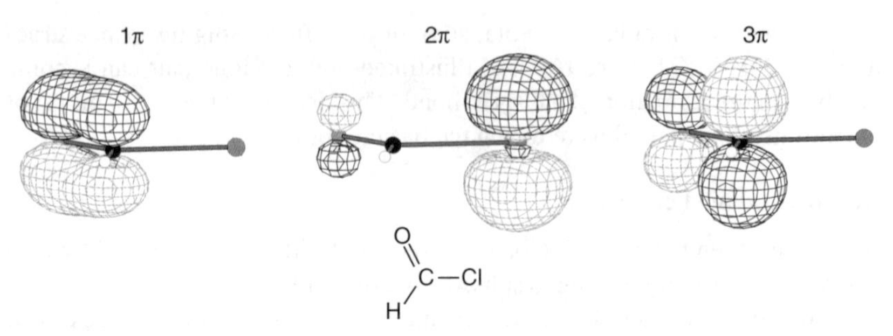

Fig. 10.27 Surface plots of the π MOs of methanoyl chloride; the molecule is viewed from the side with the C–Cl bond going to the right and the C–O bond going away from us and to the left. Note how there is little contribution from the chlorine $3p$ orbital to the 1π MO, indicating that the chlorine lone pairs are not participating in the π bonding.

We have shown the surface plots for the simplest acyl chloride, methanoyl chloride. However, it should be noted that this molecule is only stable at low temperatures and is not usually encountered as a laboratory reagent.

The surface plots of the π MOs in methanoyl chloride (HCOCl), shown in Fig. 10.27, support this explanation. The 1π MO results in bonding between the carbon and the oxygen, but this surface plot shows no contribution from the chlorine. The 2π MO is mainly just the $3p$ AO on chlorine. Compare these orbitals with those for methanamide (Fig. 10.18 on p. 161) which show the nitrogen orbital participating in the π system.

Overall, it is clear that the chlorine orbitals are not affecting the bonding in the π system and so the conjugative effect is minimal. This, then, is the explanation for the high reactivity of the acyl chloride.

In summary, there are two effects, illustrated in Fig. 10.28, of the substituent X on reactivity:

- *the inductive effect* involves the electron-withdrawing nature of X increasing the positive charge on the carbonyl carbon and hence increasing its reactivity;

- *the conjugative effect* involves the delocalization of the lone pair from X into the carbonyl system thereby raising the energy of the LUMO and so reducing the reactivity.

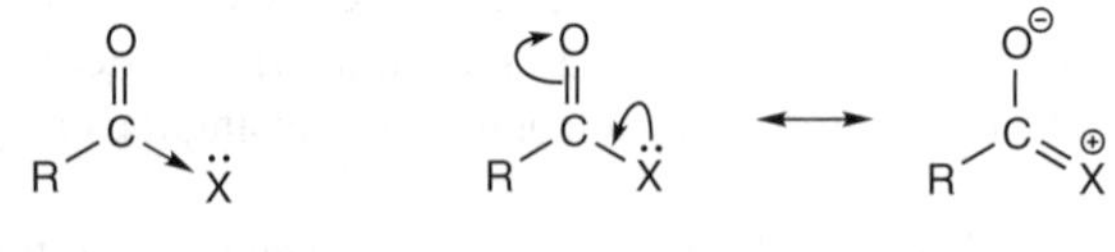

inductive effect

conjugative effect

Fig. 10.28 An electronegative substituent X has two effects: firstly, it withdraws electrons from the carbonyl carbon, thus increasing its reactivity (*inductive effect*); secondly, its lone pair is delocalized into the carbonyl π system, thus decreasing its reactivity (*conjugative effect*).

Amides and esters

The reactivity of esters falls between that of amides and acyl chlorides: an ester is readily hydrolysed by refluxing with dilute alkali, whereas the hydrolysis of an amide requires much stronger conditions, i.e. higher temperatures and longer times. Both reactions start with nucleophilic attack by hydroxide on the carbonyl, as shown in Fig. 10.29.

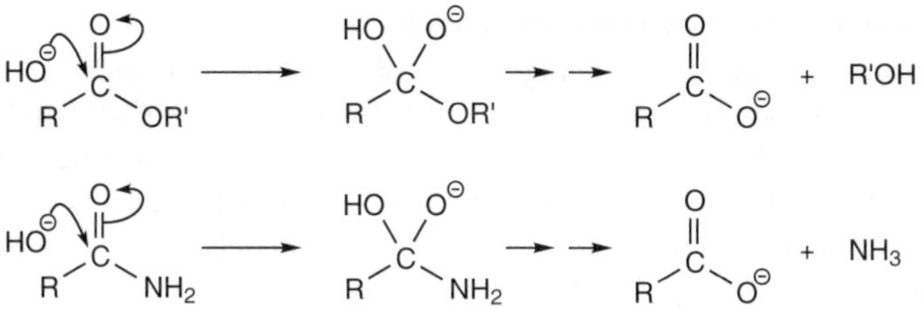

Fig. 10.29 Both esters and amides are hydrolysed by aqueous alkali and both reactions are initiated by nucleophilic attack on the carbonyl by hydroxide. However, whereas the ester is readily hydrolysed under mild conditions, hydrolysis of the amide needs 'stronger' conditions, i.e. prolonged heating.

The lone pair on the oxygen in an ester can interact with the carbonyl π orbitals to form a delocalized system, just as in the case of an amide. The result is that, as was illustrated in Fig. 10.19 on p. 161, the LUMO is raised in energy thus reducing the reactivity of the carbonyl group.

Figure 10.30 compares the effect of conjugation with a nitrogen versus an oxygen lone pair; as we have noted before, the lone pair on the oxygen is lower in energy than that on the nitrogen, as is shown in (b). Clearly, the nitrogen lone pair has a better energy match with the carbonyl $\pi^\star$ MO than does the oxygen, so we expect the 3π MO to be raised in energy more in the case of the amide than of the ester. Since it is this raising in energy of the LUMO which is responsible for the reduced reactivity of the amide, we conclude that the ester should be more reactive than the amide.

The inductive effect will also point to the ester being more reactive than the amide as the oxygen is more electron-withdrawing than the nitrogen. Thus, both the conjugative and inductive effects work together to make the ester more reactive than the amide, which is what we observe.

However, esters are still less reactive than acyl chlorides despite the inductive effect of the electronegative oxygen substituent. We therefore conclude that there must be significant delocalization of the oxygen lone pair in the ester in order to account for this reduction in reactivity.

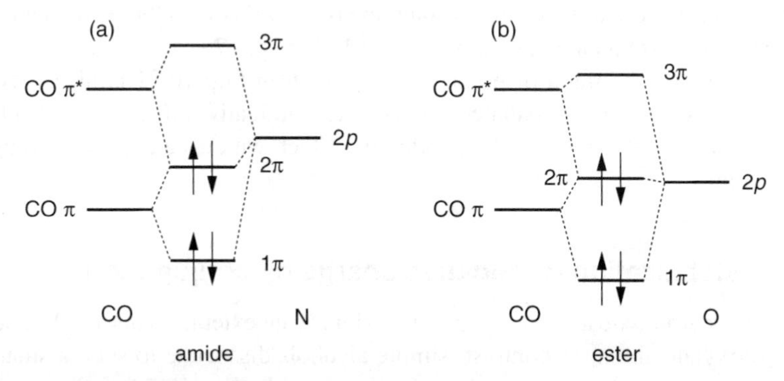

Fig. 10.30 MO diagrams showing the interaction between the lone pair on nitrogen or oxygen with the π system of the carbonyl group. Diagram (a) is identical to that of Fig. 10.19 and is appropriate for an amide; note how the interaction with the lone pair raises the energy of the LUMO (3π). Diagram (b) shows how the situation is modified for an ester; the oxygen lone pair is lower in energy than that on nitrogen, and so has a poorer energy match with the $\pi^\star$ MO, resulting in a smaller rise in the energy of the LUMO (3π). On this basis, we would predict that an ester will be more reactive than an amide.

C–O bond lengths in carbonyl compounds

When discussing the effect of conjugation of the nitrogen lone pair in amides we noted that good evidence for this was the shortening of the C–N bond length (Fig. 10.20 on p. 162). Now that we have discussed a range of other carbonyl compounds it is interesting to compare their C–O bond lengths and see how they can be rationalized in terms of the bonding models we have developed. The data are given in Fig. 10.31.

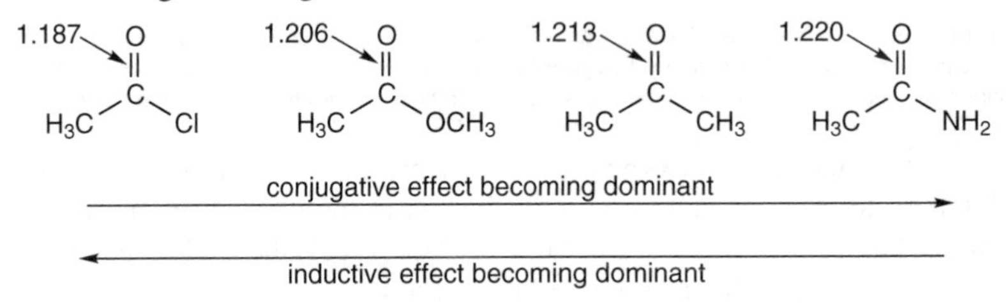

Fig. 10.31 C–O bond lengths, in Å, for a series of different carbonyl compounds.

The amide shows the longest C–O bond and we have already described (p. 162) how this can be attributed to the involvement of the nitrogen lone pair in the carbonyl π system. The ester has a shorter bond than the amide, and this is consistent with the view that the oxygen lone pair participates less in the π system than does the nitrogen lone pair, as was described on p. 167.

The acyl chloride has the shortest bond, and we can rationalize this by recalling from p. 165 that there is very little participation by the chlorine lone pair in the π system, and hence none of the weakening of the C–O bond seen in amides. The π bond is therefore just a 2c-2e bond between the carbon and the oxygen.

However, the story is a little more complicated than this, as we see that the C–O bond in the acyl chloride is actually shorter than the bond in the ketone. In this molecule there are no lone pairs to be involved in the π system, so we can take the C–O bond length in the ketone to be unaffected by conjugation. What is going on here is that the inductive effect is responsible for the shortening of the C–O bond; we will leave the explanation of why this is to the next chapter where the story is taken up again in Section 11.2 on p. 176.

In summary, the trend of bond lengths shown in Fig. 10.31 is nicely explained as resulting from a balance between the conjugative effect, which leads to lengthening of the bond, and the inductive effect, which leads to shortening of the bond.

10.4 Stabilization of negative charge by conjugation

In water, a carboxylic acid dissociates to a significant extent, giving H_3O^+ and the carboxylate anion; in contrast, simple alcohols dissociate to such a small extent that they are not thought of as even weak acids (Fig. 10.32).

In both cases it is an O–H bond which is being broken, but the crucial difference is that in the carboxylate ion the negative charge is conjugated with the carbonyl π system and is therefore delocalized on to both oxygens. No such delocalization is possible for the anion formed from the simple alcohol.

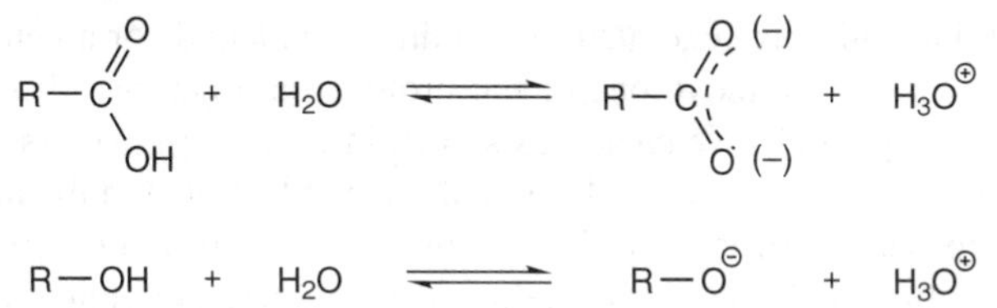

$$R{-}C({=}O)OH + H_2O \rightleftharpoons R{-}C(O^{(-)})(O^{(-)}) + H_3O^{\oplus}$$

$$R{-}OH + H_2O \rightleftharpoons R{-}O^{\ominus} + H_3O^{\oplus}$$

Fig. 10.32 Carboxylic acids dissociate quite readily in water to give the delocalized carboxylate ion (see Fig. 10.15 on p. 159). In contrast, simple alcohols hardly dissociate at all; we can attribute the difference to the extra stability of the delocalized carboxylate anion.

It seems clear, therefore, that the delocalization possible in the carboxylate ion is responsible for stabilizing this ion. In this section we will explore why this is so.

The origin of this stabilization is that the lone pair of electrons on the oxygen (formally the negative charge) becomes involved in the carbonyl π system. If we concentrate on just the p orbitals of the carbon and the two oxygens we can draw up a simple MO scheme, shown in Fig. 10.33, which explains how this stabilization comes about.

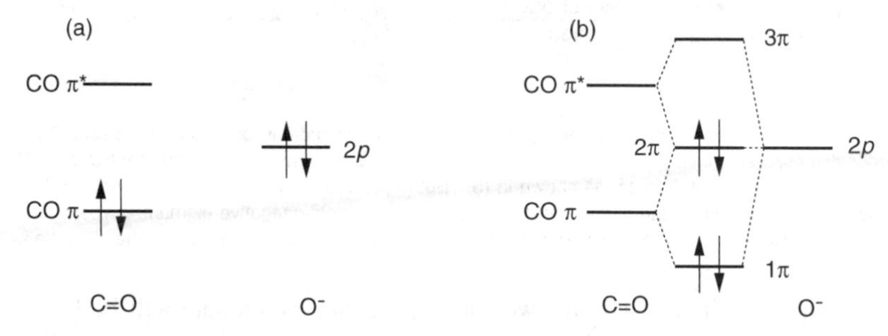

Fig. 10.33 Illustration of how the interaction of a negative charge, located in an oxygen 2*p* orbital, with an adjacent carbonyl π system results in a lowering of the energy. Diagram (a) shows the situation with no interaction and (b) shows what happens when the orbitals interact; there is a decrease in the total energy of the occupied orbitals. The exact energy of the 2π MO is somewhat uncertain but it is expected to be close in energy to the oxygen 2*p*.

In Fig. 10.33 (a) we are supposing that there is no interaction between the two π MOs and the out-of-plane $2p$ orbital. This orbital has two electrons in it, as both electrons from the O–H bond are left behind on the oxygen when the bond breaks to give H^+. In (b) the $2p$ orbital is now allowed to interact with the π MOs and this gives rise to three MOs; in fact what we have here is three $2p$ orbitals in a row, so the three MOs are similar to those given in Fig. 10.16 on p. 159.

Although the pair of electrons originally in the $2p$ orbital do not change energy very much as they move into the largely non-bonding 2π MO, the other pair of electrons are lowered in energy as they move into the 1π MO. Overall, therefore, the interaction leads to a lowering of the energy.

It is interesting to compare the carboxylate ion with two ions in which the oxygens are replaced by CH_2 groups: replacing one oxygen gives an *enolate* anion and replacing two gives an *allyl* anion which we have already described on p. 157.

The enolate anion is generated by treating an aldehyde or a ketone with a strong base; this can remove one of the protons on a carbon adjacent to the carbonyl, leaving a negative charge as shown in Fig. 10.34 (a). As in the carboxylate ion, the charge is delocalized and this is illustrated in the diagram by the use of resonance structures. Removing a proton from a carbon adjacent to a C=C double bond also gives a delocalized anion (an allyl anion) which is shown in Fig. 10.34 (b).

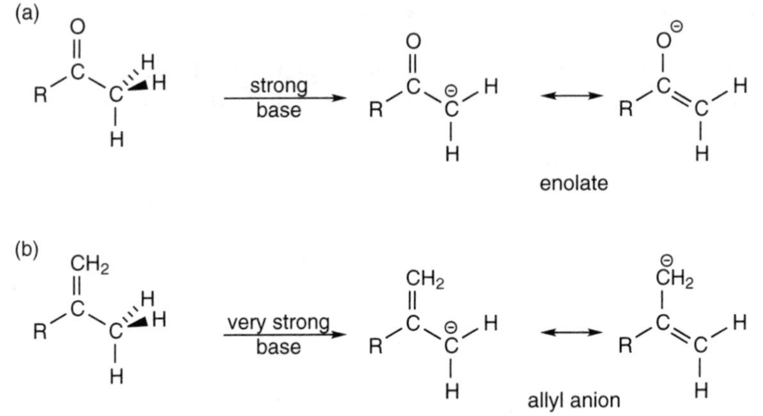

(a) strong base → enolate

(b) very strong base → allyl anion

Fig. 10.34 An enolate anion is formed by removing one of the protons on a carbon adjacent to a carbonyl group, as shown in (a). Removing one of the protons from a carbon adjacent to a C=C double bond gives an allyl anion, as shown in (b). Both anions are delocalized, but the enolate has a resonance form in which the negative charge appears on an electronegative element, oxygen. This confers extra stability and so explains why the enolate anion is much easier to form than the allyl anion.

Comparing the two anions, we can expect that the enolate will be lower in energy as the negative charge can be delocalized onto the electronegative oxygen atom. Although the allyl anion is also delocalized, there are no electronegative atoms onto which the charge can be delocalized. This explains why we need a much stronger base to form the allyl anion than is needed to form the enolate.

We have a nice progression here involving these three anions. The carboxylate, in which the negative charge can be delocalized onto two oxygens, is the most stable of the three. Next comes the enolate, in which delocalization onto one oxygen is possible, and finally the allyl anion in which there are no electronegative atoms.

Stabilization by larger π systems

If the negative charge is conjugated with a more extensive π system then the charge can be delocalized onto more atoms and this is found to confer extra stability. In general, the more delocalized the charge becomes, the greater the stabilization.

A good example of this is the anion formed by the ionization of the O–H in phenol; as shown in Fig. 10.35, the resulting negative charge can be delocalized round the benzene ring. The extra stability which this delocalization gives the anion accounts for the observation that simple alcohols (such as ethanol) are not acidic to a significant extent whereas phenol is a weak acid (indeed, it used to be known as carbolic acid). The anion formed when a simple alcohol dissociates is not delocalized, whereas that from phenol is.

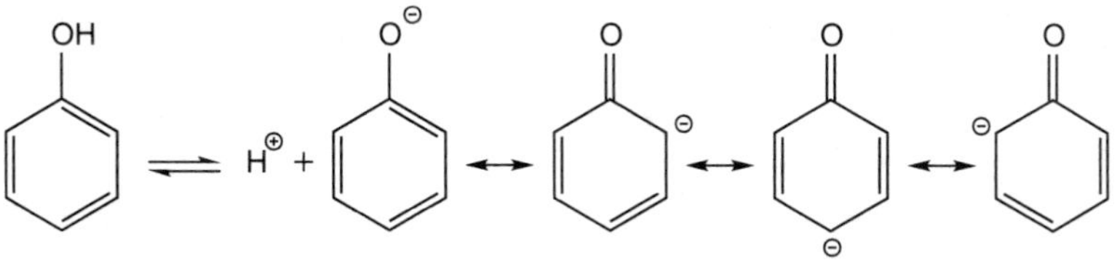

Fig. 10.35 The phenolate anion, formed by the dissociation of phenol, shows extensive delocalization of the negative charge into the benzene ring. This confers extra stabilization and so accounts for phenol being a weak acid, in contrast to simple alcohols which are not significantly acidic.

Figure 10.36 shows the HOMO of this phenolate anion. Comparing this to the resonance structures shown in Fig. 10.35 we see that the carbons on which the negative charge appears are precisely the ones whose $2p$ orbitals contribute to the HOMO.

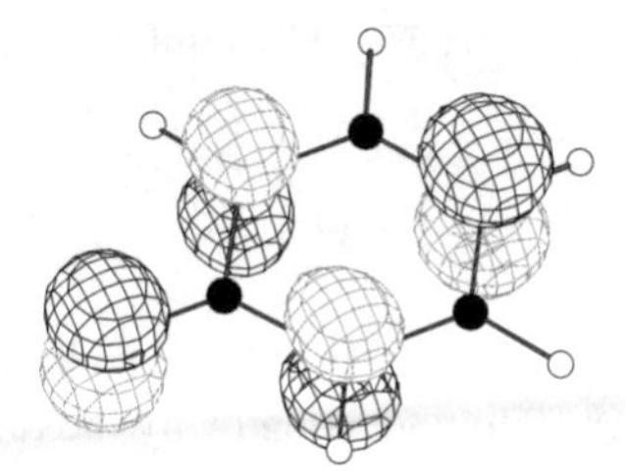

Fig. 10.36 Surface plot of the HOMO of the phenolate anion, whose resonance structures are shown in Fig. 10.35; the oxygen is on the left, coming towards us. Note how the atoms whose AOs contribute significantly to this MO are the same as those onto which the negative charge can be delocalized.

11 Substitution and elimination reactions

At the end of Chapter 9 we described that when a nucleophile replaces the bromine atom in *t*-butyl bromide, $(CH_3)_3CBr$, the rate is found by experiment to be *first* order and independent of the concentration of the nucleophile rather than second order (i.e. proportional to both the concentrations of the bromoalkane and the nucleophile) as is observed with bromomethane:

first order:

$$(CH_3)_3CBr + Nu^- \longrightarrow (CH_3)_3CNu + Br^- \quad \text{rate} = k[(CH_3)_3CBr]$$

second order:

$$CH_3Br + Nu^- \longrightarrow CH_3Nu + Br^- \qquad \text{rate} = k'[CH_3Br][Nu^-].$$

We are now going to look more closely at these and other reactions using some of the ideas from Chapter 10 to understand why the rate laws are different.

11.1 Nucleophilic substitution revisited – S_N1 and S_N2

We have seen in Sections 7.5 on p. 107 and 9.1 on p. 131 how bromomethane and hydroxide react together in a single step to form the products. For the moment let us simply assert that the reaction between a nucleophile and *t*-butyl bromide proceeds via a completely different *two-step* mechanism. In the first step, the carbon–bromine bond breaks to form a bromide ion and a reactive species, called a carbocation or *carbenium ion*, in which a carbon atom bears a positive charge. The curly arrow mechanism for this step is shown in Fig. 11.1. It is essentially the reverse of two ions reacting to form a neutral molecule as illustrated on p. 97 by the reaction of H^+ and H^- ions.

$$(CH_3)_3C\text{—}Br \longrightarrow (CH_3)_3C^{\oplus} \quad Br^{\ominus}$$

Fig. 11.1 The formation of a carbenium ion from *t*-butyl bromide.

The carbenium ion is very reactive and in the second step it quickly reacts with the nucleophile to give the product. The curly arrow mechanism for this step is shown in Fig. 11.2.

$$(CH_3)_3C^{\oplus} \quad Nu^{\ominus} \longrightarrow (CH_3)_3C\text{—}Nu$$

Fig. 11.2 The reaction between the carbenium ion and the nucleophile.

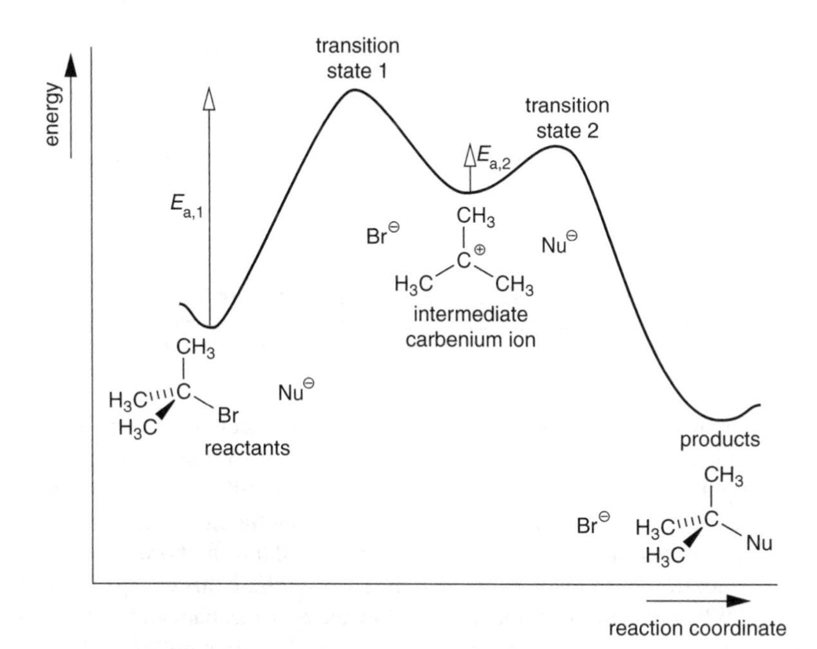

Fig. 11.3 An energy profile for the reaction between *t*-butyl bromide and a nucleophile, Nu^-. The reaction occurs in two steps: the first is the formation of the carbenium ion intermediate; the second step is the reaction between the carbenium ion and the nucleophile to give the product. The second transition state is lower in energy than the first.

Figure 11.3 shows an energy profile for the overall reaction. As the reaction between the positively charged carbenium ion and the nucleophile has such a small activation energy ($E_{a,2}$), the initial formation of the carbenium ion from the *t*-butyl bromide is the *rate-determining step* (see Section 9.6 on p. 141).

The rate of step (1), the rate at which the intermediate carbenium ion forms, depends only on the concentration of the *t*-butyl bromide – the more *t*-butyl bromide there is, the faster the carbenium ion will be formed. We can write the rate of formation of the carbenium ion as:

$$\text{rate of step (1)} = k_1[(CH_3)_3CBr].$$

As soon as it has been formed, the carbenium ion reacts with the nucleophile to give the products. Since the first step is rate-determining, it follows that the rate of formation of the product is the same as the rate of step (1):

$$\text{rate of formation of the product} = k_1[(CH_3)_3CBr].$$

This mechanism therefore explains the observed first-order kinetics of the reaction of *t*-butyl bromide.

We have two possible mechanisms for these nucleophilic substitution reactions. The reaction between bromomethane and the nucleophile depends on the concentrations of both reagents and proceeds by both reagents coming together in a single step. This mechanism is known as an S_N2 mechanism: this stands for *S*ubstitution, *N*ucleophilic with the '2' indicating that the rate-determining step involves both reagents and so the rate law is second order.

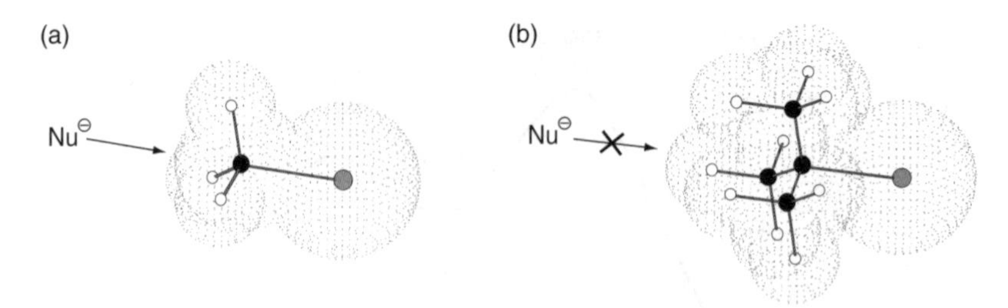

Fig. 11.4 Diagram (a) shows that the three hydrogen atoms in bromomethane do not hinder the approach of a nucleophile in the correct direction to attack the C–Br $\sigma^\star$. In contrast, (b) shows that the methyl groups in *t*-butyl bromide prevent the nucleophile approaching. The dots in each picture are used to outline the approximate atomic radii of each of the atoms.

The reaction between *t*-butyl bromide and the nucleophile proceeds in two steps – the initial formation of a carbenium ion from the *t*-butyl bromide followed by the fast reaction of this ion with the nucleophile. This mechanism is known as an S_N1 mechanism, the '1' indicating that the rate-determining step only involves one species i.e. this step is unimolecular or first order.

The question we now need to address is *why* is there this change in mechanism? Why does bromomethane not follow the S_N1 mechanism and form the cation CH_3^+? Why does *t*-butyl bromide not react directly with the nucleophile following the S_N2 mechanism?

The latter question is relatively easy to answer. We saw in Section 7.5 on p. 109 that in a reaction following the S_N2 mechanism, the attacking nucleophile must approach from directly behind the bromine leaving group. Whilst this is possible for a nucleophile attacking bromomethane (as shown in Fig. 11.4 (a)), for *t*-butyl bromide, this approach is severely hindered by the three methyl groups as shown in (b).

The question as to why bromomethane does not undergo an S_N1 reaction has a more complex answer. Put simply, the answer is that the carbenium ion that would be formed from bromomethane, H_3C^+, is so high in energy that it cannot readily form. This contrasts with the $(CH_3)_3C^+$ ion which is *much* more stable. It turns out that the methyl groups help to stabilize the positive charge in the $(CH_3)_3C^+$ ion whereas no stabilization is provided by the hydrogens in H_3C^+. Exactly how positive charges in ions may be stabilized in this and other ways is the subject of the next section.

11.2 The stabilization of positive charge

We have seen in Section 10.4 on p. 168 how negative charge can be stabilized by delocalization, i.e. where the electrons are not confined to one atom but are spread over many in a delocalized MO. Positive charges can be stabilized in a similar manner but there is one important difference: *positive charges themselves do not move.* To understand this distinction we need to think carefully about what negative and positive ions are.

A negative ion is simply a molecule (or atom) in which the number of electrons exceeds the total nuclear charge. For example, in BH_4^- the total nuclear charge is nine (five for boron, plus a total of four from the hydrogens) but there are ten electrons, giving an overall negative charge of one. Similarly, a positively charged ion is one in which there are insufficient electrons to compensate for the nuclear charge. For example, in NH_4^+ the total nuclear charge is 11, but there are just 10 electrons. In a negative ion we can imagine the electrons moving from one atom to another as a result of delocalization – for example, in the enolate anion shown in Fig. 10.34 (a) on p. 170 the negative charge can be on the carbon or the oxygen.

In a positive ion the charge may also appear on different atoms. However, this is *not* due to the positive nuclei moving but is due to the movement of the negative electrons. The positive charge only appears to move as a result of the actual movement of electrons.

As with negative charges, the formation of a positively charged species is not generally favourable unless there are special factors to stabilize it, such as delocalization; it is this stabilization which we will discuss in this section.

The structure of CH_3^+

On p. 86 we described the bonding in the planar trigonal molecule BH_3 using sp^2 hybridization of the boron. There are three 2c-2e bonds formed by these hybrids, leaving an empty $2p$ orbital pointing out of the plane. The species CH_3^+ is *isoelectronic* with BH_3, meaning that it has the same number of electrons; however, as the nuclear charge is one greater, the molecule has an overall positive charge.

We can therefore use the same description for the bonding in CH_3^+ as we used for BH_3; so the LUMO of CH_3^+ is a vacant $2p$ orbital, shown schematically in Fig. 11.5. We must not be tempted to say that the 'positive charge is in the empty $2p$ orbital'; *electrons* occupy orbitals, not positive charges.

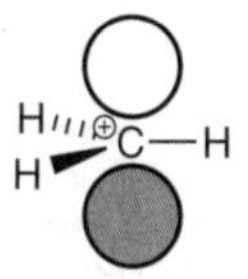

Fig. 11.5 The LUMO of CH_3^+ is the out-of-plane $2p$ orbital.

We saw in Section 7.2 starting on p. 98 that the presence of the empty $2p$ orbital in BH_3 (the LUMO) makes it susceptible to attack by a nucleophile which donates electrons from its HOMO into the BH_3 LUMO. The same is true for CH_3^+, but even more so, as the positive charge further increases the interaction with the nucleophile, especially if it is also charged. As a result, CH_3^+ is a highly reactive species which, if it could form in solution, would only have a fleeting existence.

Stabilization by adjacent lone pairs

The interaction of CH_3^+ with the HOMO of the nucleophile gives us a clue as to how the positive charge on carbon can be stabilized. Suppose that we make a pair of electrons available not from the HOMO of a nucleophile but from an atom attached to the carbon; to be specific, let us replace one of the hydrogen atoms with O^-. Such a replacement gives us the rather odd looking species $^+CH_2–O^-$ depicted in Fig. 11.6 (a). We can imagine that the negative charge on oxygen is located in a $2p$ orbital, which is of course adjacent to the empty $2p$ orbital on carbon. These two orbitals will interact to give a π MO – shown in (b) – which is occupied by the two electrons. In other words, a π bond is formed. The two representations (a) and (b) are simply resonance structures of the same molecule, methanal, $CH_2{=}O$ as shown in (c).

The MO picture of this interaction is shown in Fig. 11.7; as we have noted before, the lone pair on oxygen is lower in energy than the carbon orbital. Interaction with the carbon orbital results in a π bonding MO which is lower in

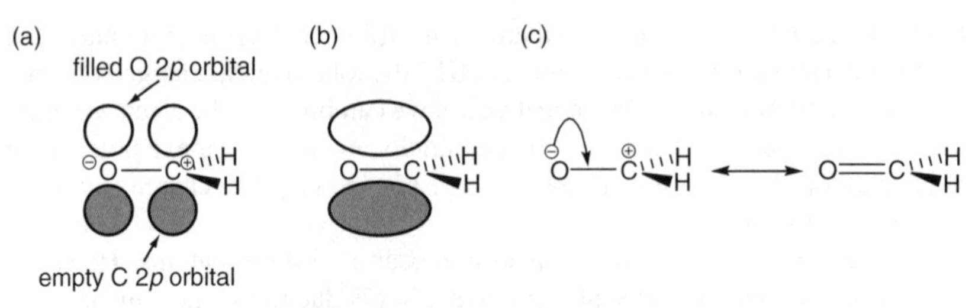

Fig. 11.6 Replacing one of the hydrogen atoms in CH_3^+ with an O^- gives the odd-looking species shown in (a). The two $2p$ orbitals interact to form the π MO shown in (b); this is occupied by the two electrons to give a π bond. The two forms are actually resonance structures of the same molecule, methanal, $H_2C{=}O$, as shown in (c).

energy than the original orbital on oxygen. The lone pair of electrons move into this orbital, thus leading to an overall lowering of the energy. We see, therefore, how having an adjacent lone pair leads to stabilization of the positively charged carbon.

Using a lone pair from a negatively charged oxygen is an odd way to stabilize a positive charge on carbon – as we have seen, the net result is the formation of a neutral molecule with no positive charge at all. However, this example highlights that the stabilization results in the positive charge being no longer localized on the carbon, an idea which we will continue to explore.

The stabilizing lone pair does not need to come from a negatively charged atom; exactly the same effect is observed if an –OH group is used in place of the O^-. At first glance, it might be tempting to say that the electronegative oxygen would *destabilize* an adjacent positive charge; this is not the case. Rather it is because the oxygen has relatively high-energy lone pairs which can interact with an empty orbital in such a way that the charge is stabilized.

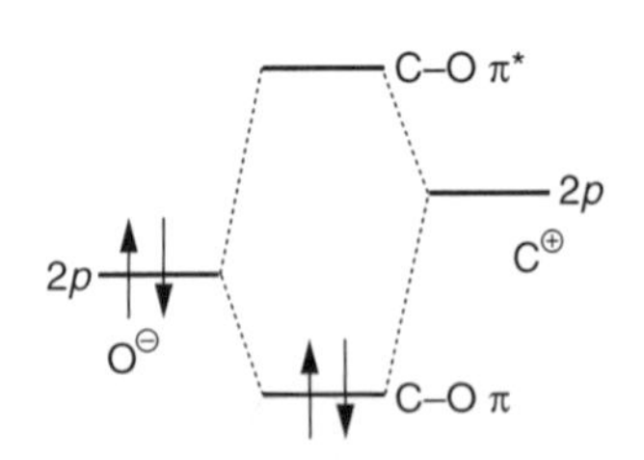

Fig. 11.7 MO picture showing how the energy of the system is lowered by the interaction of a lone pair on oxygen (shown on the left) with an empty orbital on carbon (shown on the right).

As Fig. 11.8 shows, a better way of drawing this species is as a protonated carbonyl. Whilst the unstabilized carbenium ion CH_3^+ is far too reactive to exist in aqueous solution, protonated carbonyl cations are formed when acid is added to a solution of the carbonyl compound.

Experimental evidence confirms that the better representation of the cation is as the protonated carbonyl compound with the positive charge on the oxygen rather than on carbon. In forming the new C–O π bond, the donation of the oxygen electrons lowers the energy of the system just as we saw in Fig. 11.7.

Fig. 11.8 An OH group stabilizes an adjacent carbenium ion. A better representation of this ion is as a protonated carbonyl.

The acylium ion

We are now in a position to understand why the C=O bond in an acyl chloride, such as ethanoyl chloride, is so short, as mentioned on page 168. If ethanoyl chloride is mixed with silver tetrafluoroborate in an inert solvent such as nitromethane (CH_3NO_2) at low temperatures, silver chloride precipitates out leaving a solution of the salt $[CH_3CO]^+BF_4^-$:

$$CH_3COCl + AgBF_4 \longrightarrow [CH_3CO]^+BF_4^- + AgCl(s).$$

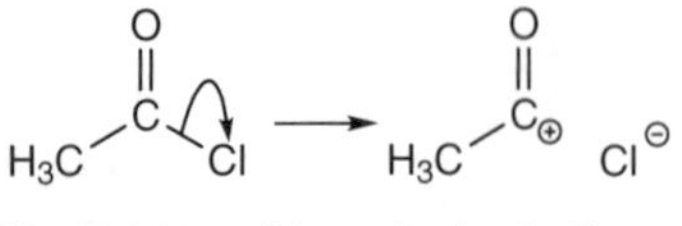

The reaction is driven by the precipitation of the insoluble silver chloride. It might be tempting to think that the structure of the cation is just like ethanoyl chloride minus the chloride ion as shown in Fig. 11.9. However, the crystal structure of the salt has been determined and the cation turns out to be *linear*

Fig. 11.9 A possible mechanism for the formation of the acylium cation.

with a C–O bond length of 1.108 Å, notably shorter than the C–O bond length in ethanoyl chloride at 1.187 Å.

Once the chloride ion has departed, there are only two atoms or groups around the carbonyl carbon in $[CH_3CO]^+$, the methyl group and the oxygen; it should therefore be no surprise that these are at 180° to each other. We now have just the situation described above with a positive charge, nominally on the carbon atom, and an adjacent oxygen atom. The oxygen atom can stabilize the positive charge by donating a pair of electrons. This is shown in Fig. 11.10.

$$:\ddot{O}{=}\overset{\oplus}{C}{-}CH_3 \longleftrightarrow :\overset{\oplus}{O}{\equiv}C{-}CH_3$$

Fig. 11.10 The positive charge on the carbon is stabilized by the donation of a pair of electrons from the adjacent oxygen atom.

The donation of the lone pair from the oxygen formally creates a triple bond between the oxygen and carbon which explains the shortening of the bond length. The acylium ion plays an important role in a number of reactions such as the acylation of aromatic rings.

The way in which the bond length is shortened in the acylium ion allows us to understand why the CO bond in ethanoyl chloride is shorter than in other carbonyl compounds. Whilst ethanoyl chloride does not exist as the separate acylium and chloride ions, the chlorine is electronegative and withdraws some electron density towards itself, increasing the partial positive charge on the carbonyl carbon as mentioned on p. 168. This in turn brings down some electron density from the oxygen lone pairs, strengthening the CO π bond as shown in Fig. 11.11.

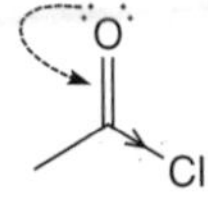

Fig. 11.11 The electronegative chlorine withdraws electrons from the carbon, thereby making it slightly more positively charged and this, in turn, brings down more electron density from the oxygen, thereby strengthening the C=O π bond.

Comparisons of the CO bond lengths in propanone, ethanoyl chloride and fluoride and the acylium cation are shown in Fig. 11.12. We have already described how the electron-withdrawing chlorine leads to a shortening of the CO bond. Changing chlorine to the even more electronegative fluorine results in further shortening of the bond. Finally, in the acylium cation, there is essentially a triple bond between the carbon and oxygen, giving a further reduction in bond length.

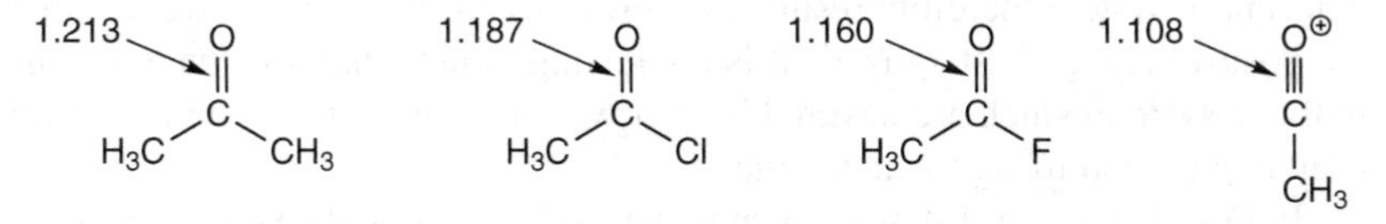

Fig. 11.12 Along the series propanone → ethanoyl chloride → ethanoyl fluoride the partial positive charge on the carbonyl carbon increases leading to a decrease in the CO bond length (shown in Å). The acylium cation with its full positive charge has one of the shortest CO bond lengths known.

Stabilization by adjacent π systems

Other atoms with lone pairs, such as nitrogen, can stabilize positive charges in the same way that oxygen does but if no lone pair is available, then other filled orbitals may help to stabilize the positive charge. We saw on p. 96 that generally after a lone pair, the next highest energy orbital, and hence the next best for stabilizing a positive charge, is a π MO. An example of stabilization by such an MO is the allyl cation shown in Fig. 11.13 where the positive charge is stabilized by a neighbouring C=C.

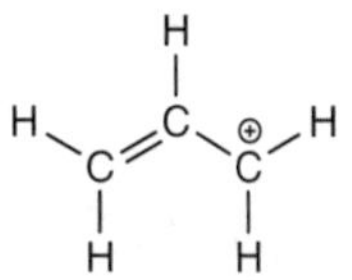

Fig. 11.13 The allyl cation. The positive charge is stabilized by the adjacent π bond.

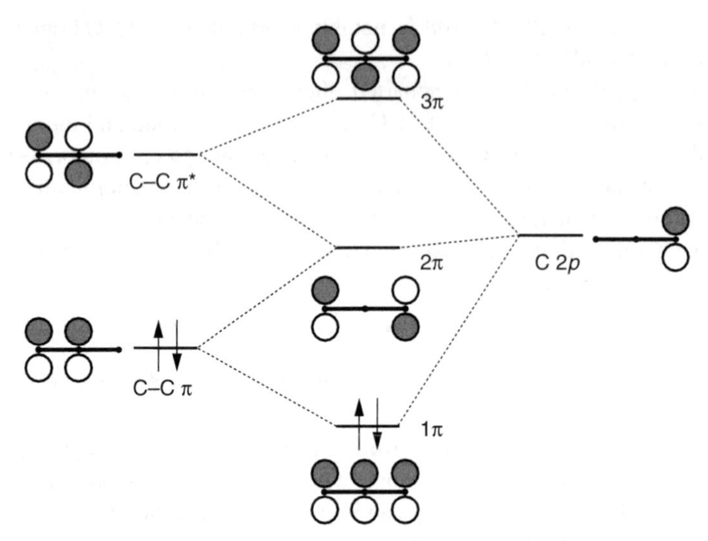

Fig. 11.14 MO diagram showing how the positive charge on carbon in the allyl cation is stabilized by interaction with the adjacent π system. The C=C π and π^* MOs interact with the vacant p orbital on the adjacent carbon to give three MOs. The lowest, 1π, is lower in energy than the combining orbitals. Since this is the only orbital occupied, the energy of the system has been lowered. MO 2π is non-bonding and 3π is anti-bonding.

In the allyl cation we have a π system formed from three p orbitals: two can be thought of as originating from the C=C π bond and one is the vacant orbital on the adjacent carbon. We have already described in Fig. 10.11 on p. 158 the three MOs which arise from the interaction of these three $2p$ orbitals; the MOs are depicted once more in Fig. 11.14. In this figure we imagine that the three MOs arise from the interaction of the C=C π and π^* MOs with the $2p$ AO on the adjacent carbon. For example, the 1π MO can be thought of as arising from an in-phase interaction between the $2p$ and the C=C π MO. Overall though, the final result is just the same as for three p orbitals overlapping in a line.

Of these MOs, only the 1π is occupied and this clearly results in a lowering of energy as the electrons move from the C=C π MO. The interaction with the adjacent π system therefore results in a lowering of the energy – we say that the positive charge is stabilized. It is interesting to note that it is the electrons in the π system which are lowered in energy by the interaction with an empty orbital, thus stabilizing the molecule.

In Fig. 10.14 on p. 159 we saw how the allyl anion could be represented by two different resonance structures or a delocalized structure. The allyl cation may be represented in an analogous manner as shown in Fig. 11.15.

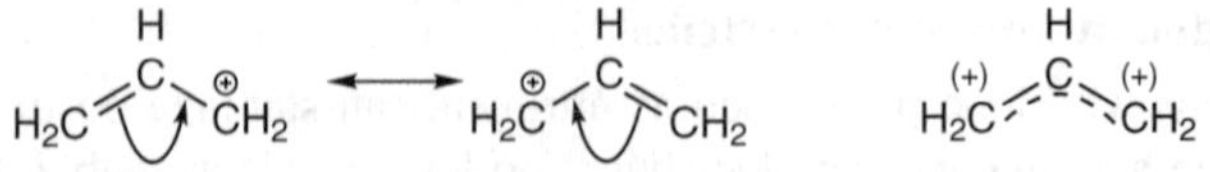

$H_2C{=}CH{-}CH_2^{\oplus}$ ⟷ $^{\oplus}H_2C{-}CH{=}CH_2$ $\quad$ (+)$H_2C{\cdots}CH{\cdots}CH_2$(+)

Fig. 11.15 On the left, connected by a double-headed arrow, are two resonance structures for the allyl cation. On the right is an alternative representation in which the dashed line shows the partial π bond across all three atoms; the atoms onto which the positive charge can be delocalized are indicated by (+).

Stabilization by adjacent σ bonds – σ conjugation

If there are no lone pairs or π bonds adjacent to the positive charge, the cation can still be stabilized by the electrons in neighbouring σ bonds. This type of stabilization is called *σ conjugation.*

In order for the σ bond to help stabilize the cation, it must be in the correct orientation. We saw how lone pairs and π bonds stabilize the positive charge when their occupied MOs overlap with the vacant p orbital on the carbon; effective overlap can only occur if the orbitals involved are in the same plane. The same is true for σ conjugation – the σ MO must be in the same plane as the vacant p orbital as shown in Fig. 11.16 (a).

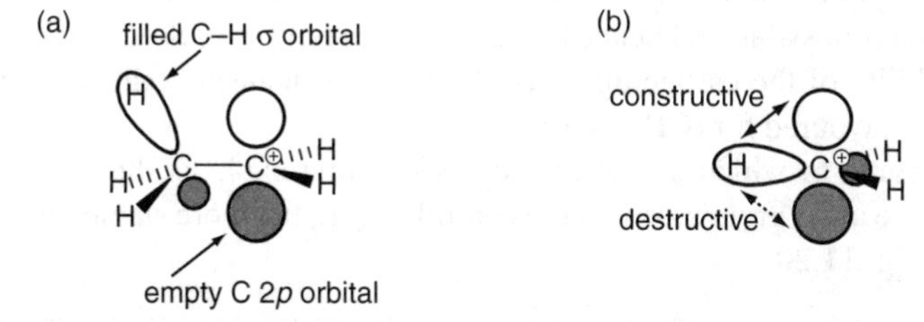

Fig. 11.16 In order for an adjacent σ bond to help stabilize a cation, the σ MO must be able to get into the same plane as the vacant p orbital. This has been achieved in (a) where the C–H σ MO involves the carbon atom next to the one with the positive charge. In the case of CH_3^+, shown in (b), the only σ MOs are at right angles to the vacant p orbital and so no net overlap is possible.

If the orbitals can overlap, they interact as shown in Fig. 11.17. The electrons in the σ MO are now more delocalized and hence lowered in energy. It is this lowering in energy of the electrons in the σ MO that stabilizes the cation. Figure 11.18 shows the delocalized MO formed essentially from the interaction between the C–H σ MOs and the vacant p orbital. It clearly shows that some electron density is shared between the two carbon atoms.

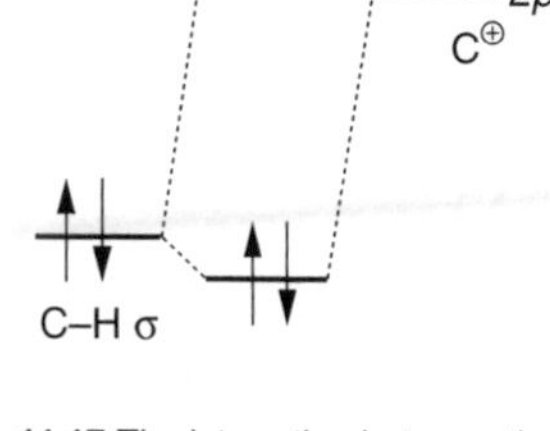

Fig. 11.17 The interaction between the σ bonding MO and the vacant p orbital lowers the energy of the electrons in the σ MO, thereby stabilizing the cation.

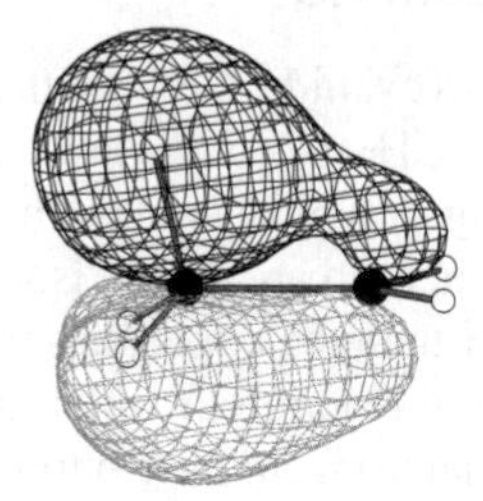

Fig. 11.18 A delocalized MO formed from the interaction between the C–H σ MOs and the vacant p orbital; the electrons can be found in the region between the two carbon atoms. This delocalization lowers the energy of the cation.

If the σ MOs cannot get into the same plane as the vacant p orbital then no stabilization is possible. This is the case in the CH_3^+ cation; the C–H bonds are at right angles to the vacant p orbital and consequently no net overlap is possible as illustrated in Fig. 11.16 (b).

The σ MO need not be from a C–H bond; a C–C bond, for example, can also stabilize the cation. The degree of stabilization by σ conjugation is not as significant as conjugation involving a lone pair or π bond. The reason for this is that σ MOs are lower in energy than both lone pairs and π MOs and so the degree of overlap between a σ MO and a vacant p orbital will be less than between the lone pair or π MO and the vacant p orbital. However, the more σ conjugation that is possible, the more stabilized the cation will be.

This finally explains the mystery we started this chapter with: why *t*-butyl bromide undergoes a substitution reaction by first forming the carbenium ion whereas bromomethane follows the S_N2 mechanism instead. In the $(CH_3)_3C^+$ ion, C–H σ MOs from each of the three methyl groups are able to conjugate with the vacant p orbital. Figure 11.19 shows a delocalized MO where the p orbital on the central carbon is σ conjugated to all of the C–H MOs, spreading electron density over the whole molecule.

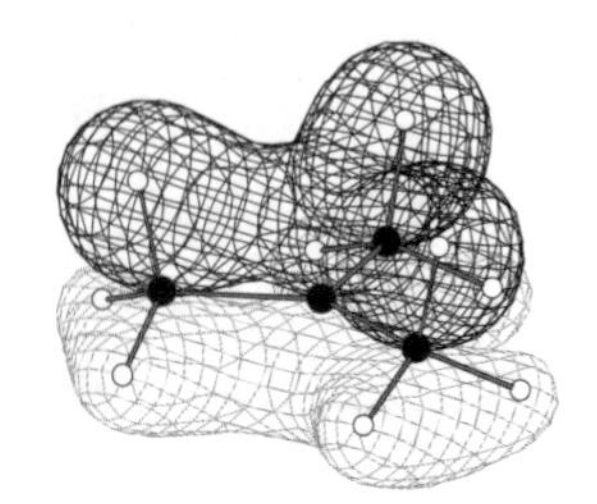

Fig. 11.19 A MO in the carbenium ion, $(CH_3)_3C^+$, showing the extensive delocalization due to the σ conjugation.

The large amount of σ conjugation possible in this ion is sufficient for it to be relatively stable and so be formed readily. In contrast, for CH_3^+ no σ conjugation is possible and hence this cation does not form. So on the grounds of the stability of the carbenium ions, the S_N1 mechanism is disfavoured for CH_3Br and favoured for $(CH_3)_3CBr$.

The degree to which σ conjugation is possible explains the observation that the more alkyl groups around the central cation, the more stable the ion, as shown in Fig. 11.20.

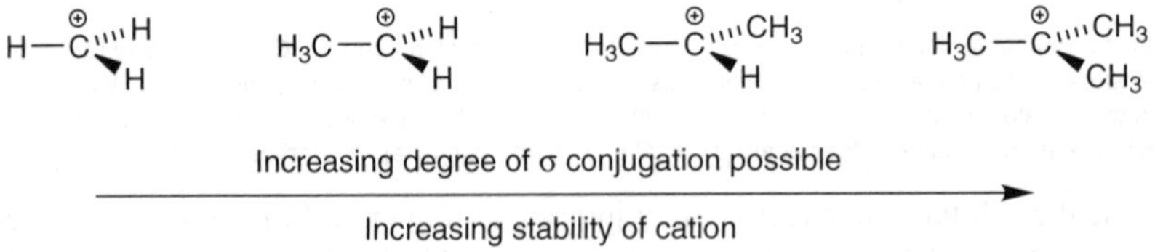

CH_3^+ $\quad$ $H_3C{-}CH_2^+$ $\quad$ $(H_3C)_2CH^+$ $\quad$ $(H_3C)_3C^+$

Increasing degree of σ conjugation possible

Increasing stability of cation

Fig. 11.20 The more alkyl groups attached to the positive carbon, the more σ conjugation is possible and hence the more stable the carbenium ion.

11.3 Elimination reactions

Mixing a nucleophile, such as cyanide, CN^-, with *t*-butyl bromide gives the expected substitution product $(CH_3)_3C{-}CN$. However, adding hydroxide (which we might think of as being a typical nucleophile) to *t*-butyl bromide does *not* yield $(CH_3)_3C{-}OH$ but instead an alkene, isobutene (2-methylpropene), as shown in Fig. 11.21. Rather than the expected substitution reaction occurring, what we see is an *elimination* reaction in which the net result is that HBr has effectively been lost (eliminated) from the *t*-butyl bromide (no HBr appears in the products since under basic conditions water and bromide ion are formed).

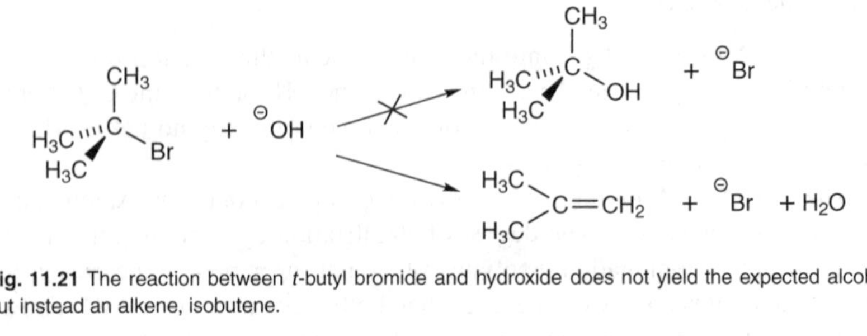

$(CH_3)_3C{-}Br + {}^{\ominus}OH \not\rightarrow (CH_3)_3C{-}OH + {}^{\ominus}Br$

$(CH_3)_3C{-}Br + {}^{\ominus}OH \rightarrow (H_3C)_2C{=}CH_2 + {}^{\ominus}Br + H_2O$

Fig. 11.21 The reaction between *t*-butyl bromide and hydroxide does not yield the expected alcohol but instead an alkene, isobutene.

Let us suppose for the moment that in the elimination reaction the initial step is as we outlined above, namely the formation of the relatively stable trimethylcarbenium ion, $(CH_3)_3C^+$. Why does hydroxide not react with this to form the substitution product $(CH_3)_3C$–OH? We can gain a clue to understanding this by looking again at the delocalized MO of the trimethylcarbenium ion shown in Fig. 11.19. The orbital shows that there is electron density spread out over the whole of the ion; this delocalization has a number of important consequences.

The delocalization removes some electron density from the C–H σ MOs and shares it between the two carbons. Removing electron density from the C–H σ MOs means that these σ bonds are weakened. However, the C–C bond is being strengthened since it gains some partial π character. This accounts for the fact that the C–C bond is shorter in the cation than in the neutral hydrocarbon as illustrated in Fig. 11.22.

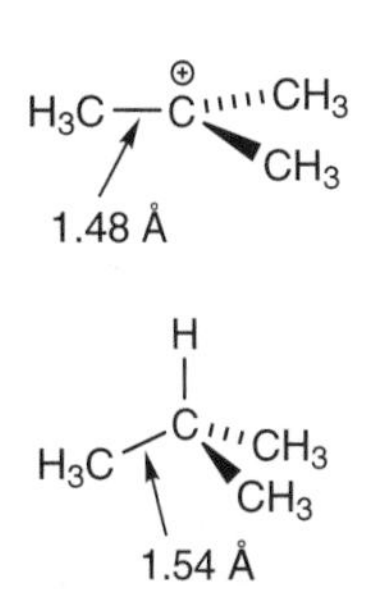

Fig. 11.22 The delocalization in the trimethylcarbenium ion shortens the C–C bond length relative to that in the hydrocarbon.

If we continued this withdrawal of electrons from one of the C–H σ bonds to the extreme, eventually *all* of the electron density from the σ bond would end up being shared between the two carbon atoms. In other words, eventually there would be a full C–C π bond and a free proton, H^+. This process is illustrated in Fig. 11.23 where the stabilizing interaction between the C–H σ bond and the vacant p orbital is shown in (a) and the result of taking this to the extreme is shown in (b). The corresponding curly arrow mechanism is shown in the lower part of the diagram.

The mechanism forming the alkene in Fig. 11.23 is not realistic – the carbenium ion does *not* spontaneously form the alkene and a proton in solution. Indeed, as we shall see later, the reverse is true – a proton will attack the alkene to form the cation. Nonetheless, the σ conjugation does weaken the C–H bond and this means that the hydrogen becomes more *acidic*. This gives us the clue as to what happens when a base (such as hydroxide) is added.

In order to form the *substitution* product, the incoming hydroxide has to reach the central carbon. Although this is possible, the approach is crowded by the hydrogens – see Fig. 11.24 (a) – and in fact the hydroxide reacts preferentially with these. Remember that these hydrogens are, as a result of the

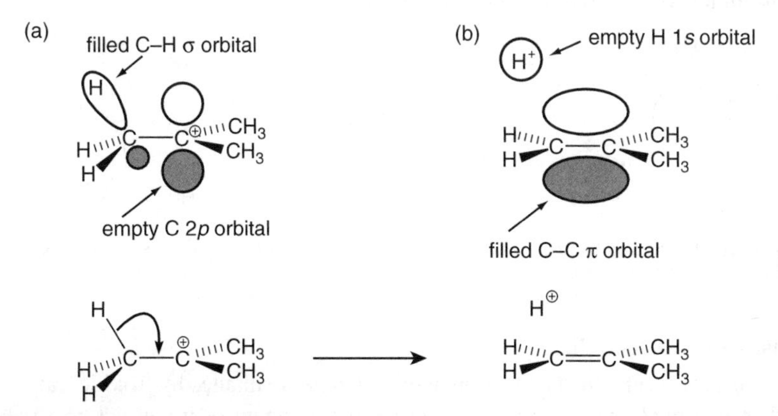

Fig. 11.23 The interaction between the C–H σ MO and the vacant p orbital is shown in (a). If carried to the extreme, this would eventually lead to the complete formation of a C–C π bond and a free proton shown in (b). The lower part of the diagram shows the curly arrow representation of the same process.

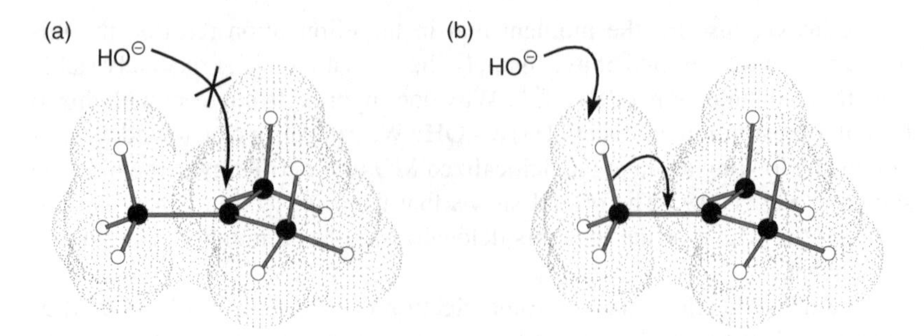

Fig. 11.24 Rather than squeeze past the hydrogens to attack the vacant *p* orbital on the carbon as shown in (a), the hydroxide just helps to take off the already acidic hydrogen as shown in (b).

σ conjugation, somewhat acidic and so all that happens is that the OH^- abstracts one of these hydrogens (as H^+) to give water and the alkene as shown in Fig. 11.24 (b). The corresponding curly arrow mechanism for this reaction is shown in Fig. 11.25.

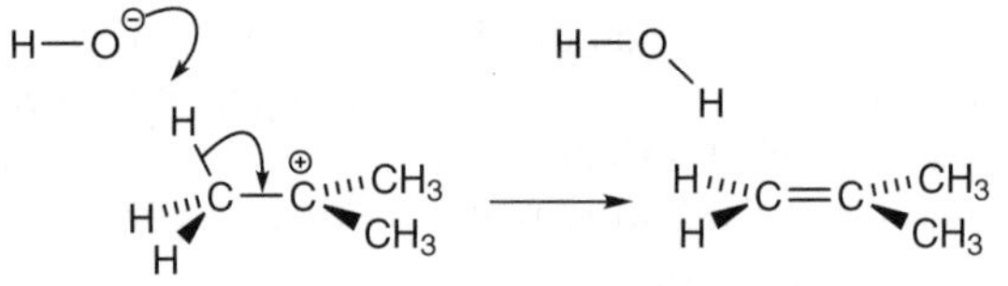

Fig. 11.25 The reaction between the trimethylcarbenium ion and the strong base hydroxide to form the elimination product, isobutene.

We have seen that, instead of forming *t*-butyl alcohol, hydroxide ion reacts with *t*-butyl bromide to give isobutene. It *is* possible to form *t*-butyl alcohol from the bromide but we cannot use hydroxide since it is too strong a base. Using water instead of hydroxide will form more of the desired substitution product, although some of the alkene will still form. Water is less basic than hydroxide and therefore not so good at removing the acidic hydrogens from the cation. However, it is still perfectly able to react with the reactive trimethylcarbenium ion. The curly arrow mechanism for the reaction of the trimethylcarbenium ion with water is shown in Fig. 11.26.

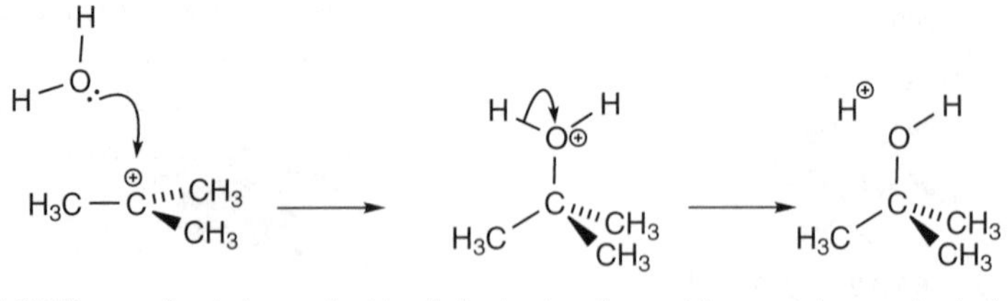

Fig. 11.26 The reaction between the trimethylcarbenium ion and the weak base water to form the substitution product, *t*-butyl alcohol.

Bases and nucleophiles

As we have seen, in its reaction with a halogenoalkane, hydroxide can act either as a nucleophile and substitute for the halogen, or it can act as a base by removing a hydrogen ion. In a way, it is unfortunate that we have two separate words for this behaviour since it encourages us to think that there is a fundamental difference in the way the hydroxide reacts; there is not. When acting as a base, hydroxide (or any other base) is carrying out a nucleophilic attack but specifically on an H–X bond rather than on a C–X bond. All that is needed is for the base/nucleophile to possess a high-energy pair of electrons, usually a lone pair. The base/nucleophile could be charged, like OH^- or NH_2^-, or neutral, like NH_3.

The idea that a base is a hydrogen ion acceptor and that an acid is a hydrogen ion donor, is known as the Brønsted-Lowry theory, named after Johannes Brønsted and Thomas Lowry who independently proposed these ideas in the 1920s. A more refined definition of acids and bases was proposed by Gilbert N. Lewis: a Lewis base is a species which is a good donor of electrons whilst a Lewis acid is a species which can act as an electron acceptor. The Lewis definition of a base removes any distinction between something acting as a base or as a nucleophile!

It is true that some substances are better at removing hydrogen ions than attacking at other centres; an example being inorganic amides such as sodium amide, $Na^+NH_2^-$ or lithium diisopropylamide, $Li^+N[CH(CH_3)_2]_2^-$. Other substances, such as iodide, I^-, are good nucleophiles but poor bases. Most substances can (and do) act as both nucleophiles and bases.

Exactly *why* something prefers to attack an H–X bond and remove a hydrogen ion, rather than attack any other centre, depends on a number of factors; these may include the energy of its lone-pair electrons, which other groups or counter-ions are present, what solvent is used and, of course, exactly what it is reacting with. These ideas will be explored in the future chapters.

11.4 Addition reactions – elimination in reverse

We mentioned above that whilst it is fairly easy for a base such as hydroxide to remove a hydrogen from the trimethylcarbenium cation, the H^+ will not spontaneously fall off to give an alkene. We are now going to look at the *reverse* reaction, adding H^+ to an alkene to form a carbenium ion, specifically we shall look at the protonation of isobutene to form the trimethylcarbenium ion.

On p. 101 we saw that in trying to understand how two reagents might react, we should first identify the highest energy occupied MO of the whole system and then the very lowest energy unoccupied MO. Treating our acid as a source of H^+ means that the LUMO must be from this species as it has no electrons; we will assume that this orbital is a 1*s* AO. The highest energy electrons must therefore come from the alkene. There are no lone pairs in the alkene – the HOMO is the C–C π bonding MO.

The problem is, for an unsymmetrical alkene like isobutene, there are two positions in which the incoming H^+ could end up being bound, as shown in Fig. 11.27.

In route (a), the proton adds to the *least* substituted carbon atom, meaning that the positive charge ends up on the carbon atom with the *most* substituents. Protonation on the other carbon, as shown in route (b), means that the positive charge ends up on the least substituted carbon. We have seen on p. 180 that the

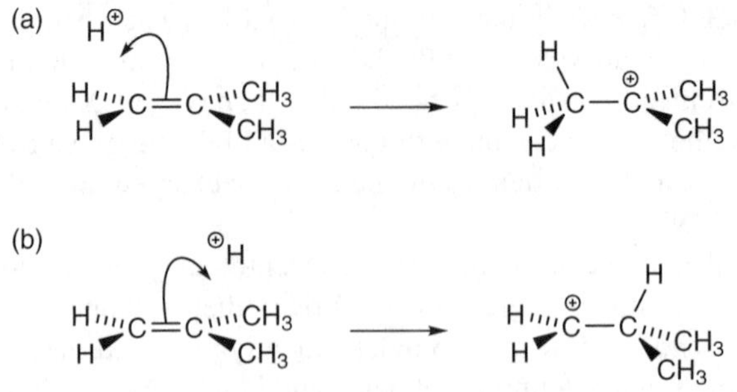

Fig. 11.27 An unsymmetrical alkene, such as isobutene, may protonate in either of two positions. In (a) the proton adds to the carbon atom with the most hydrogens and results in the positive charge forming on a carbon which has three methyl groups attached. In (b) the proton adds to the carbon with two methyl groups and results in the positive charge forming on a carbon with just one alkyl group attached. Route (a) is the preferred choice since this forms the most stable carbenium ion.

most stable cation will be the one where the most σ conjugation is possible. Since hydrogens attached to a positively charged carbon atom cannot stabilize the cation but adjacent C–H or C–C σ bonds can, the most stable cation will be the one with least hydrogens attached to the positive carbon. Looking at it the other way round, the most stable carbenium ion is the one in which the positive carbon has the greatest number of stabilizing alkyl groups attached to it.

Once the carbenium ion has been formed, it will react with any nucleophile present, as shown in Fig. 11.28. This is exactly the same mechanism we saw in our discussion of the S_N1 mechanism.

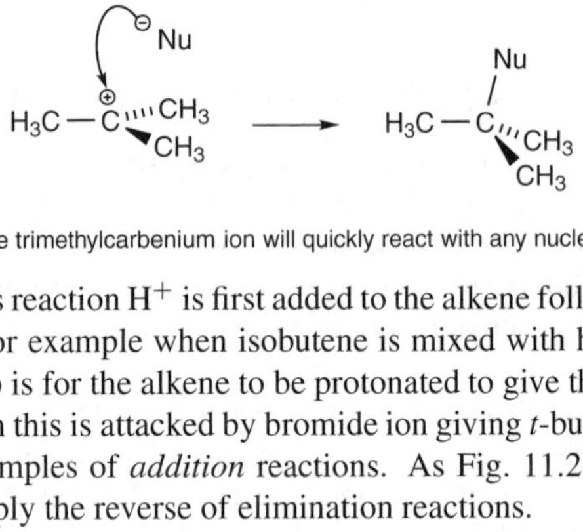

Fig. 11.28 The trimethylcarbenium ion will quickly react with any nucleophile present.

Overall in this reaction H^+ is first added to the alkene followed by attack by a nucleophile. For example when isobutene is mixed with hydrobromic acid, HBr, the first step is for the alkene to be protonated to give the trimethylcarbenium ion and then this is attacked by bromide ion giving *t*-butyl bromide. Such reactions are examples of *addition* reactions. As Fig. 11.29 shows, addition reactions are simply the reverse of elimination reactions.

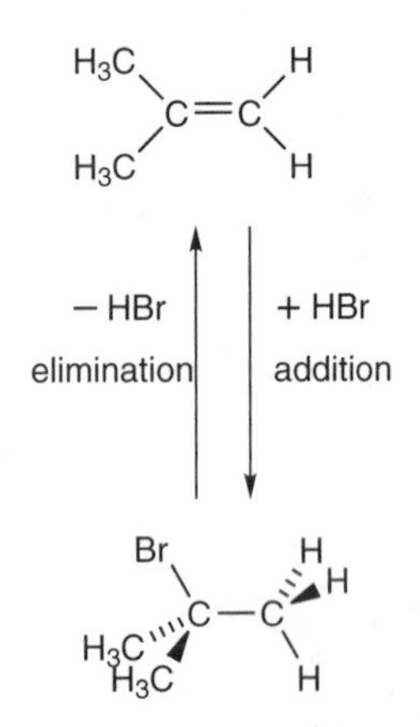

Fig. 11.29 Addition of HBr to isobutene is the reverse of the elimination of HBr from *t*-butyl bromide.

Markovnikov's Rule

The observation that the proton adds to the carbon with the greater number of hydrogens attached, is sometimes known as *Markovnikov's rule* after the Russian chemist, Vladimir Markovnikov, who first formulated the rule. However, sometimes the rule breaks down. A better version of the rule would be to say that the proton adds to the carbon atom of the alkene that gives rise to the most stable cation. The reason for this slight modification is that, as we have seen above, other groups are better at stabilizing positively charged carbon atoms than simple alkyl groups. An example of the application of this modified rule is shown in Fig. 11.30.

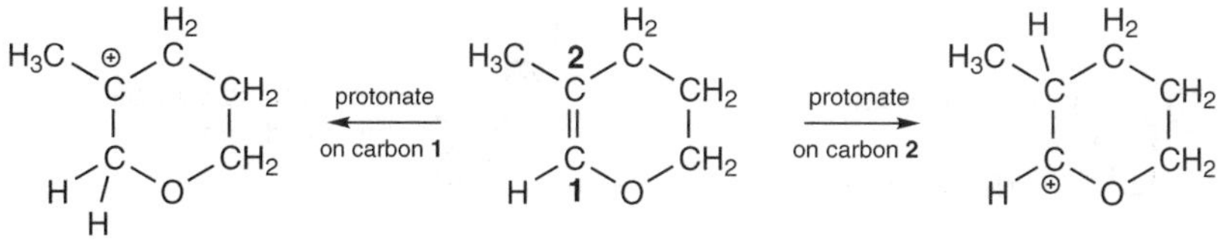

Fig. 11.30 The structure in the centre can be protonated on either C_1 or C_2. Protonating C_1 gives a carbenium ion which is stabilized by σ conjugation alone; protonating C_2 gives a carbenium ion which is stabilized by the neighbouring oxygen atom. The oxygen is much better at stabilizing the charge and so the molecule is protonated preferentially on C_2.

The structure in the centre can protonate on either C_1 or C_2. C_1 has more hydrogens attached than C_2 which has none and so Markovnikov's rule would suggest that protonation occurs on C_1. Protonating C_1 does give a carbenium ion which can be stabilized by σ conjugation, but protonating C_2 gives a carbenium ion with a neighbouring oxygen atom. Remember (see p. 176) that an adjacent oxygen atom stabilizes a positive charge very effectively – much more so than just σ conjugation alone – so this molecule does protonate preferentially on C_2. A better way of representing the protonated product is shown in Fig. 11.31 where the oxygen has donated one of its lone pairs and formed a π bond between the C and O. So in this example the molecule does not protonate on the carbon with the most hydrogens attached but so as to give the most stable cation. Understanding the rule is always better than just applying it blindly!

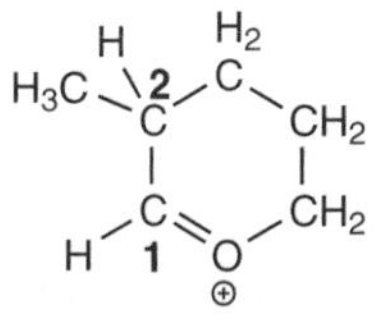

Fig. 11.31 After protonating on C_2 the carbenium ion is stabilized by the oxygen as shown in this structure.

11.5 E2 elimination

We have seen how *t*-butyl bromide can eliminate HBr by first losing a bromide ion to form the stable carbenium ion and then having a proton removed to form the alkene. HBr can be eliminated from the other bromoalkanes to give alkenes; for example, 1-bromopropane forms propene when mixed with hydroxide as shown in Fig. 11.32.

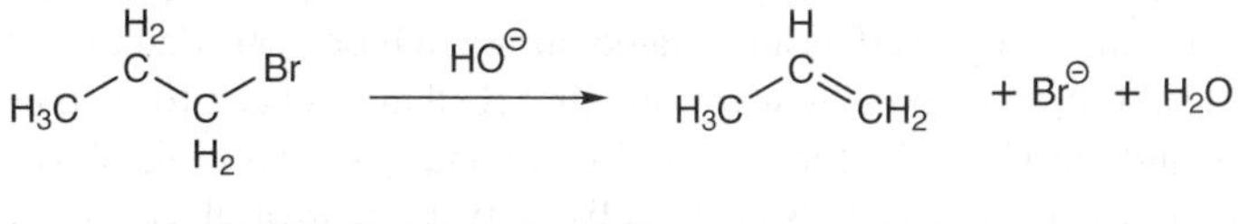

Fig. 11.32 Hydroxide reacts with 1-bromopropane to give propene.

However, a carbenium ion could not form from 1-bromopropane since there would be insufficient σ-conjugation in the $CH_3CH_2CH_2^+$ ion for it to be stable (see Fig. 11.20 on p. 180). We therefore need to find an alternative mechanism for the elimination which does not involve the formation of this ion.

We saw on p. 181 that the reason why a base can remove one of the protons from the trimethylcarbenium ion was that the interaction of the C–H σ MOs with the vacant p orbital weakens the C–H bonds making the hydrogens more acidic. In 1-bromopropane, there is no vacant p orbital for the C–H σ MOs to interact with. However, the vacant C–Br $\sigma^\star$ MO can interact to *some* degree with the C–H σ MOs.

The C–Br $\sigma^\star$ is not as low in energy as a vacant p orbital so the energy match between this and the filled C–H σ MO is not so good. Nonetheless, if

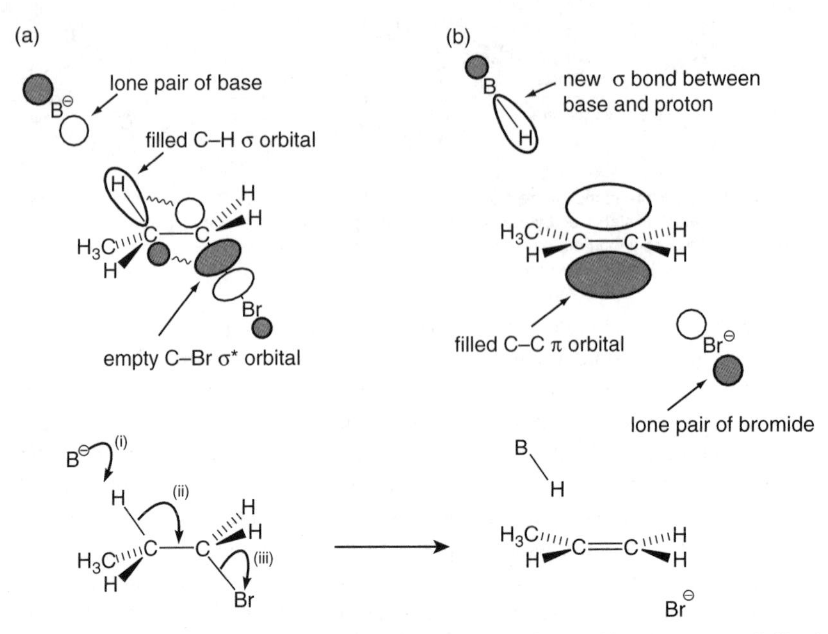

Fig. 11.33 Diagram (a) shows the interaction between the filled C–H σ MO and the vacant C–Br $\sigma^\star$ MO in 1-bromopropane as a base approaches. The products are shown in (b); a new bond has formed between the base and the proton, a new C–C π bond has formed and the bromide ion has been lost. The corresponding curly arrow mechanism for the process is shown below the structures.

the filled C–H σ MO and the vacant C–Br $\sigma^\star$ can be aligned in the same plane, they can have an interaction similar to that between the C–H σ MO and the vacant p orbital (see Fig. 11.23 on p. 181).

The interaction between the C–H σ MO and the vacant C–Br $\sigma^\star$ MO in 1-bromopropane as a base approaches is shown in Fig. 11.33; the curly arrow mechanism for the reaction is also shown in the lower part of Fig. 11.33.

There are three points to note in this reaction. The first is the formation of a new bond between the proton and the base. As we saw before, the hydrogen cannot just fall off by itself as an isolated proton; a base is needed to remove it. In Fig. 11.33 (b) we see the new σ bonding MO that has been formed between the base and the H^+. The formation of this bond is shown in the curly arrow mechanism by arrow (i). This arrow tells us that a pair of electrons from the base is ultimately going to be bonding the base to the hydrogen.

The second point to note is the formation of a new π bond between the two central carbon atoms. We can understand this as being a result of the gradual overlap of the filled C–H σ MO and the vacant C–Br $\sigma^\star$ MO in the starting material. The bonding interaction between these two orbitals is shown by the wavy lines in Fig. 11.33 (a). The new π bond that has formed from this interaction is shown in (b). Curly arrow (ii) tells us two things: that electron density is moving from the C–H bond as it is breaking and that it builds up in between the carbons forming the new π bond.

The final point to note is the breaking of the C–Br bond. We could think of this as arising from the electrons from the C–H σ bond moving into the anti-bonding C–Br $\sigma^\star$ orbital. As we saw in Chapter 5, filling an anti-bonding MO cancels out the effects of a filled bonding MO. So the interaction between

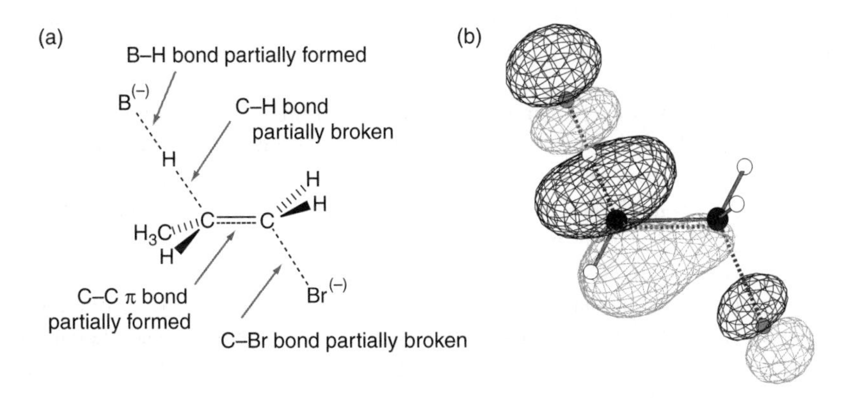

Fig. 11.34 The transition state of the reaction between 1-bromopropane and a base is shown in (a). The new base–proton and C–C π bonds have partially formed and the old C–H and C–Br bonds have partially broken. Charge is beginning to develop on the bromine atom and is being lost from the base. Diagram (b) shows a delocalized MO for a transition state analogous to that shown in (a). This shows that electron density is spread over the whole molecule. Its shape can be understood by comparing with the orbitals in Fig. 11.33.

the C–H σ bonding MO and the C–Br $\sigma^\star$ simultaneously forms a new C–C π bond and breaks the C–Br bond. The breaking of the C–Br bond is represented by curly arrow (iii) which tells us that the electrons from this bond end up on the bromide ion as a lone pair, as shown in (b).

The important point is that these three steps are all occurring at the same time. The mechanism is said to be *concerted*. Fig. 11.34 (a) shows a transition state for this process. It shows the new bond forming between the base and the proton, the C–H bond beginning to break, the C–C π bond forming and the C–Br bond beginning to break. The negative charge that was initially on the base is being spread over the whole molecule – still partially on the base but also beginning to form on the bromide.

The full MO picture for this reaction is rather involved but the HOMO for the transition state is rather revealing and is shown in Fig. 11.34 (b). This is one, delocalized, MO which can hold just two electrons. It clearly shows that electrons are being transferred from the base to the bromide leaving group with the form of the orbital resembling the interactions shown in Fig. 11.33.

Elimination kinetics

Since this elimination mechanism involves the base and the halogenoalkane coming together in a single encounter, it should not surprise us that this reaction rate is first order with respect to the concentrations of both the base and the halogenoalkane, i.e. second order overall:

$$\text{rate of elimination} = k[\text{1-bromopropane}][\text{base}].$$

This mechanism is known as an 'E2' mechanism, i.e. an *E*limination reaction in which the rate-determining step involves *two* species i.e. is bimolecular. The first elimination mechanism we saw in Section 11.3, where the bromide ion first fell off the *t*-butyl bromide to leave the relatively stable trimethylcarbenium ion which was then quickly attacked by the base, would be termed an

'E1' mechanism. Like the S_N1 mechanism, this is a first-order reaction since the rate-determining step, the initial loss of the bromide ion, involves only the *t*-butyl bromide.

The E1 and E2 mechanisms are really the extremes of a continuum of possibilities. In the E1, the halide or other leaving group leaves first before the proton is abstracted. In the E2 mechanism, both events happen together, so in the transition state the halide ion has partly left and the proton has partly been abstracted. If the halide is nearer to having left than the proton is, then the reaction is beginning to look more E1-like.

A pure E1 mechanism is rather less common than the E2. An elimination reaction will only take place by an E1 mechanism if a relatively stable cation can form following the loss of the leaving group and if the base that is present is not too powerful. If the base is strong, it will get involved and help the hydrogen come off before the leaving group gets a chance to fall off by itself.

12 The effects of the solvent

Up to now we have neglected a crucial component of reactions – the solvent. Almost all reactions are carried out in solution, including all the ones going on in our bodies, but it is all too easy to overlook how much the solvent influences reactions. A good example is the reaction of *t*-butyl chloride shown in Fig. 12.1. As was discussed in Chapter 11, this molecule reacts by first forming the carbenium ion $(CH_3)_3C^+$ which then goes on to react with either a nucleophile to form a substitution product, or with a base to form the elimination product, isobutene.

Fig. 12.1 *t*-butyl chloride reacts first by forming the carbenium ion and then forming either the substitution product or the elimination product. In each case, the rate-determining step for the reaction is the initial formation of the carbenium ion.

The rate of formation of either the substitution or elimination product depends only on the rate of the initial step in which the carbenium ion is formed; the reaction is therefore first-order. The table below gives values of this first-order rate constant, k_1, at 298 K for the reaction of *t*-butyl chloride in a number of different solvents. The table also gives the half-life, $t_{1/2}$, of the reaction which is the time needed for the reactant concentration to halve.

	water	ethanol	acetone	benzene	pentane
k_1 / s^{-1}	2.9×10^{-2}	8.5×10^{-8}	1.3×10^{-10}	6.9×10^{-13}	10^{-16}
$t_{1/2}$	24 sec	94 days	169 years	30 000 years	220 million years

The table shows that whilst the reaction of *t*-butyl chloride in water proceeds quickly with half of the reagent being used up in 24 seconds, the reaction in pentane is so slow that it essentially does not occur at all. Even changing the solvent from water to ethanol results in a considerable change in the rate. It is clear that the solvent has an enormous effect on the rate of this reaction.

The reaction between CH_3Br and Cl^-, shown in Fig. 12.2, proceeds by the S_N2 mechanism and it is found that the influence of the solvent on the rate of this reaction is opposite to the case of *t*-butyl chloride. For example, changing the solvent from water to propanone (acetone) results in a 10^4 fold *increase* in the rate. Moving to the gas phase, where there is no solvent at all, results in a rate increase by a factor of 10^{15}.

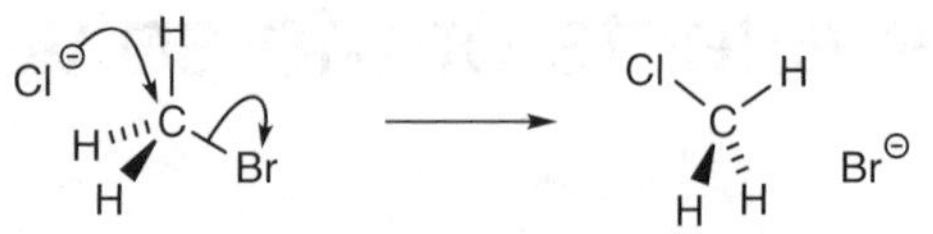

Fig. 12.2 The substitution reaction between bromomethane and chloride ion proceeds 10^4 times faster in propanone (acetone) than water, and 10^{15} times faster when no solvent is present at all.

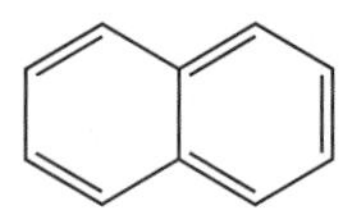

Fig. 12.3 The structure of the aromatic hydrocarbon naphthalene. At room temperature it is a white solid which sublimes readily giving off a vapour which repels many insects – hence the use of naphthalene in moth balls.

A further example of the importance of the solvent is the solubility of one substance in another. Solid NaCl dissolves in water but not in hexane, whereas the reverse is true for a substance such as naphthalene (Fig. 12.3). We attribute the ready solubility of ionic compounds such as NaCl in water as being due to the ability of water to solvate ions – something which hexane is very much less effective at doing.

The rate-determining step of the reaction of *t*-butyl chloride involves the formation of ions. The fact that the reaction goes so much faster in water than in pentane is clearly related to the excellent solvation of ions by water.

In order to appreciate fully how the solvent can affect the reaction we first need to understand the nature of the different solvents and how they can interact with ions. It is to this topic that the next few sections are devoted and then, towards the end of the chapter, we will return to answer the question as to why the solvent has such a large influence on the rate of a reaction.

12.1 Different types of solvent

Solvents may usefully be classified according to how polar they are, but trying to quantify this is not particularly straightforward. A useful measure is the *relative permittivity* (or *dielectric constant*) of the solvent. In Section 3.2 on p. 21 we saw that the attractive force between two oppositely charged ions with charges z_+ and z_- varies with the separation between the two ions, r, according to:

$$\text{force} = \frac{z_+ z_- e^2}{4\pi \varepsilon_0 r^2}$$

where e is the charge on the electron (the fundamental unit of charge) and ε_0 is the vacuum permittivity.

In a solvent, the force between the two ions is reduced by a factor called the *relative permittivity*, ε_r, of the medium:

$$\text{force} = \frac{z_+ z_- e^2}{4\pi \varepsilon_0 \varepsilon_r r^2}. \qquad (12.1)$$

Thus, moving to a solvent with a larger relative permittivity *decreases* the interaction between the ions, as is illustrated schematically in Fig. 12.4.

Typical values of ε_r are around 2 for hydrocarbon solvents such as hexane to over 100 for primary and secondary amides. Water – the solvent used in nature – has a relative permittivity of 80.

Solvents with a relative permittivity greater than about 15 are usually described as *polar* whereas those with smaller values of the relative permittivity are described as *non-polar* or *apolar*. As we shall see, polar solvents are much better at dissolving ionic salts than are non-polar solvents.

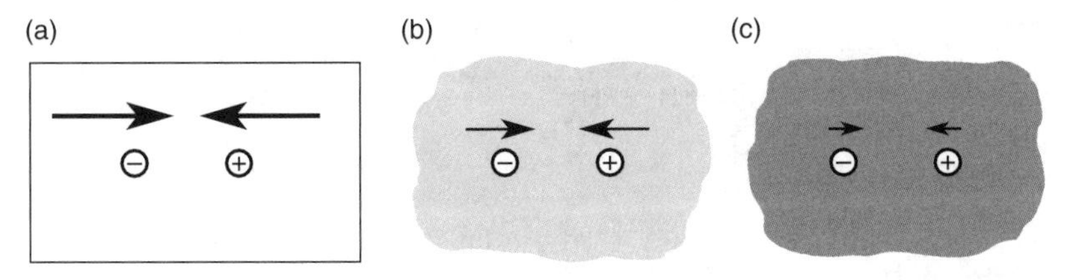

Fig. 12.4 The force between two ions (represented by the size of the arrows in the diagrams) depends on the relative permittivity of the solvent. The force is greatest when the ions have nothing in between them as in a vacuum, represented in (a), and decreases on moving to solvents with progressively increasing relative permittivities as represented in (b) and (c).

The relative permittivity can easily be determined from simple physical measurements but rationalizing the particular value for a given solvent is not so straightforward. The value of ε_r depends on the dipole moment of the solvent molecules and also the extent to which they can interact with one another. Particularly important in this regard is the ability of solvent molecules to form *hydrogen bonds* with one another.

Hydrogen bonds

In order for a hydrogen atom in a molecule to form a hydrogen bond, it first needs to be slightly acidic – usually, this simply means that it is bound to an electronegative element such as fluorine, oxygen or nitrogen. Anything with a high energy lone pair can then interact with this acidic hydrogen, resulting in the lone pair being shared to some extent between the two atoms – in other words, a partial bond is formed to the hydrogen.

Figure 12.5 illustrates the formation of a hydrogen bond in water, which has both the required acidic hydrogens (attached to electronegative oxygen) and the high energy lone pairs on oxygen. Liquid HF is another example of a solvent in which there is extensive hydrogen bonding between solvent molecules.

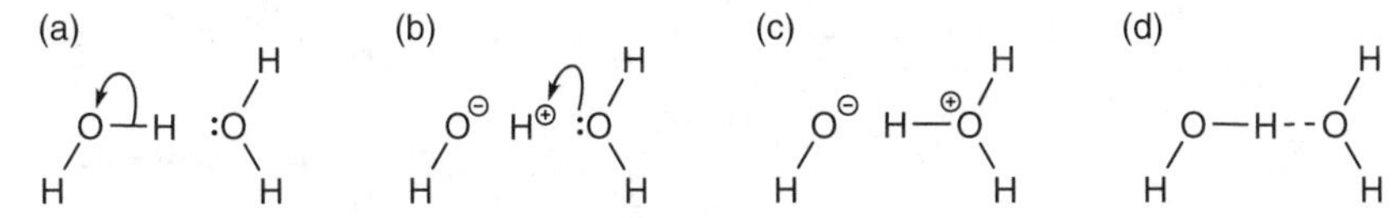

Fig. 12.5 Illustration of the formation of a hydrogen bond between two water molecules. The electronegative oxygen withdraws electron density from the O–H bond as shown in (a). In the extreme case, this would give a hydroxide ion and a proton, H^+, shown in (b). Any neighbouring water would be protonated at the oxygen to give the products shown in (c). In reality the process is not so extreme; there is still almost a full bond in the water molecule itself and a partial bond – a hydrogen bond – between neighbouring water molecules. This is usually indicated by a dashed line as shown in (d).

Some approximate strengths of hydrogen bonds are shown in Fig. 12.6; for comparison, the strengths of the 'full' bonds in these molecules are also shown. In each case it can be seen that the hydrogen bonds are very much weaker than the full bonds. The strengths of the hydrogen bonds increase with the electronegativity of the element to which the hydrogen is attached, so the hydrogen bonding increases along the series H_2S, NH_3, H_2O, HF.

The final point to note from Fig. 12.6 is that the hydrogen bond to an *ion* is much stronger than that to a neutral molecule. The strongest hydrogen bond in the diagram is that between HF and F^- shown in (h). In this species both

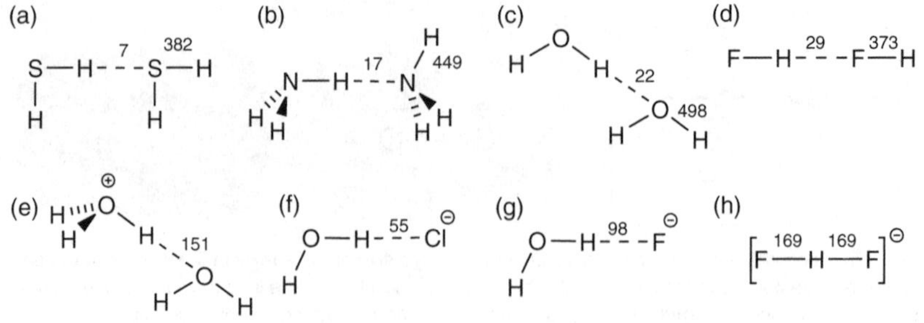

Fig. 12.6 Approximate bond strengths and hydrogen bond strengths (in kJ mol^{-1}) (a) in hydrogen sulfide; (b) in ammonia; (c) in water; (d) in hydrogen fluoride; (e) between H_3O^+ and water; (f) between chloride ion and water; (g) between fluoride ion and water and (h) between fluoride ion and hydrogen fluoride.

H–F bond lengths and bond strengths are the same. The resulting symmetrical ion, $[F–H–F]^-$, is so stable that it is readily obtainable as the solid sodium salt, sodium hydrogenfluoride, $NaHF_2$.

Hydrogen bonds are often responsible for holding molecules in one particular shape. This is important in proteins since it is the three-dimensional shape of such molecules which gives them their special properties. Another example is DNA in which the two strands which are twisted together into the famous double-helix are held together by hydrogen bonds between the individual base pairs. As is shown in Fig. 12.7, the base *adenine* hydrogen bonds with *thymine*, and *guanine* with *cytosine*.

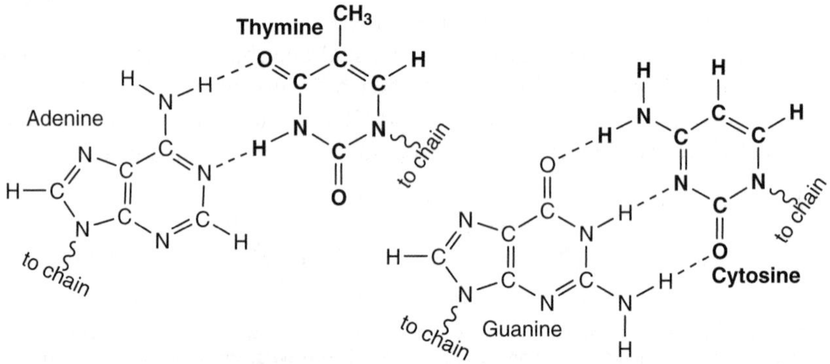

Fig. 12.7 DNA exists as a double-helix due to hydrogen bonding. The adenine subunits from one strand hydrogen bond with the thymine subunits from the other; similarly guanine and cytosine are hydrogen bonded to one another.

Protic and aprotic solvents

Solvents which have weakly acidic hydrogens capable of forming hydrogen bonds are called *protic* solvents. In order for the hydrogens to be acidic they must be bonded to an electronegative atom, which makes the solvent polar. Polar solvents which do not contain acidic hydrogens are called *polar aprotic*.

Figure 12.8 lists several common solvents in general order of polarity and classifies them as polar protic, polar aprotic and non-polar. Also included in the table are data on the relative permittivities of the solvent and the dipole moments of the solvent molecules.

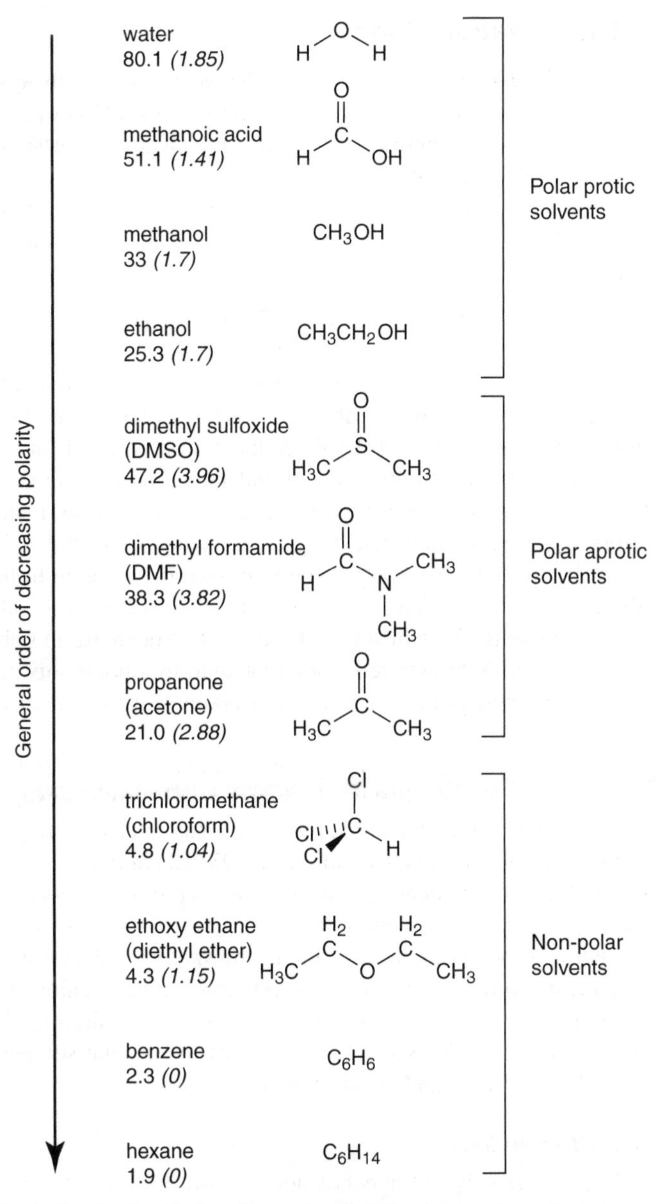

Fig. 12.8 Table showing the classification of some common solvents into polar protic, polar aprotic and non-polar. The number immediately following the name is the relative permittivity at 293 K; the number italicized in brackets is the dipole moment (in Debye) for the molecule in the gas phase.

For polar protic solvents, hydrogen bonding is possible between solvent molecules. This group of solvents tends to have large relative permittivities but small dipole moments. The polar aprotic solvents are not able to form hydrogen bonds between solvent molecules. This group tends to have mid-range relative permittivities but larger dipole moments.

Non-polar solvents cannot form any hydrogen bonds and have small relative permittivities and small or zero dipole moments.

12.2 The solvation of ions

We saw on p. 190 that the force of attraction between two ions always decreases when the ions are transferred from a vacuum to a solvent. For essentially the same reasons, the Gibbs energy of an isolated ion always decreases on moving from a vacuum into a solvent.

This change in Gibbs energy on taking an ion from the gas phase to a solvent (the Gibbs energy of solvation, $\Delta G^\circ_{\mathrm{solv}}$) can be estimated using the Born equation:

$$\Delta G^\circ_{\mathrm{solv}} = -\frac{z^2 e^2 N_{\mathrm{A}}}{8\pi\varepsilon_0 r}\left(1 - \frac{1}{\varepsilon_r}\right). \tag{12.2}$$

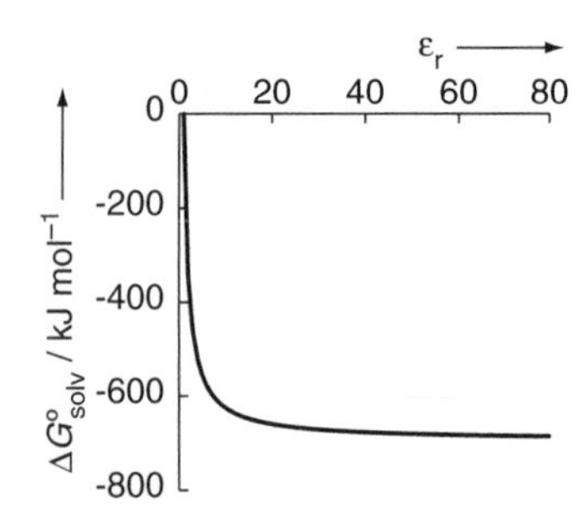

Fig. 12.9 Graph showing the prediction of the Born equation, Eq. 12.2, for how the Gibbs energy of solvation varies with the relative permittivity of a solvent. Note how $\Delta G^\circ_{\mathrm{solv}}$ is always negative and becomes more so as the relative permittivity increases. However, once ε_{r} is greater than about 20, there is little further change on $\Delta G^\circ_{\mathrm{solv}}$. The radius of the ion has been taken as 1 Å and $z = 1$.

In this relationship, z is the charge on the ion, ε_{r} is the relative permittivity of the solvent, N_{A} is Avogadro's number and r is the radius of the ion.

Figure 12.9 shows a plot of the Born equation prediction for $\Delta G^\circ_{\mathrm{solv}}$ as a function of relative permittivity; we can understand its form in the following way. Since ε_{r} is always greater than 1, the term in the bracket is positive and so the Born equation predicts that $\Delta G^\circ_{\mathrm{solv}}$ will always be negative, i.e. there is always a decrease in Gibbs energy on going from the vacuum to the solvent. Also, the term in the bracket is largest (closest to 1) for large values of ε_r, i.e. for polar solvents. This means that the greatest decrease in Gibbs energy occurs when an ion is transferred from a vacuum to a polar solvent; in other words, the Gibbs energy of an ion is lower in a polar solvent than in a non-polar solvent.

The specific interactions between a charged ion and the solvent depend on the type of solvent. Polar solvents – both protic and aprotic – usually have an oxygen or nitrogen atom with a lone pair which can interact with positive ions as shown for the solvent dimethyl sulfoxide (DMSO) in Fig. 12.10.

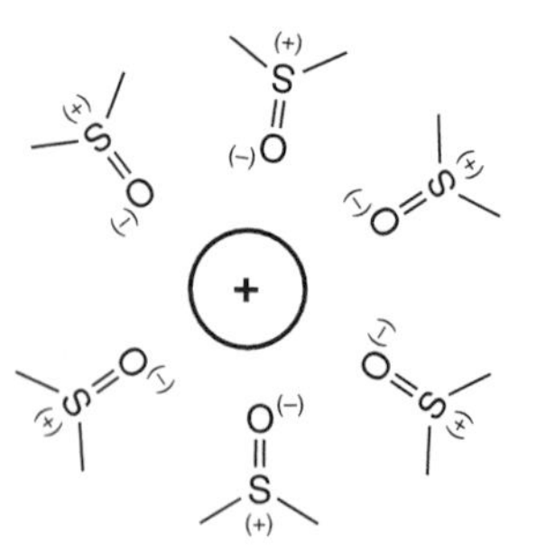

Fig. 12.10 A polar solvent such as DMSO solvates a cation by using its lone pairs to interact with the ion.

The big difference between polar protic and aprotic solvents arises in the solvation of *anions*. As we saw in Fig. 12.6 (f)–(h), protic solvents are able to form hydrogen bonds to anions; this is shown for methanol in Fig. 12.11 (a). In contrast, no hydrogen bonds are possible between anions and aprotic polar solvents – these can only solvate anions poorly by aligning their dipole moments as shown for DMSO in Fig. 12.11 (b). Non-polar solvents are poor at solvating both positive and negative ions.

Solvating different ions

How well an ion is solvated depends not only on the solvent, but on the ion itself. The Born equation (Eq. 12.2) tells us that for a given solvent, the Gibbs energy of solvation is proportional to z^2/r, i.e. the charge of the ion squared divided by its radius. This means that the Gibbs energy of solvation of an ion becomes more negative as the ion gets smaller and as the charge on the ion increases.

It turns out that experimental values of the *enthalpy* and *entropy* of solvation also correlate quite well with the value of z^2/r. For example, Fig. 12.12 shows a plot of the enthalpies of solvation in water against z^2/r for some common cations.

The enthalpies are clearly grouped depending on the charge on the cation – the ions with the greatest charge having the most negative enthalpies of solva-

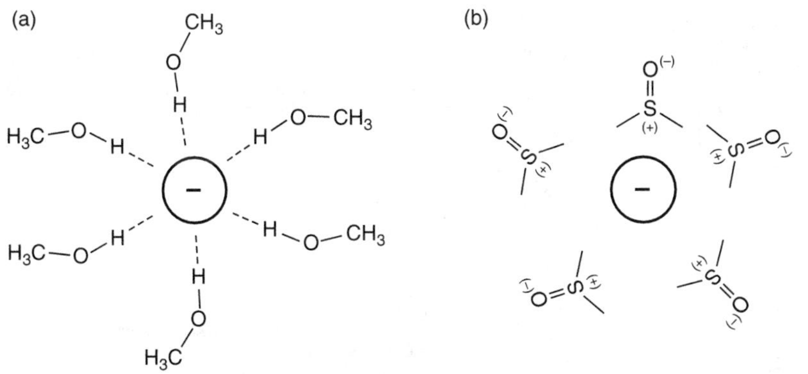

(a) (b)

Fig. 12.11 Protic solvents, such as methanol, solvate anions by forming hydrogen bonds as in (a). Aprotic solvents, such as DMSO, only solvate anions weakly by orienting their dipole moments with the positive end towards the anion as shown in (b).

tion. Within each group, the enthalpy of solvation is inversely proportional to the radius of the ion: for example, the enthalpy of solvation for the large Ba^{2+} ion is less negative than that for the smaller Mg^{2+} ion.

Figure 12.13 shows that the entropy of solvation in water correlates quite well with z^2/r, although the correlation is not as good as it is for the enthalpy of solvation. For the majority of ions the entropy of solvation is negative, which we can attribute to the ordering of the water molecules as they become coordinated to the ions. The smaller the ion and the more highly charged it is the greater restriction it causes and hence the more negative the entropy of solvation.

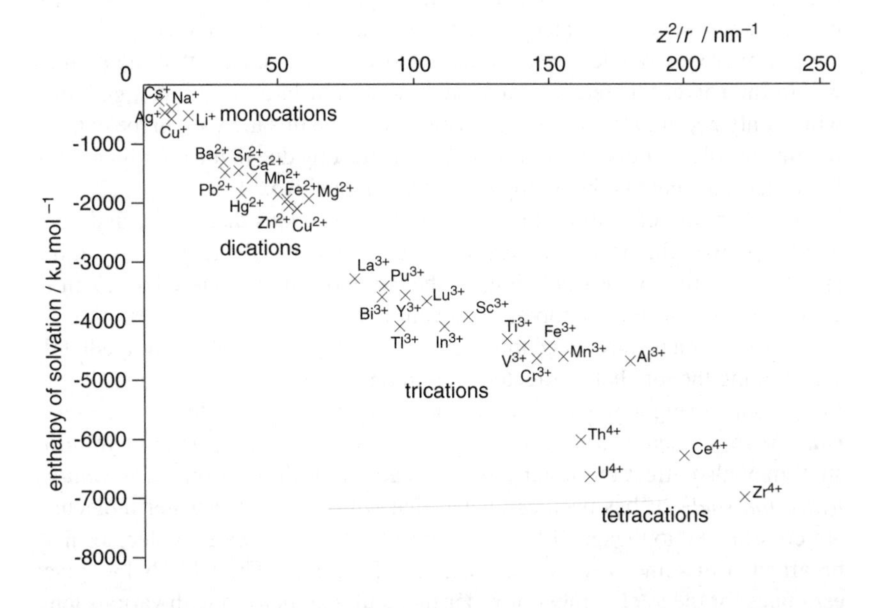

Fig. 12.12 A plot of the enthalpy of solvation of various cations in water against z^2/r shows that there is a good correlation between these two quantities. The plot also shows that the ions fall into groups depending on their formal charge. Within each group, the smaller the radius of the ion, the greater the enthalpy of solvation; thus Mg^{2+} has a more negative enthalpy of solvation than Ba^{2+}.

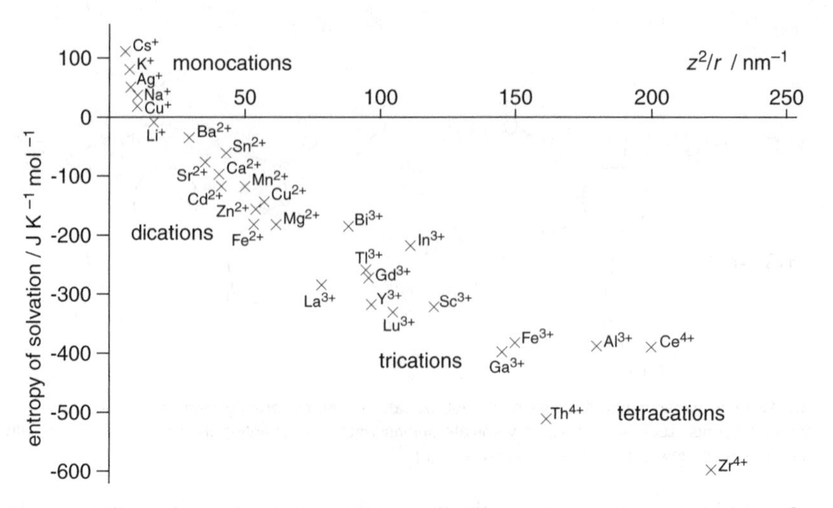

Fig. 12.13 Plot showing how the entropy of solvation of various cations in water correlates with z^2/r; the correlation is not as good as in the case of the enthalpy of solvation (Fig. 12.12). However, the ions are still in groups depending on their formal charge and, within each group, the smaller the radius of the ion, the more negative the entropy of solvation. The entropy of solvation for the large singly charged cations are positive, implying that they disrupt the ordering of the water molecules to some extent.

What is surprising is that the entropies of solvation for some of the larger singly charged cations are actually *positive*; this means that transferring such an ion from the gas phase to water results in an *increase* in the disorder in the solvent. To work out what is going on here we need to recognize that in pure water there is an extensive network of hydrogen bonding. When an ion is introduced, some of the water molecules interact with this ion thereby disrupting the hydrogen bonded network in the solvent: this means that the entropy within the solvent increases to a small extent. For large singly charged ions, which only coordinate the water molecules weakly, this increase in the entropy within the solvent outweighs the decrease in entropy due to the coordination of the water thus making the entropy of solvation positive.

Another way of putting this is to say that these ions disrupt the hydrogen bonded network in the solvent, leading to an increase in the entropy. A similar effect is seen for anions (including all the common singly charged ones), most of which have positive entropies of solvation in water.

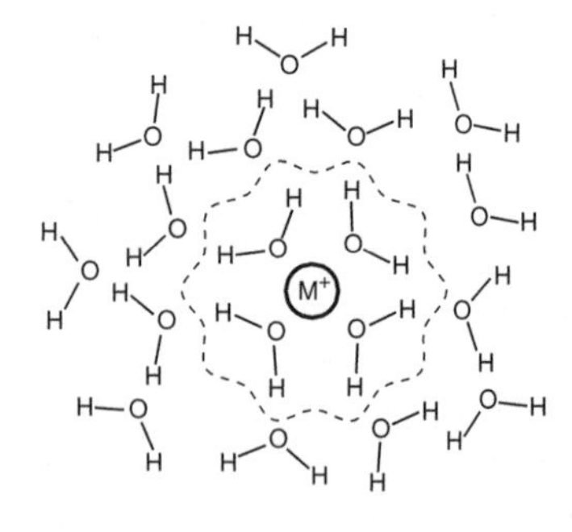

Fig. 12.14 The water molecules directly coordinated to the ion form the *primary hydration shell*, shown within the dotted line. These water molecules affect their neighbours, binding to them more tightly than they would in the free solvent forming a *secondary hydration shell*.

When an ion is introduced into water, it is not only the solvent immediately surrounding the ion that is affected. A number of water molecules, usually between four and eight, will be directly coordinated to the ion in the so-called *primary hydration shell* as shown in Fig. 12.14. However, the primary hydration shell may also affect the water molecules around it thus forming a *secondary hydration shell*. It has been estimated that whilst only four water molecules are coordinated to the small Li^+ ion, up to a total of 22 water molecules may be affected in some way by the presence of the ion. The table below gives estimates for the total number of water molecules associated with various ions, sometimes called the *hydration number*.

ion	Cs^+	K^+	Na^+	Li^+	Ca^{2+}	Mg^{2+}	Zn^{2+}
hydration number	6	7	13	22	29	36	44

The hydration shell around a given ion often results in the ion appearing much larger in solution than might be expected. For example, while the actual ionic radii of the gaseous ions of the Group I metals increase as we go down the Group from Li^+ to Cs^+, the effective radii of the ions and their coordinated water molecules actually decrease as we go down the Group.

The coordination of water molecules with a given ion may also play an important part in how soluble a compound is in water, as we shall see in the next section.

12.3 Solubilities of salts in water

On p. 14 in Chapter 2 we saw that by considering the entropy changes of *both* the system and the surroundings we could understand why a salt such as ammonium nitrate dissolves in water despite the process being endothermic. We are now going to revisit this whole question of why some salts are soluble and some are not – not surprisingly the solvent plays a crucial role in this and we will use the ideas developed so far in this chapter to understand the factors which affect solubility.

Rather than looking at the entropy change of the system and the surroundings, and hence determining the entropy change of the Universe, it is easier just to consider the Gibbs energy change of the system – we saw in Section 2.7 starting on p. 16 that these two approaches are entirely equivalent. So, when trying to predict the degree to which a salt, MX, is soluble, we simply need to look at the Gibbs energy of solution, $\Delta G^\circ_{\text{solution}}$, for the process

$$\text{MX}(s) \longrightarrow \text{M}^+(aq) + \text{X}^-(aq). \tag{12.3}$$

The crucial thing is the sign of $\Delta G^\circ_{\text{solution}}$: if it is negative then the products are favoured at equilibrium, i.e. the salt is soluble; if it is positive then the solid salt is favoured and so it is insoluble (or, more accurately, the salt will be sparingly soluble).

We can also assess how soluble a salt is by simply looking at the equilibrium constant for the reaction of Eq. 12.3. This equilibrium constant is known as the *solubility product*, K_{sp}, and is given by

$$K_{\text{sp}} = [\text{M}^+(aq)]_{\text{eq}}[\text{X}^-(aq)]_{\text{eq}},$$

where you need to recall that solid species do not contribute terms to the expression for an equilibrium constant, so only the ions from Eq. 12.3 appear in the expression.

As we saw in Section 8.3 on p. 118, for any equilibrium the standard Gibbs energy change and the equilibrium constant are related; in this case $\Delta G^\circ_{\text{solution}}$ and K_{sp} are related by

$$\Delta G^\circ_{\text{solution}} = -RT \ln K_{\text{sp}}.$$

In trying to rationalize *why* a particular salt is either soluble or insoluble, it is often helpful to break down $\Delta G^{\circ}_{\text{solution}}$ for the salt into the enthalpy and entropy contributions, $\Delta H^{\circ}_{\text{solution}}$ and $\Delta S^{\circ}_{\text{solution}}$:

$$\Delta G^{\circ}_{\text{solution}} = \Delta H^{\circ}_{\text{solution}} - T\Delta S^{\circ}_{\text{solution}}.$$

It is also helpful to analyse the values of each of these using the cycle shown in Fig. 12.15. In this cycle the solid salt is first dissociated to give gas-phase ions; the enthalpy change for this process is the (standard) lattice enthalpy, $\Delta H^{\circ}_{\text{lattice}}$, discussed on p. 26. Then the gaseous ions are solvated for which the enthalpy change is the (standard) enthalpy of solvation of M^+ plus that for X^-.

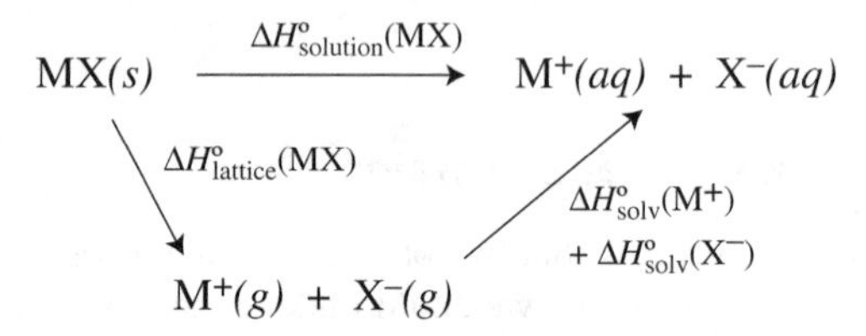

Fig. 12.15 The enthalpy of solution of a salt may usefully be broken down into the lattice enthalpy of the solid plus the enthalpies of solvation of the gaseous ions. A similar cycle can be constructed for the entropy of solution.

Soluble salts

We shall first look at some salts which are readily soluble in water; some data (at 298 K) for three such examples are given in the following table.

salt	K_{sp}	$\Delta G^{\circ}_{\text{solution}}$ / kJ mol^{-1}	$\Delta H^{\circ}_{\text{solution}}$ / kJ mol^{-1}	$\Delta S^{\circ}_{\text{solution}}$ / J K^{-1} mol^{-1}	$-T\Delta S^{\circ}_{\text{solution}}$ / kJ mol^{-1}
$Ca(NO_3)_2$	410 000	−32	−19	+45	−13
$MgSO_4$	16 000	−24	−87	−210	+63
$NaNO_3$	15	−6.7	+20	+90	−27

All three of these salts are soluble in water as shown by the large values of K_{sp} and the negative values of $\Delta G^{\circ}_{\text{solution}}$. The contributions to $\Delta G^{\circ}_{\text{solution}}$ made by $\Delta H^{\circ}_{\text{solution}}$ and $\Delta S^{\circ}_{\text{solution}}$ are visualized in Fig. 12.16; as it is most directly relevant, $-T\Delta S^{\circ}_{\text{solution}}$ has been plotted, rather than $\Delta S^{\circ}_{\text{solution}}$. Let us look at each salt in turn and work out why it is soluble.

Why is calcium nitrate soluble in water?

The reaction is exothermic, which must mean the enthalpy of solvation is greater in magnitude than the lattice enthalpy. The entropy also increases, which we can attribute to the breaking up of the ordered lattice. Both $\Delta H^{\circ}_{\text{solution}}$ and $-T\Delta S^{\circ}_{\text{solution}}$ are therefore negative, contributing to a negative $\Delta G^{\circ}_{\text{solution}}$.

Why is magnesium sulfate soluble in water?

For this salt the reaction is much more exothermic than for calcium nitrate. The enthalpy of solvation is now much greater in magnitude than the lattice enthalpy, something we can attribute to the fact that there are *two* doubly charged

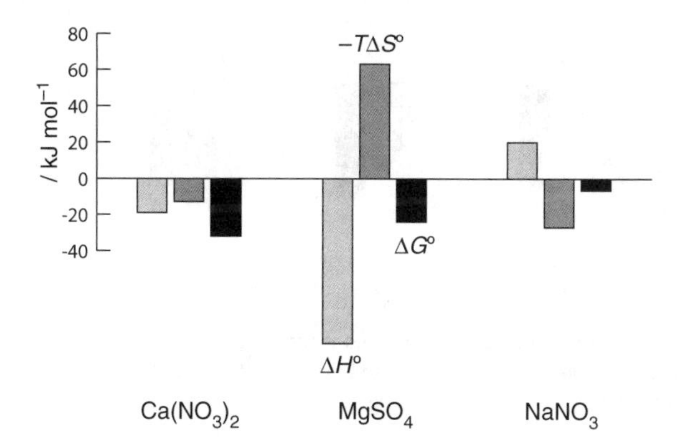

Fig. 12.16 Visualization of the separate contributions made by $\Delta H^\circ_{solution}$ (light grey bars) and $\Delta S^\circ_{solution}$ (darker grey bars) to the values for $\Delta G^\circ_{solution}$ (black bars) for three different salts dissolving in water. The entropy term has been plotted as $-T\Delta S^\circ_{solution}$ as this is most directly relevant. In all three cases $\Delta G^\circ_{solution}$ is negative, indicating that the salt is soluble, but this negative value is achieved in different ways for each salt.

ions being solvated. However, there is a cost for this strong solvation – the entropy of solvation is negative because many water molecules are being tied up solvating the ions. However $\Delta G^\circ_{solution}$ is negative as the large enthalpy term outweighs the unfavourable positive $-T\Delta S^\circ_{solution}$ term.

We can say that this salt is soluble because the favourable enthalpy term outweighs the unfavourable entropy term.

Why is sodium nitrate soluble in water?

For this salt the reaction is *endothermic*, which must mean that the enthalpy of solvation is smaller in magnitude than the lattice enthalpy. This should not be too surprising – both ions are singly charged and as we have seen these are not strongly solvated in water.

There is a large increase in the entropy when the salt dissolves. We can attribute this to the loss of the ordered lattice and the weak solvation of these singly charged ions. Overall, the negative $-T\Delta S^\circ_{solution}$ term outweighs the positive enthalpy term, resulting in a negative value for $\Delta G^\circ_{solution}$.

We can say that this salt is soluble because the favourable entropy term outweighs the unfavourable enthalpy term. These three examples show how a negative ΔG°_{solv} can be achieved using different combinations of ΔH°_{solv} and ΔS°_{solv}.

Sparingly soluble salts

We shall now consider some salts that are only sparingly soluble in water. These salts will all readily be precipitated out of solution when their ions are mixed. Since the precipitation reaction *does* occur whereas the salts do not dissolve to a significant degree, rather than looking at the Gibbs energy for the salt dissolving we shall instead look at the Gibbs energy of precipitation, ΔG°_{ppt}, which is $\Delta_r G^\circ$ for the reaction

$$M^+(aq) + X^-(aq) \longrightarrow MX(s);$$

of course, $\Delta G^\circ_{ppt} = -\,\Delta G^\circ_{solution}$.

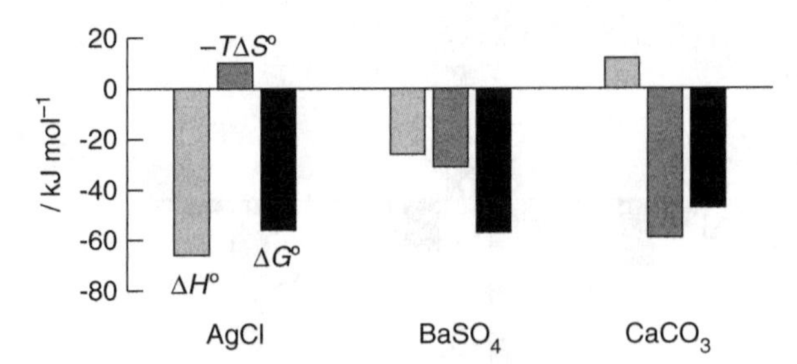

Fig. 12.17 Visualization of the separate contributions made by ΔH°_{ppt} (light grey bars) and ΔS°_{ppt} (darker grey bars) to the values for ΔG°_{ppt} (black bars) for three different salts being precipitated from solution in water. The entropy term has been plotted as $-T\Delta S^\circ_{ppt}$.

As before, the Gibbs energy change may also be broken down into the enthalpy and entropy of precipitation, ΔH°_{ppt} and ΔS°_{ppt}:

$$\Delta G^\circ_{ppt} = \Delta H^\circ_{ppt} - T\Delta S^\circ_{ppt}.$$

Some data (at 298 K) for three sparingly soluble salts are given in the following table and, as before, the data are visualized in Fig. 12.17.

salt	K_{sp}	ΔG°_{ppt} / kJ mol^{-1}	ΔH°_{ppt} / kJ mol^{-1}	ΔS°_{ppt} / J K^{-1} mol^{-1}	$-T\Delta S^\circ_{ppt}$ / kJ mol^{-1}
AgCl	1.5×10^{-10}	−56	−66	−34	+10
$BaSO_4$	1.1×10^{-9}	−51	−19	+105	−31
$CaCO_3$	5.8×10^{-9}	−47	+12	+200	−59

The fact that these three salts are all only sparingly soluble in water is indicated by the very small values for K_{sp}. Each ΔG°_{ppt} is negative, indicating that all the salts will readily precipitate out of solution. We shall look more closely at each reaction to see why the precipitation occurs.

Why does silver chloride precipitate out of solution?

The formation of the precipitate is an exothermic reaction which means that the lattice energy is greater than the magnitude of the solvation energy; we can rationalize this by noting that these singly charged ions are not strongly solvated. There is a decrease in the entropy as the ions go from the solution to the lattice. This is perhaps to be expected since the free ions are only weakly solvated and so there is a large loss of disorder when the lattice forms. However, the resulting positive $-T\Delta S^\circ_{ppt}$ term is not large enough to overcome the negative ΔH°_{ppt} term, and so ΔG°_{ppt} is negative.

We can say that precipitation takes place because the favourable enthalpy term is dominant. Looking at the process the other way round, AgCl is insoluble because of the unfavourable enthalpy term.

Why does barium sulfate precipitate out of solution?

This is an exothermic reaction, although less so than for the precipitation of silver chloride; we can attribute this to the stronger solvation of the doubly charged ions. Precipitation is accompanied by an *increase* in the entropy, leading to a negative $-T\Delta S^{\circ}_{\text{ppt}}$ term. So, both the enthalpy and the entropy terms favour the reaction. At first it seems odd that the entropy should increase as the ions form a lattice, however this increase is due to the liberation of the water molecules that were strongly complexed to the doubly charged ions.

The precipitation of barium sulfate is favoured by both the enthalpy and entropy change. The opposite reaction – barium sulfate dissolving – is disfavoured by both.

Why does calcium carbonate precipitate out of solution?

This is perhaps the most interesting of all the reactions as the formation of the precipitate is actually *endothermic*. We can attribute this to the strong solvation of the doubly charged ions resulting in the solvation energy being larger in magnitude than the lattice energy. As with barium sulfate, and for the same reasons, the entropy change on precipitation is positive, resulting in a negative $-T\Delta S^{\circ}_{\text{ppt}}$ term which outweighs the unfavourable positive $\Delta H^{\circ}_{\text{ppt}}$ term.

So, calcium carbonate precipitates from solution on account of the large increase in the entropy due to the liberation of the solvent from the ions; this favourable entropy change outweighs the unfavourable enthalpy change. To have a reaction in which the formation of a solid is accompanied by an increase in entropy seems at first sight rather odd until we recognize the crucial role which the solvent plays.

Solubilities of salts in other solvents

The last section illustrated how a number of factors contribute to determining how soluble a given salt is in water. The same ideas can be applied to solvents other than water but generally these are not as good as water at solvating inorganic salts. The table below shows $\Delta H^{\circ}_{\text{solv}}$ and $\Delta S^{\circ}_{\text{solv}}$ (at 298 K) for Na^+ in a number of polar solvents.

	water	methanamide (formamide)	methanol	DMF
$\Delta H^{\circ}_{\text{solv}}$/ kJ mol^{-1}	−418	−434	−438	−450
$\Delta S^{\circ}_{\text{solv}}$/ J K^{-1} mol^{-1}	+5.9	−18	−44	−59

It is initially surprising that the enthalpy of solvation becomes more negative on moving from water to dimethylformamide (DMF). Each of these solvents has a lone pair on oxygen which can solvate the cation in an exothermic process. However, when the ion is in water it disrupts the hydrogen bonding between water molecules; the breaking of these hydrogen bonds is an endothermic process and this makes the enthalpy of solvation of the ion less negative than it might otherwise be.

The amount of hydrogen bonding is greatest in water, and reduces as we go to formamide and then methanol; there is no hydrogen bonding between DMF molecules. The trend in the enthalpies of solvation thus follows the extent of hydrogen bonding in the solvent.

The trend in $\Delta S^{\circ}_{\text{solv}}$ supports this interpretation. Whilst there is actually a small increase in entropy on adding Na^+ to water due to the disruption of the hydrogen bonding, for all the other solvents the entropy change is negative because the solvent molecules become coordinated to the positive ion. Since there is no hydrogen bonding to disrupt between DMF molecules, placing a sodium ion in this solvent gives the largest decrease in entropy.

The greater decrease in the entropy on moving to an aprotic solvent to some extent offsets the fact that the reaction becomes more exothermic. The net result is that it is often difficult to predict how soluble an ion will be in a given solvent.

The reduced solubility of ionic salts in organic polar solvents is often due to the large unfavourable negative $\Delta S^\circ_{\text{solv}}$ term. For example, whilst the solubility of NaCl decreases as we go from water to methanoic acid to methanol, the process is actually endothermic in water, has almost no enthalpy change in methanoic acid and is exothermic in methanol.

12.4 Acid strengths and the role of the solvent

An acid, which we will denote HA, dissociates in water to give H^+ and A^- according to the equilibrium:

$$\text{HA}(aq) \rightleftharpoons \text{H}^+(aq) + \text{A}^-(aq).$$

H^+ exists in water as the solvated hydronium ion, $H_3O^+(aq)$, so the equilibrium is more properly written

$$\text{HA}(aq) + \text{H}_2\text{O}(l) \rightleftharpoons \text{H}_3\text{O}^+(aq) + \text{A}^-(aq). \quad (12.4)$$

This reaction involves breaking a bond (to give ions), which is certainly an endothermic process, but the solvation of the resulting ions is exothermic and so can compensate for energy needed to break the bond. The whole process has many similarities to the case of an ionic solid dissolving in water and so, on the basis of the previous discussion, we can expect that entropy terms will play an important role.

In this section we are going to look at two examples of how we can understand trends in acidity by recognising the important role of the solvent. In the first example we will see how it is possible to rationalize the observation that whereas HCl is a very strong acid, HF is rather weak. In the second example, we will look at the effect that substitution of H by Cl has on the acidity of CH_3COOH; we will find, perhaps rather surprisingly, that the origin of the observed increase in acidity is entirely entropic.

The equilibrium constant for the dissociation of an acid (Eq. 12.4) is called the *acid dissociation constant*, K_a, which is given by:

$$K_a = \frac{[\text{H}_3\text{O}^+][\text{A}^-]}{[\text{HA}]}.$$

As it is the solvent, the fraction of the water present which is actually used to form H_3O^+ is negligible, so the concentration of water is essentially constant. It is therefore not included in the expression for the equilibrium constant, K_a.

The value of K_a can be related to $\Delta_r G^\circ$, and $\Delta_r G^\circ$ is computed from $\Delta_r H^\circ$ and $\Delta_r S^\circ$ in the usual way:

$$\Delta_r G^\circ = -RT \ln K_a$$
$$\Delta_r G^\circ = \Delta_r H^\circ - T\Delta_r S^\circ.$$

The acidity of HCl and HF

We will discuss the acidity of these two acids in water by analysing their dissociation using the following Hess' Law cycle:

$$\begin{array}{ccc} \mathrm{HA}(g) & \xrightarrow{\Delta_r G^\circ(2)} & \mathrm{H}^+(g) + \mathrm{A}^-(g) \\ \Delta_r G^\circ(1) \uparrow & & \downarrow \Delta_r G^\circ(3) \\ \mathrm{HA}(aq) & \xrightarrow{\Delta_r G^\circ(4)} & \mathrm{H}^+(aq) + \mathrm{A}^-(aq) \end{array}$$

The acid is first taken to the gas phase (step 1), where it then dissociates into gas phase ions (step 2) and then finally these ions are hydrated (step 3); for simplicity we have written the dissociation as giving H^+ rather than H_3O^+. Using this cycle we can compute $\Delta_r G^\circ$ for step 4 by knowing the values for steps 1–3: $\Delta_r G^\circ(4) = \Delta_r G^\circ(1) + \Delta_r G^\circ(2) + \Delta_r G^\circ(3)$.

In the case of HCl the values of $\Delta_r G^\circ$ (in kJ mol^{-1}) for each step are:

$$\begin{array}{ccc} \mathrm{HCl}(g) & \xrightarrow{1354} & \mathrm{H}^+(g) + \mathrm{Cl}^-(g) \\ -4 \uparrow & & \downarrow -1392 \\ \mathrm{HCl}(aq) & \xrightarrow{-42} & \mathrm{H}^+(aq) + \mathrm{Cl}^-(aq) \end{array}$$

The first thing that we note is that the dissociation of HCl in water is very favourable ($\Delta_r G^\circ(4) = -42$ kJ mol^{-1}), which is as expected as we know that HCl is a strong acid and so dissociates fully in water. Why this is so can be rationalized by looking at the enthalpy and entropy terms of the separate steps in the cycle. These are shown in the table:

$\Delta_r H^\circ$/ kJ mol^{-1}	$\Delta_r S^\circ$/ J K^{-1} mol^{-1}
$\begin{array}{ccc} \mathrm{HCl}(g) & \xrightarrow{1384} & \mathrm{H}^+(g) + \mathrm{Cl}^-(g) \\ 17 \uparrow & & \downarrow -1459 \\ \mathrm{HCl}(aq) & \xrightarrow{-58} & \mathrm{H}^+(aq) + \mathrm{Cl}^-(aq) \end{array}$	$\begin{array}{ccc} \mathrm{HCl}(g) & \xrightarrow{96} & \mathrm{H}^+(g) + \mathrm{Cl}^-(g) \\ 75 \uparrow & & \downarrow -226 \\ \mathrm{HCl}(aq) & \xrightarrow{-55} & \mathrm{H}^+(aq) + \mathrm{Cl}^-(aq) \end{array}$

First, consider the enthalpy terms: $\Delta_r H^\circ(2)$, the enthalpy of dissociation to ions in the gas phase, is very large and positive. An explanation for this is that dissociating a molecule into ions requires a lot more energy than dissociating into neutral atoms on account of the strong attraction between the ions which must be overcome in order to separate them.

However, $\Delta_r H^\circ(3)$, the enthalpy of solvation of the ions, is very large and negative; we can attribute this to the favourable interactions between the ions and the polar solvent. Overall, this term wins, making $\Delta_r H^\circ(4)$, the enthalpy of dissociation in solution, negative. The relative sizes of the four terms are visualized in Fig. 12.18; what is striking is how $\Delta_r H^\circ(2)$ and $\Delta_r H^\circ(3)$ are almost equal and opposite, leading to a value for $\Delta_r H^\circ(4)$ which is very much smaller than either $\Delta_r H^\circ(2)$ or $\Delta_r H^\circ(3)$.

Of the entropy terms, the most important is the large *negative* value for $\Delta_r S^\circ(3)$, the entropy of solvation of the ions. As we have already noted, the ordering effect that an ion has on the solvent is the origin of this reduction in entropy. The smaller the ion, the greater the effect, and so we should not be surprised to find that the solvation of $H^+(g)$ results in an entropy change which is significantly negative. As we can see, $\Delta_r S^\circ(3)$ is so negative that it dominates the other two terms and makes $\Delta_r S^\circ(4)$ negative; this is visualized in Fig. 12.18.

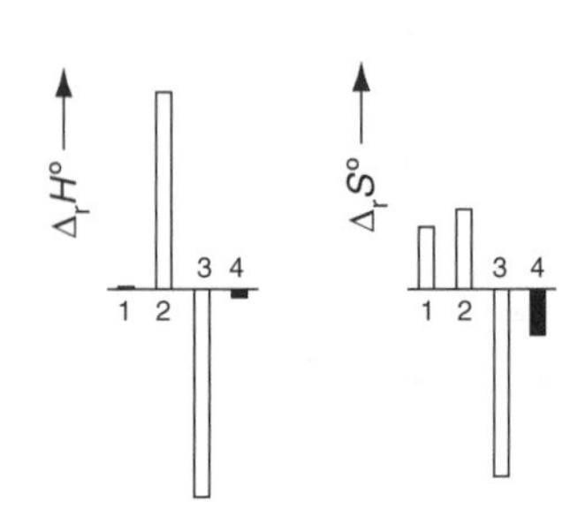

Fig. 12.18 Visualization of the relative sizes of the $\Delta_r H^\circ$ (on the left) and $\Delta_r S^\circ$ (on the right) values of the steps involved in the dissociation of HCl in water. The numbering on the columns refers to the steps in the table above. $\Delta_r H^\circ(1)$ and $\Delta_r H^\circ(4)$ are all but invisible on this scale; this highlights how the small value of $\Delta_r H^\circ(4)$ is determined by the difference in the two large terms, $\Delta_r H^\circ(2)$ and $\Delta_r H^\circ(3)$.

The dissociation of HCl in water has a (negative) favourable $\Delta_r H^\circ$ and a (negative) unfavourable $\Delta_r S^\circ$. From these values we can compute $\Delta_r G^\circ$ (at 298 K) as:

$$\begin{aligned}\Delta_r G^\circ &= \Delta_r H^\circ - T\Delta_r S^\circ \\ &= -58 - 298 \times (-55 \times 10^{-3}) \\ &= -58 + 16 \\ &= -42 \text{ kJ mol}^{-1}.\end{aligned}$$

Overall the $\Delta_r H^\circ$ term dominates, making $\Delta_r G^\circ$ negative, but note that there is rather a fine balance between $\Delta_r H^\circ$ and the $-T\Delta_r S^\circ$ term – it would not take much of a change in either to make $\Delta_r G^\circ$ positive.

It is perhaps surprising at first that HF turns out to be rather a weak acid (only 7% of the HF molecules are dissociated in a 0.1 mol dm^{-3} solution). However, we can rationalize why this is by using a cycle in the same way as we did for HCl and comparing the $\Delta_r H^\circ$ and $\Delta_r S^\circ$ values:

$\Delta_r H^\circ$/kJ mol^{-1}			$\Delta_r S^\circ$/J K^{-1} mol^{-1}		
HF(g)	$\xrightarrow{1534}$	$H^+(g) + F^-(g)$	HF(g)	$\xrightarrow{96}$	$H^+(g) + F^-(g)$
50 ↑		↓ −1593	96 ↑		↓ −263
HF(aq)	$\xrightarrow{-9}$	$H^+(aq) + F^-(aq)$	HF(aq)	$\xrightarrow{-71}$	$H^+(aq) + F^-(aq)$

Comparing these values to the ones for HCl we see that $\Delta_r H^\circ(2)$ is larger for HF than HCl which just tells us that the HF bond is stronger than the HCl bond. $\Delta_r H^\circ(3)$ is more negative for HF than HCl, and this can be rationalized by noting that F^- is smaller than Cl^-; as is illustrated in Fig. 12.12 on p. 195, the smaller an ion, the more negative the enthalpy of hydration. The bond strength increases by a little more than the amount by which the solvation energy becomes more negative and as a result $\Delta_r H^\circ(4)$, although still negative, is much less so than for HCl.

When it comes to the entropy values we have a larger negative entropy of solvation ($\Delta_r S^\circ(3)$) for $H^+ + F^-$ than for $H^+ + Cl^-$; again, we can attribute this to the smaller size of the F^- ion. $\Delta_r S^\circ(4)$ is thus more negative for HF than for HCl.

The value of $\Delta_r G^\circ$ at 298 K now turns out to be positive:

$$\begin{aligned}\Delta_r G^\circ &= \Delta_r H^\circ - T\Delta_r S^\circ \\ &= -9 - 298 \times (-71 \times 10^{-3}) \\ &= -9 + 21 \\ &= +12 \text{ kJ mol}^{-1}.\end{aligned}$$

In contrast to the case of HCl, the $-T\Delta_r S^\circ$ term now dominates, making $\Delta_r G^\circ$ positive and so HF is a weak acid.

The comparison with HCl shows how relatively small changes in the values of $\Delta_r H^\circ$ or $\Delta_r S^\circ$ for the individual steps can swing us from a strong acid (HCl) to a weak acid (HF). It also shows the importance of the entropy terms which in the end are the deciding factors in making HF a weak acid.

Chloroethanoic acids

Ethanoic acid itself is rather a weak acid, but as we substitute chlorine for the hydrogens in the methyl group the acid strength increases to the point where trichloroethanoic acid is classified as strong. It is interesting to see how this trend can be rationalized with the aid of $\Delta_r H^\circ$ and $\Delta_r S^\circ$ values.

The equilibrium in question is:

$$CH_{3-n}Cl_nCOOH(aq) + H_2O(l) \rightleftharpoons CH_{3-n}Cl_nCOO^-(aq) + H_3O^+(aq)$$

where n can be 0, ..., 3. We define the acid dissociation constant, K_a, in the usual way as

$$K_a = \frac{[CH_{3-n}Cl_nCOO^-][H_3O^+]}{[CH_{3-n}Cl_nCOOH]}.$$

When discussing acid strengths it is usual not to give the value of K_a but the quantity pK_a defined as

$$pK_a = -\log K_a.$$

Note that the logarithm is to the base 10, as opposed to the natural logarithm (ln). With this definition a weak acid with a value of K_a which is less than 1 will have a positive value for the pK_a. For example, K_a for ethanoic acid (at 298 K) is 1.7×10^{-5} which is $10^{-4.8}$, so the pK_a is +4.8.

On the other hand, strong acids which have large values of K_a have smaller pK_a values, and indeed for the very strongest acids the values are negative. For example, K_a for HCl (at 298 K) is 2.3×10^6 which is $10^{+6.4}$ so the pK_a is −6.4. The logarithmic scale compresses a wide range of K_a values into a much smaller scale.

The table gives thermodynamic data for the dissociation of chloroethanoic acids, all at 298 K.

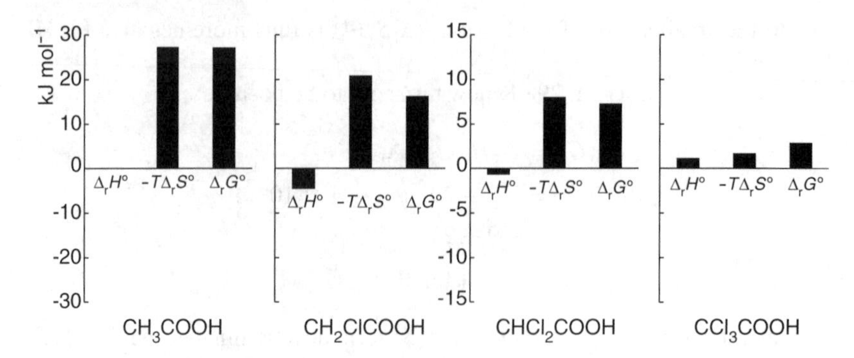

Fig. 12.19 Diagrammatic representation of the relative contributions that the $\Delta_r H^\circ$ and $-T\Delta_r S^\circ$ terms make to $\Delta_r G^\circ$ for the dissociation of a series of chloroethanoic acids at 298 K. Note that the vertical scale has been expanded two-fold for the di- and trichloro acids. The diagram clearly indicates that the value of $\Delta_r G^\circ$ is dominated by the $-T\Delta_r S^\circ$ term.

	CH_3COOH	$CH_2ClCOOH$	$CHCl_2COOH$	CCl_3COOH
$\Delta_r H^\circ$ / kJ mol^{-1}	-0.08	-4.6	-0.7	$+1.2$
$\Delta_r S^\circ$ / J K^{-1} mol^{-1}	-91.6	-70.2	-27.0	-5.8
$-T\Delta_r S^\circ$ / kJ mol^{-1}	$+27.3$	$+20.9$	$+8.1$	$+1.7$
$\Delta_r G^\circ$ / kJ mol^{-1}	$+27.2$	$+16.3$	$+7.3$	$+2.9$
pK_a	4.76	2.86	1.28	0.51

In addition to giving the values for $\Delta_r H^\circ$ and $\Delta_r S^\circ$ we have also given the value of $-T\Delta_r S^\circ$; the reason for this is that as $\Delta_r G^\circ = \Delta_r H^\circ - T\Delta_r S^\circ$ the two terms which contribute directly to $\Delta_r G^\circ$ are $\Delta_r H^\circ$ and $-T\Delta_r S^\circ$.

The first thing we notice is that adding each successive chlorine increases the strength of the acid, as evidenced by the *decreasing* pK_a values; indeed the effect is rather dramatic when we recall that the pK_a values are on a logarithmic scale – the K_a values in fact change by four orders of magnitude.

A simple explanation for what is happening is to say the electron-withdrawing chlorine atoms are polarizing the O–H bond in the carboxyl group, thus making it easier to dissociate. We could also argue that the electronegative chlorine atoms are helping to stabilize the anion formed on dissociation. Tempting though both of these arguments are, the data in the table reveal that they are entirely incorrect.

The crucial observation to make on the data is that for all but the trichloroethanoic acid the $\Delta_r H^\circ$ terms are *negligible* compared to the $-T\Delta_r S^\circ$ terms; the variation in pK_a values is therefore entirely an *entropy effect*, and has nothing to do with the strength or otherwise of the O–H bond. Figure 12.19 illustrates in a graphical way the contributions which the $\Delta_r H^\circ$ and $-T\Delta_r S^\circ$ terms make for each acid; it is clear that the enthalpy terms play an insignificant role.

Our explanation for the increase in acidity must therefore centre on the variation in the entropy terms. The first thing to note is that $\Delta_r S^\circ$ is negative; we have seen this before for the dissociation of HCl and HF. As we noted there, the origin of this reduction in entropy is the ordering effect which the ions have on the water.

The $\Delta_r S^\circ$ values become *less* negative as the methyl hydrogens are successively replaced by chlorine; we can argue that the electron-withdrawing chlorine atoms are leading to a spreading out of the negative charge (formally residing on the oxygen) across the whole molecule. The charge is therefore less concentrated and so the solvent water molecules are less tightly held, leading to a smaller reduction in the entropy. Figure 12.20 illustrates the idea.

The case of trichloroethanoic acid is perhaps a little different from the others, as for this molecule the $\Delta_r H^\circ$ and $-T\Delta_r S^\circ$ values are comparable so that we cannot say that the entropy term is dominant. However, it is still clear that the high acidity of this species compared to all of the others is attributable to the fact that it has the least negative $\Delta_r S^\circ$ term. In fact, the $\Delta_r H^\circ$ for this acid is the only one of the series which is positive (i.e. unfavourable), but despite this the trichloro derivative is the most acidic; the less unfavourable entropy term is clearly dominating over the enthalpy term.

This example illustrates two things. The first is that we really need to look at the enthalpy and entropy terms to see what is going on in a series of equilibria – just looking at the equilibrium constants can be rather misleading. The second is that when ions are involved in an equilibrium the contribution that the entropy of solvation makes is often decisive.

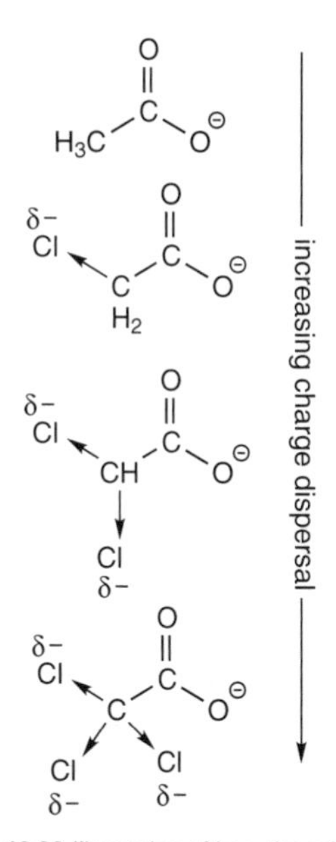

Fig. 12.20 Illustration of how the addition of successive electron-withdrawing chlorine atoms leads to a dispersal of negative charge for the carboxylate anion; as a result the solvating water molecules are less tightly held and so the $\Delta_r S^\circ$ values for dissociation become less negative.

12.5 How the solvent affects the rate of reaction

We are now in a position to understand the effects of changing the solvent on the rates of the reactions which we discussed at the start of the chapter.

For the reaction of *t*-butyl chloride, the first step is the formation of the trimethylcarbenium and chloride ions. These ions are clearly far more polar than the starting material and hence are preferentially stabilized by polar solvents. The energy profiles shown in Fig. 12.21 illustrate this point: the solid line shows a profile for the reaction in a non-polar solvent and the dotted line for the reaction in a polar solvent.

The intermediate ions are stabilized by the more polar solvent and so are lower in energy. More importantly, the transition state for the reaction in which the carbenium ion is formed is also lower in energy since it too is more polar than the starting material. The starting material, *t*-butyl chloride, is not charged and so is little affected by changing the polarity of the solvent.

The net effect is that the activation energy for the reaction is *decreased* on moving to the *more polar* solvent and hence the rate of the reaction increases. This is exactly what is observed, as was discussed on p. 189.

In contrast, the reaction between CH_3Br and Cl^- (Fig. 12.2 on p. 190) goes faster on moving from water to acetone and is fastest when no solvent is present. Since this reaction proceeds via the S_N2 mechanism, the rate-determining step is the reaction between the bromoalkane and the incoming chloride ion.

The key to understanding the difference in rate is the effect of the solvent on the chloride anion. In a protic solvent like water or methanol, this anion is strongly solvated by hydrogen bonding whereas in an aprotic solvent like acetone, it is less strongly solvated (see Section 12.2 on p. 194). In the S_N2 mechanism, the nucleophile must approach the bromomethane in a specific geometry (see p. 109) – the hydrogen bonded solvent hinders this approach and hence the reaction rate increases on moving from the protic solvent water to the aprotic solvent acetone.

An energy profile for this reaction is shown in Fig. 12.22. In a protic solvent, the negatively charged nucleophile is strongly solvated through hydrogen bonding. Thus the energy of the starting materials (chloride ion and bromomethane) are lower in the protic solvent than in the aprotic solvent.

In the transition state, a bond is beginning to form between the chlorine and carbon and the bond between the bromine and the carbon is beginning to break. The negative charge is spread over the molecule (mainly on the chlorine and bromine) rather than being concentrated on just one ion as it is in the starting materials or products. In other words, the charge is becoming dispersed. As a result, the solvent is less firmly associated with the transition state than with the incoming nucleophile before the attack.

The important point is that while the chloride ion and the transition state are all lowered in energy due to solvation by the protic solvent, the transition state is lowered less than the chloride ion due to the charge being more dispersed in the transition state. This means that, as shown in Fig. 12.22, the activation energy is greater in the protic solvent than in the aprotic solvent. Thus, the reaction proceeds more quickly in the aprotic solvent, which is exactly what is found experimentally.

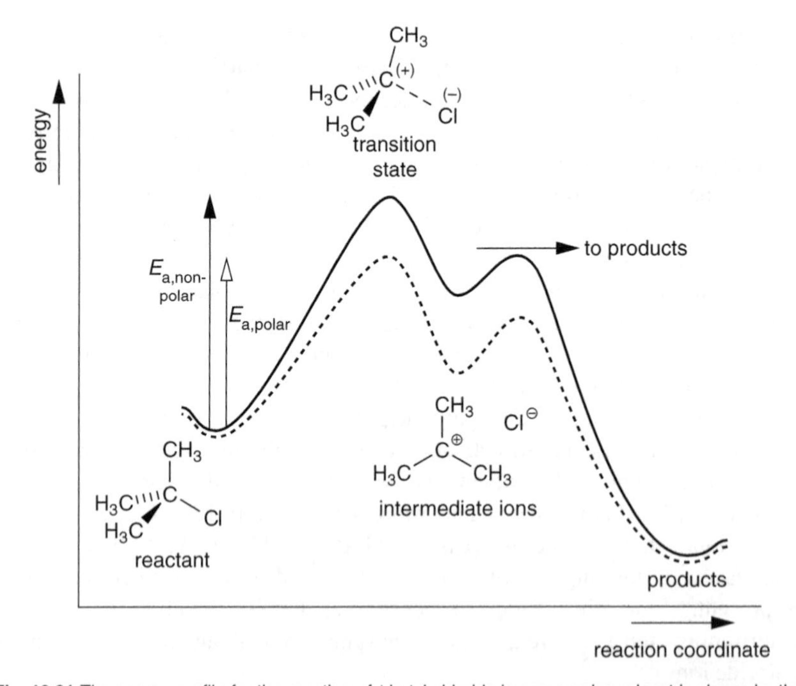

Fig. 12.21 The energy profile for the reaction of *t*-butyl chloride in a non-polar solvent is shown by the solid line; the dotted line shows the profile in a more polar solvent. Whilst the change in solvent has little effect on the energy of the uncharged starting material, the charged intermediates are lower in energy in the more polar solvent. As charges are building up in the transition state for the first reaction (which leads to the formation of the ions), this too is lowered in energy in the more polar solvent. The net effect is that the activation energy for the reaction in the polar solvent is less than in the non-polar solvent and hence the reaction rate is greater in the polar solvent.

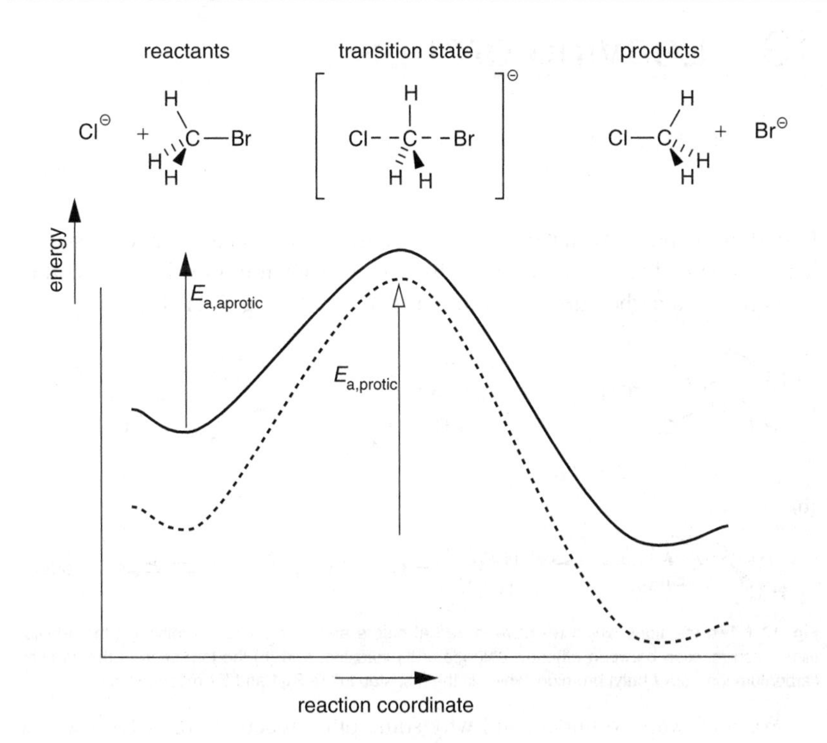

Fig. 12.22 The energy profile for the reaction between chloride ion and bromomethane in a polar aprotic solvent is shown by the solid line and for the same reaction in a protic solvent by the dotted line. In the protic solvent the free chloride ion is more strongly solvated than the transition state due to the charge dispersal in the latter. The result is that on moving to a protic solvent the energy of the transition state is lowered by less than is the energy of the chloride ion. Therefore the activation energy for the substitution reaction is greater in the protic solvent than in the aprotic solvent.

13 Leaving groups

In earlier chapters we have looked at a number of simple reactions, such as that between ethanoyl chloride and hydroxide, shown in Fig. 13.1 (a), and the formation of a carbenium ion from *t*-butyl bromide, Fig. 13.1 (b).

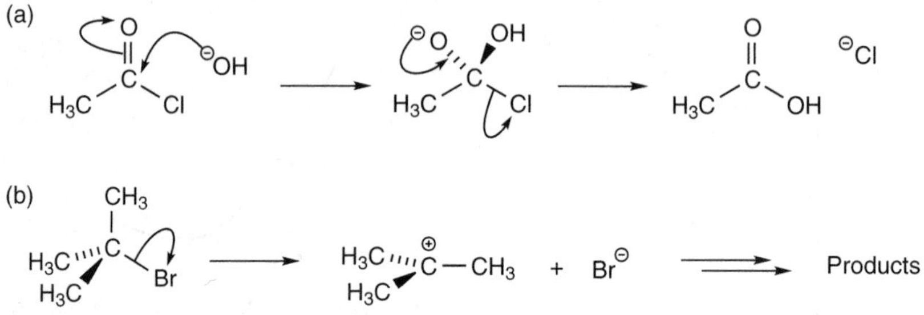

(a) H_3C–C(=O)–Cl + $^{\ominus}OH$ ⟶ H_3C–C($O^{\ominus}$)(OH)–Cl ⟶ H_3C–C(=O)–OH + $^{\ominus}Cl$

(b) $(H_3C)_3C$–Br ⟶ $(H_3C)_2\overset{\oplus}{C}$–$CH_3$ + $Br^{\ominus}$ ⟶ Products

Fig. 13.1 Two reactions which we have looked at before and which occur readily: (a) the addition-elimination reaction between ethanoyl chloride and hydroxide; and (b) the formation of the trimethyl-carbenium ion from *t*-butyl bromide, which is the first step in the S_N1 and E1 mechanisms.

We now want to understand why some other reactions for which we can write plausible looking mechanisms do *not* occur. For example, why does Cl^- not attack ethanoic acid to form ethanoyl chloride, as shown in Fig. 13.2 (a)? This reaction is simply the reverse of that shown in Fig. 13.1 (a), but it does not take place. Similarly, the attack on a ketone by Cl^- to give an acyl chloride, as shown in Fig. 13.2 (b), does not take place, neither does the formation of a carbenium ion by loss of H^-, shown in (c). We know that the *same* carbenium ion is formed readily by loss of Br^- from *t*-butyl bromide, so why is it that this ion cannot be formed by loss of H^- from a hydrocarbon?

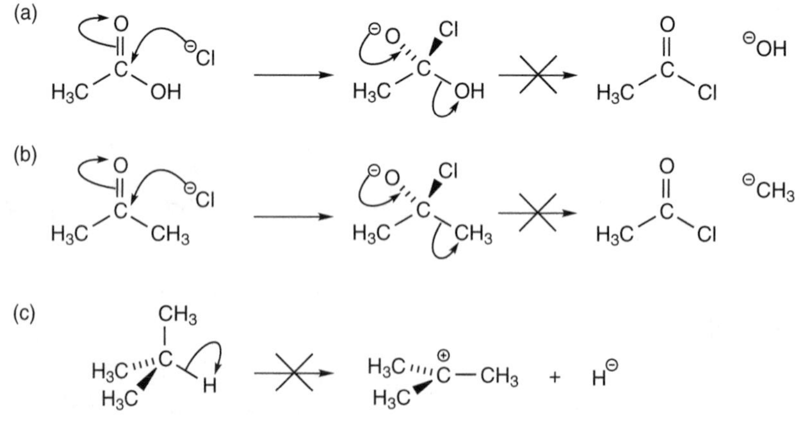

(a) H_3C–C(=O)–OH + $^{\ominus}Cl$ ⟶ H_3C–C($O^{\ominus}$)(Cl)–OH ⟶✕ H_3C–C(=O)–Cl + $^{\ominus}OH$

(b) H_3C–C(=O)–CH_3 + $^{\ominus}Cl$ ⟶ H_3C–C($O^{\ominus}$)(Cl)–CH_3 ⟶✕ H_3C–C(=O)–Cl + $^{\ominus}CH_3$

(c) $(H_3C)_3C$–H ⟶✕ $(H_3C)_2\overset{\oplus}{C}$–$CH_3$ + $H^{\ominus}$

Fig. 13.2 Three reactions which do not occur: (a) the formation of ethanoyl chloride from ethanoic acid and chloride ion; (b) the formation of ethanoyl chloride from propanone and chloride ion and (c) the formation of trimethylcarbenium ion from 2-methylpropane.

We will see that we can answer these questions simply by considering the relative energies of the species involved. This will lead to the development of a general set of ideas for predicting which reactions will go readily and which will not.

13.1 Energy profiles and leaving groups

An energy profile diagram for the reaction between ethanoyl chloride and hydroxide is shown in Fig. 13.3. Since we know that the products are favoured at equilibrium, we have drawn the products (chloride ion and ethanoic acid) lower in energy than the reactants (ethanoyl chloride and hydroxide ion) by an amount labelled ΔE .

To a rough approximation, the only difference between the ethanoyl chloride and the ethanoic acid is that the C–Cl bond is replaced by a C–OH bond. At 450 kJ mol^{-1}, the C–OH bond in the carboxylic acid is stronger than the C–Cl bond in the acyl chloride (350 kJ mol^{-1}) and so this change contributes to ΔE, the lowering of energy as the reaction proceeds to the products. However, the relative energies of the OH^- and Cl^- ions will also contribute to the value of ΔE as these too are reactants and products.

The key species on the energy profile is the tetrahedral intermediate. It is perhaps not too surprising that it is higher in energy than both the reactants and the products as the intermediate has four groups crowded around the carbon whereas the reactants and products have just three. Once formed, the intermediate can either lose Cl^- to form the products or lose OH^- to re-form the reactants. Since it is harder to break a C–O bond than to break a C–Cl bond,

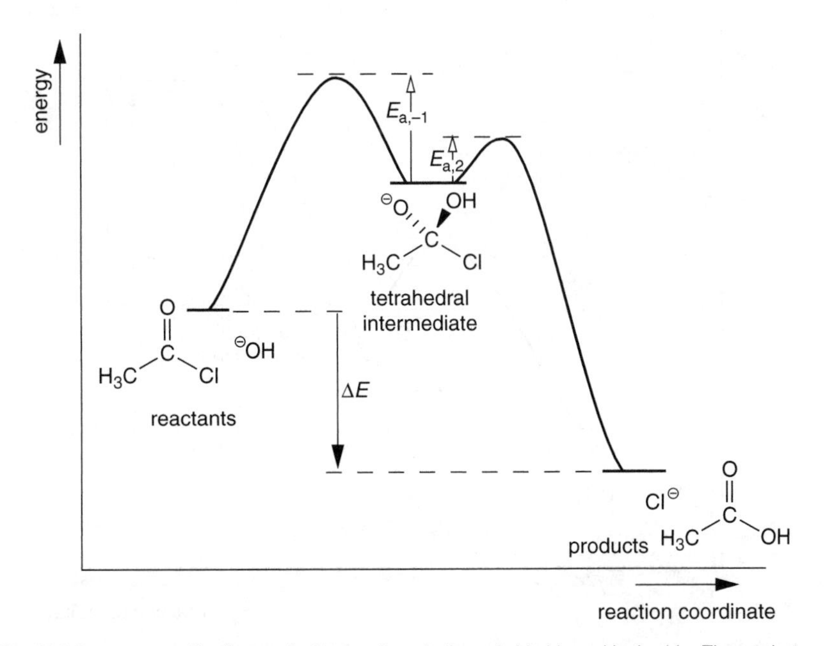

Fig. 13.3 An energy profile diagram for the reaction of ethanoyl chloride and hydroxide. The products are lower in energy than the reactants by an amount labelled ΔE.

it is easier for the intermediate to lose the Cl^- and form the products. Consequently, we have shown the activation energy for the tetrahedral intermediate to from the products, ($E_{a,2}$ in Fig. 13.3) as being lower less than the activation energy required for the tetrahedral intermediate to return to the reactants ($E_{a,-1}$).

In these two possible reactions of the tetrahedral intermediate, Cl^- and OH^- are called *leaving groups*. The fact that loss of Cl^- occurs more readily than loss of OH^- leads us to describe Cl^- as a *better leaving group* than OH^-.

We can now answer the question as to why Cl^- does not attack ethanoic acid to form ethanoyl chloride. If the Cl^- were to attack, as in Fig. 13.2 (a), the tetrahedral intermediate so formed would be the same as that in Fig. 13.3. As we have seen, for this intermediate the Cl^- is far more likely to leave than the OH^-, and this loss of Cl^- will simply regenerate the ethanoic acid. No ethanoyl chloride is produced as the intermediate preferentially collapses back to the starting materials.

A similar line of argument can be used to explain why Cl^- does not react with propanone to form ethanoyl chloride and CH_3^-, as shown in Fig. 13.2 (b). The products from this hypothetical reaction are *much* higher in energy than the reactants, as shown in the energy profile diagram in Fig. 13.4. We have also shown the products being even higher in energy than the tetrahedral intermediate.

The reason the products are so high in energy is mainly due to the formation of the CH_3^- ion; it is *much* harder to break a C–C bond and liberate CH_3^- than it is to break a C–Cl bond to form Cl^-. We describe the Cl^- ion as being more stable, meaning lower in energy, than the CH_3^- ion. This is because the

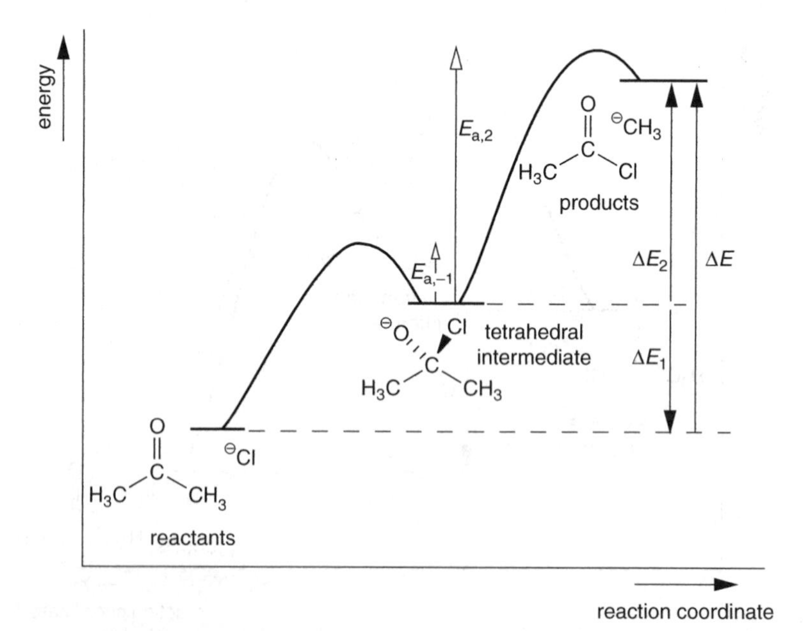

Fig. 13.4 Energy profile diagram for the reaction between Cl^- and propanone to form ethanoyl chloride and CH_3^-; the reaction is shown in Fig. 13.2 (b).

electrons in the Cl^- ion experience a greater effective nuclear charge than those on carbon and hence are lower in energy (see Section 4.8 on p. 57).

Even if the chloride ion did attack propanone to form the tetrahedral intermediate (which is entirely possible), this species would eliminate Cl^- and return to the starting materials in preference to eliminating CH_3^- to form ethanoyl chloride. We say that Cl^- is a much better leaving group than CH_3^-.

Whilst the forward reaction shown in Fig. 13.4 does not take place, the reverse reaction can. If care is taken to prevent further reactions, ketones can be prepared from acyl chlorides by reaction with organometallic compounds, which essentially act as a source of the alkyl anion – see Fig. 13.5. Once the alkyl anion has attacked the acyl chloride to form the tetrahedral intermediate, the loss of chloride ion inevitably follows yielding the ketone. The driving force for this reaction is essentially the high-energy organometallic starting material – both the intermediate and products are lower in energy than this.

(+) (–)
M—R

Fig. 13.5 Organometallic compounds are strongly polarized with the metal positive and the alkyl group (such as a methyl group) negative. They essentially behave as a source of the R^- anion. Organocadmium compounds such as $(CH_3)_2Cd$ are particularly good for forming ketones from acyl chlorides.

Finally we turn to the question as to why the trimethylcarbenium ion can form from *t*-butylbromide, as shown in Fig. 13.1 (b), but not from 2-methylpropane, as shown in Fig. 13.2 (c).

Since the same carbenium ion is being produced in each reaction, the difference must be because of the anion formed. We can therefore say that hydride ion, H^-, is a *much worse leaving group* than Br^-. The reasons are twofold: firstly, it is harder to break a C–H bond than a C–Br bond and secondly, the greater effective nuclear charge on Br means that its valence orbitals are lower in energy than those of hydrogen. This is often expressed by saying that Br^- is 'more stable' than the H^- – but we have to remember to be careful about what we mean by 'more stable'. As we have seen in the previous chapter, the stability of an anion is often strongly influenced by its interactions with the solvent.

13.2 Leaving group ability

In the previous section we saw that Cl^- is a better leaving group than both OH^- and CH_3^-, and that Br^- is a better leaving group than H^-. From the study of many reactions, a general order for *leaving group ability* has been established; Fig. 13.6 lists some common leaving groups in order of the ease with which they can leave a compound.

How good something is as a leaving group depends on two things: (a) the bond strength between the leaving group and the atom with which it was initially bonded (usually carbon) and (b) the 'stability' of the leaving group itself once it has been formed.

The very best leaving group is neutral nitrogen, N_2. This is a very stable molecule – once it has been eliminated, there is very little chance that it will add back again. Further, when bonded in a molecule the N_2 group bears a positive charge which draws electrons towards it; this results in a very weak C–N bond which may easily be broken, further enhancing the leaving group ability.

The halide ions appear in the list in the order $I^- > Br^- > Cl^- > F^-$: of them iodide is the best leaving group and fluoride the worst. This order reflects the halogen–carbon bond strengths as exemplified by the series CH_3I,

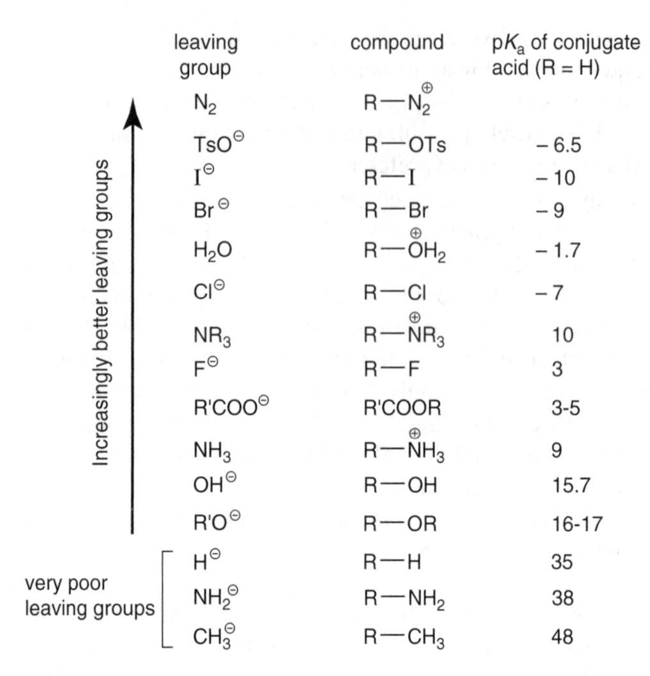

	leaving group	compound	pK_a of conjugate acid (R = H)
Increasingly better leaving groups ↑	N_2	$R-\overset{\oplus}{N_2}$	
	$TsO^{\ominus}$	R—OTs	– 6.5
	$I^{\ominus}$	R—I	– 10
	$Br^{\ominus}$	R—Br	– 9
	H_2O	$R-\overset{\oplus}{O}H_2$	– 1.7
	$Cl^{\ominus}$	R—Cl	– 7
	NR_3	$R-\overset{\oplus}{N}R_3$	10
	$F^{\ominus}$	R—F	3
	$R'COO^{\ominus}$	R'COOR	3-5
	NH_3	$R-\overset{\oplus}{N}H_3$	9
	$OH^{\ominus}$	R—OH	15.7
	$R'O^{\ominus}$	R—OR	16-17
very poor leaving groups	$H^{\ominus}$	R—H	35
	$NH_2^{\ominus}$	$R-NH_2$	38
	$CH_3^{\ominus}$	$R-CH_3$	48

Fig. 13.6 The left-hand column lists some ions or molecules in order of their ability to act as a leaving group from an alkyl carbon or a carbonyl group. The middle column shows how each leaving group occurs in a compound; R represents any alkyl group. TsO^- is the tosylate group, shown in Fig. 13.7. The right-hand column gives the approximate pK_a value of the conjugate acid of the leaving group, which is the acid formed when R=H.

CH_3Br, CH_3Cl and CH_3F in which the bond strengths are 237, 293, 352 and 472 kJ mol^{-1} respectively.

The order $F^- > OH^- > NH_2^- > CH_3^-$ reflects the stability of the anions themselves. The elements formally bearing the negative charge are all in the same row in the Periodic Table and so the effective nuclear charge experienced by the valence electrons in these atoms increases in the order F > O > N > C. Consequently the lowest energy orbitals are in F^- and the highest in CH_3^-, which means that F^- is lower in energy (i.e. more stable) than O^- and so on for the others.

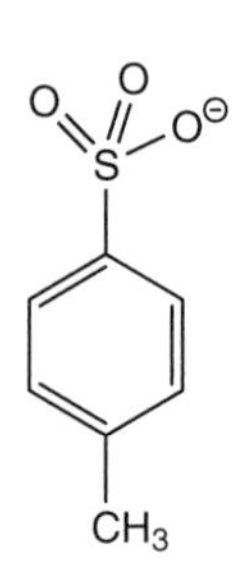

Fig. 13.7 The structure of the *tosylate* group, abbreviated to TsO^-.

The order $TsO^- > RCOO^- > RO^-$ also reflects the stability of the anion which is related to the degree of delocalization that is possible in these ions (see Section 10.4 beginning on p. 168). The negative charge in the tosylate ion can be delocalized over three oxygen atoms, which confers greater stability when compared to the carboxylate anion in which the charge can only be delocalized over two oxygen atoms. Both of these ions are lower in energy than an alkoxide, RO^- (e.g. $CH_3CH_2O^-$), in which the charge essentially remains on the single oxygen atom.

From the list we see that it is always harder for an anion to act as a leaving group than the protonated form of the same group. For example, OH^- is a much poorer leaving group than neutral H_2O, and NH_2^- is a much poorer leaving group than neutral NH_3. Thus one way to transform an –OH group into a much better leaving group is to protonate it to form $-OH_2^+$. Now, rather than the anion OH^- leaving, it is the neutral species H_2O which leaves.

13.3 Leaving groups and pK_a

How well an ion (X^-) or molecule (Y) can act as a leaving group is to some extent reflected by the acidity of HX or HY^+. HX is called the *conjugate acid* of X^-, similarly HY^+ is the conjugate acid of Y; these conjugate acids are formed simply by adding H^+ to the relevant leaving group.

The connection between acidity of the conjugate acid and leaving group ability comes about because similar factors affect both. Acidity is determined by the H–X bond strength and the stability of X^-. As we have seen, the same factors determine how good X^- is as a leaving group, although we should note that for a leaving group it is the C–X, rather than H–X, bond strength which is relevant.

As we saw on p. 205, we usually quantify the strength of an acid using pK_a values. A negative pK_a corresponds to an acid which readily dissociates in water, whereas a positive pK_a corresponds to a weak acid which only partly dissociates, if at all.

The pK_a values for the conjugate acids of the leaving groups are given in Fig. 13.6. It can be seen that the conjugate acids of good leaving groups, such as TsO^-, I^- and H_2O, have negative pK_a values whereas the conjugate acids of groups which do not ordinarily act as leaving groups, such as H^-, NH_2^- and CH_3^-, have large positive pK_a values.

The same trends that we see for leaving group ability are found in the strengths of the conjugate acids. For example, the acidity of the hydrogen halides increases along the series HF, HCl, HBr, HI – which mirrors exactly the trend in leaving group ability.

The acidity of hydrides increases as we move across the Periodic Table: CH_4 is not at all acidic and hence has an extremely high pK_a (around 48), NH_3 (pK_a = 38) can only be deprotonated by very strong bases, H_2O is a very weak acid (pK_a = 15.7) and HF is acidic, but still only weakly so (pK_a = 3). As we saw for leaving groups, this order reflects the stability of the ions formed on dissociation of the acid, with F^- being the most stable due to the large effective nuclear charge on fluorine, and CH_3^- being the least stable as a result of the smaller effective nuclear charge of carbon.

Whilst we see the same general trends in both leaving group ability and acid strength, the order of leaving group ability from carbon does not match exactly the order of the acidity of the conjugate acids (see Fig. 13.6). For example, H_2O is a better leaving group than Cl^-, yet HCl is a stronger acid than H_3O^+. Such discrepancies are due to the fact that how good something is as a leaving group depends on its bond strength to *carbon*, whereas how strong its conjugate acid is depends on its bond strength to *hydrogen*; the two are not necessarily the same.

Leaving group ability is a good guide to working out which reactions will go and which will not, but often it is not as clear cut as the examples given here. This brings us to the last topic which is the more complex one of what happens when there are *feasible* alternative reactions open to a given set of reactants.

14 Competing reactions

In this final chapter we will look at what happens when there are a number of possible outcomes for a given set of reactants i.e. when there are alternative pathways available. To understand and rationalize which of these pathways is the dominant one and – more importantly – to understand how to alter the balance between different pathways, will need us to bring together *all* the concepts and ideas we have been developing in the previous chapters.

14.1 Reactions of carbonyls with hydroxide

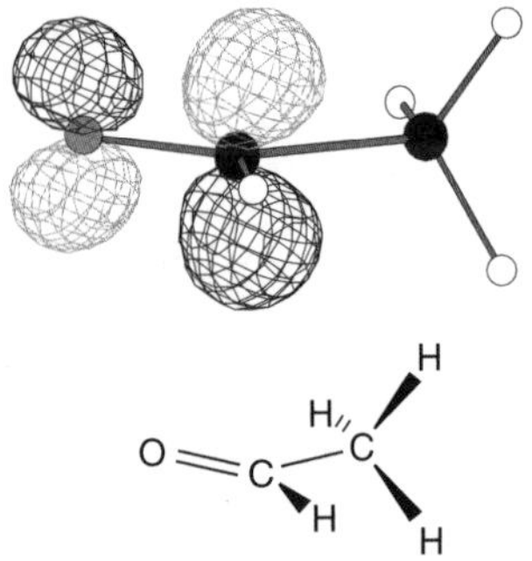

Fig. 14.1 The LUMO of ethanal is the $\pi^\star$ MO which has a greater contribution from the carbon $2p$ AO than from the oxygen $2p$.

We will start by thinking about the initial reaction between hydroxide ion and an aldehyde or ketone. To keep the structures and orbitals as simple as possible, we shall look at the reaction between hydroxide and ethanal (acetaldehyde). As we saw in Section 7.3 on p. 102, the hydroxide ion could act as a nucleophile and attack the C=O $\pi^\star$ MO to form a tetrahedral intermediate. The $\pi^\star$ MO is shown in Fig. 14.1. As we have seen before, this MO is essentially just the out-of-phase interaction between the carbon $2p$ and oxygen $2p$ AOs; the MO has a greater contribution from carbon than from oxygen.

The hydroxide attacks at the carbonyl carbon for two reasons: (i) due to the effect of the electron-withdrawing oxygen, the carbonyl carbon is slightly positively charged; and (ii) the $\pi^\star$ LUMO has a greater contribution from the carbon AO than from the oxygen. The curly arrow description of the attack by OH^- is shown in Fig. 14.2 (a).

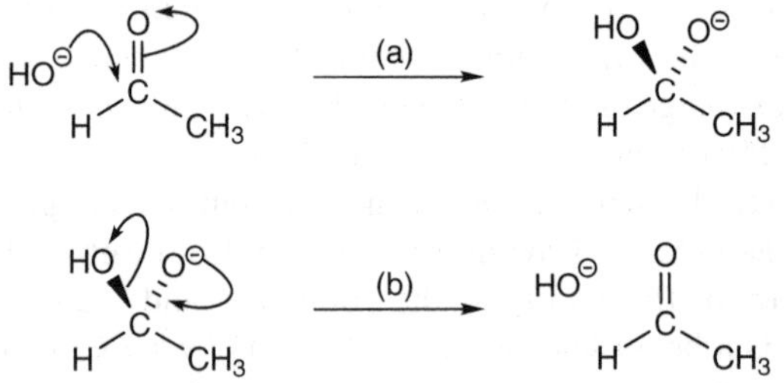

Fig. 14.2 In (a) hydroxide attacks the carbonyl carbon, simultaneously breaking the C=O π bond to form a tetrahedral intermediate. This intermediate breaks down, as shown in (b), by re-forming the π bond and eliminating the only realistic leaving group, which is the hydroxide ion which attacked to form the tetrahedral intermediate in the first place.

Once formed, the only thing the tetrahedral intermediate can do is re-form the ketone by eliminating the only realistic leaving group – the hydroxide ion that just attacked; CH_3^- and H^- are simply not good enough leaving groups to compete with the OH^-. The curly arrow description of the collapse of the tetrahedral intermediate is shown in Fig. 14.2 (b).

An equilibrium will be established with both the intermediate and the aldehyde present. For an aldehyde such as ethanal, significant amounts of the intermediate will be present, but for a ketone, the equilibrium lies very much in favour of the carbonyl.

Formation of enolates

The other way the hydroxide could react with the carbonyl compound is as a base; deprotonation leads to an *enolate anion* of the type shown in Fig. 10.34 on p. 170. The enolate anion formed from ethanal is shown in Fig. 14.3.

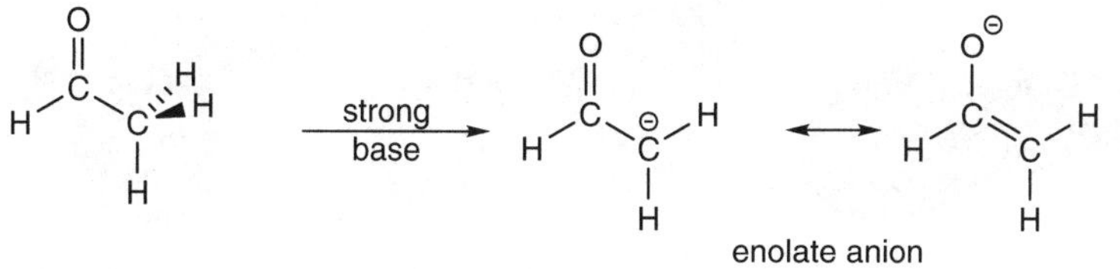

Fig. 14.3 A strong base may remove one of the hydrogens from the methyl group in ethanal to form an enolate anion. This can be drawn with the charge on either carbon or oxygen; the best representation is probably that with the charge on the more electronegative oxygen.

We shall now look more closely to see *why* hydroxide can do this. What we will find is that the C–H bonds on the carbon adjacent to the carbonyl are weakened as a result of interaction between their orbitals and the π system. This, combined with the fact that the negative charge on the enolate can be delocalized on to oxygen, results in these hydrogens being significantly more acidic than in simple alkanes.

We saw on p. 160 that in an amide there is an interaction between the nitrogen $2p$ orbital and the C=O π system which leads to the formation of π MOs which are delocalized over the carbon, nitrogen and oxygen. The 'lone pair' on nitrogen is not, therefore, localized on that atom; rather, it participates in a delocalized system (see Fig. 10.18 on p. 161).

In an analogous fashion, whilst we might like to think of the C–H and C=O π bonds in ethanal as being separate, the computed MOs show there is some degree of interaction between the C–H σ bonding orbitals of the methyl group and the π system. Figure 14.4 (a)(i) shows a surface plot of a π MO which contributes to the bonding between the carbon and hydrogens in the CH_3 group in ethanal. The MO may be thought of as arising from an interaction between the methyl carbon $2p$ AO and the $1s$ AOs of the hydrogens above and below the plane of the σ-framework; this interaction is illustrated in Fig. 14.5.

Figure 14.4 (a)(ii) is a surface plot of the *same* MO but at a lower value of the wavefunction than in (a)(i). Whilst (a)(i) shows that this MO has most electron density on the methyl carbon and the attached hydrogens, (a)(ii) shows that electron density is also, to a lesser extent, delocalized over the π system.

Similarly, Fig. 14.4 (b)(i) shows the C=O π bonding MO resulting from the in-phase combination of the C and O $2p$ AOs. As usual, there is a greater contribution to this MO from the oxygen AO than from the carbon AO. However, there is also a small contribution from the C–H σ bonding MOs on the methyl carbon, as shown in Fig. 14.4 (b)(ii) which is plotted at a lower value of the wavefunction.

The $\pi^\star$ anti-bonding MO, shown in Fig. 14.4 (c)(i), is made up largely from the out-of-phase combination of the C and O $2p$ AOs, with a greater contribution from the carbon than the oxygen. As can be seen from (c)(ii),

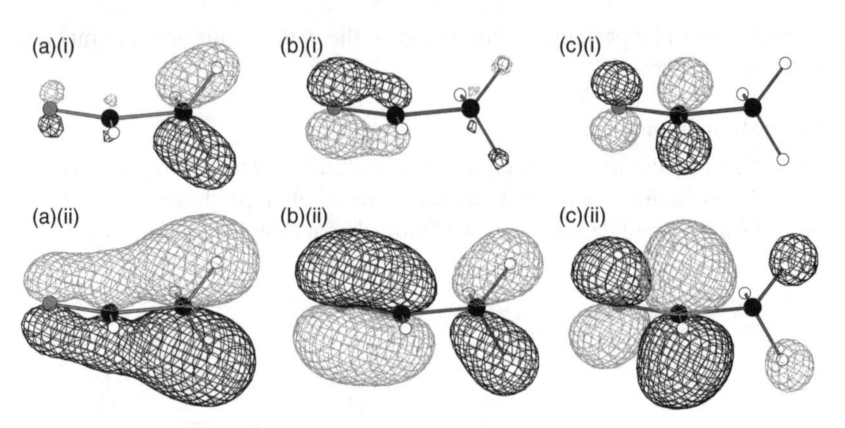

Fig. 14.4 Surface plots of the π MOs of ethanal, i.e. the orbitals for which the plane of the molecule is a nodal plane. The plots in the upper half of the figure are made for a higher value of the wavefunction than those in the lower half. Thus the plots in the upper half just show where the electron density is most concentrated, whereas those in the lower half show the atoms on which smaller amounts of electron density are found. Orbital (a) essentially just contributes to the bonding between the methyl carbon and its attached hydrogens. Orbital (b) is essentially the C=O π bonding MO and (c) is the π^* anti-bonding MO. All the orbitals are to some extent delocalized over the oxygen atom, both carbon atoms and the two out-of-plane hydrogen atoms.

even this MO has a small contribution from the C–H σ bonding MOs.

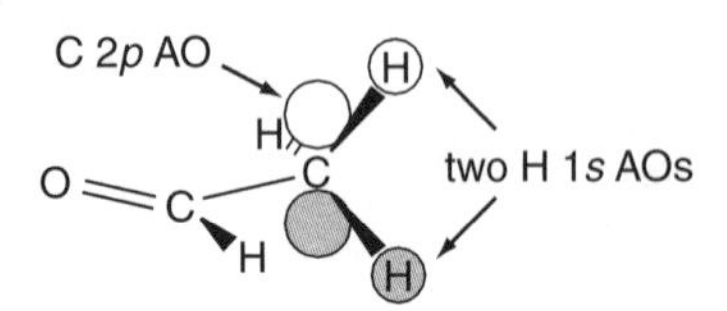

Fig. 14.5 The *p* orbital on the methyl carbon can contribute to the bonding of the methyl hydrogens by overlapping with the 1*s* orbitals on the two out-of-plane hydrogens.

What these three MOs tell us is that there is some weak σ conjugation between the C–H σ bonding MOs of the methyl group and the π system; this is shown schematically in Fig. 14.6. Note that this conjugation is not possible for the C–H bond between the carbonyl carbon and the attached hydrogen as the σ orbitals for this bond lie in the plane and so are not able to interact with the π system. In contrast, the orbitals from the C–H bonds of the methyl group lie out of the plane so that interaction with the π system is possible.

The effect of this σ conjugation is that some electron density is withdrawn from the methyl C–H bonds towards the oxygen. This weakens the bonds and makes the hydrogens more acidic than they would be in a simple alkane. We saw a similar example on p. 179 where neighbouring C–H σ bonds helped to stabilize a carbenium ion. We also saw how the resulting withdrawal of electrons from the C–H bonds made the hydrogens sufficiently acidic that they could be removed by a base, such as hydroxide, and how this led to elimination reactions (see Section 11.3 on p. 180).

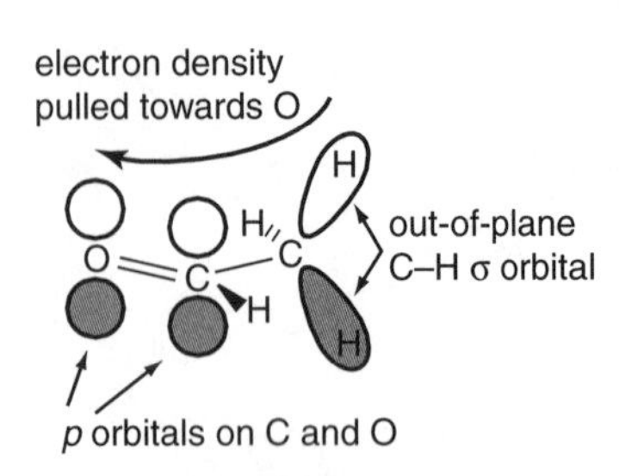

Fig. 14.6 The C–H σ bonding MOs of the methyl group and the 2*p* orbitals on the carbonyl carbon and the oxygen can all interact to form a delocalized set of π MOs. Note that for this interaction to be successful the C–H bonds must lie out of the plane of the molecule. Electron density is pulled towards the oxygen due to its greater effective nuclear charge.

In ethanal, this σ conjugation increases the acidity of the hydrogens attached to the carbon adjacent to the carbonyl. In addition, the enolate anion which results from the removal by a base of one of these hydrogens is delocalized. Taken together, these two effects account for the fact that it is much easier to remove H^+ from either an aldehyde or a ketone than from a simple alkane. Typically, these carbonyl compounds have pK_a values of around 20, in contrast to simple alkanes which have values of between 40 and 50.

Figure 14.7 shows two curly arrow descriptions of the formation of an enolate. Although both mechanisms are acceptable, (b) is preferred since it emphasizes the role of the C=O bond withdrawing electrons from the C–H bond, making the hydrogen acidic in the first place. Also, as we saw in Fig. 10.34 on p. 170, when representing the enolate ion it is better to have the negative charge

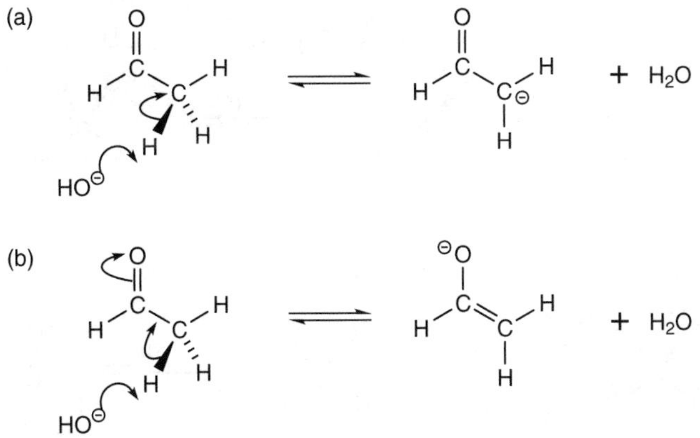

Fig. 14.7 Two curly-arrow representations of the formation of an enolate. Version (a) shows the deprotonation by hydroxide ion to form the enolate; this version does not emphasize the electron-withdrawing nature of the C=O bond which is what makes the hydrogen acidic in the first place. The effect of the carbonyl is represented in version (b); in addition, this also suggests the enolate formed has a greater charge on the oxygen than on the carbon.

on the more electronegative oxygen atom. However, the structure which has the charge on the carbon does make a contribution, so we should think of the enolate as having significant charge on both the oxygen *and* the carbon.

In summary there are two possible reactions that could occur between hydroxide and ethanal: the first is the direct attack of the hydroxide into the $\pi^\star$ MO of the ethanal to form the tetrahedral intermediate; the second is the deprotonation of one of the methyl hydrogens to form the enolate. The question is, which reaction actually occurs? The answer to this is simple: they *both* do. Both of these reactions are reversible so at any one time the reaction mixture will contain the unreacted hydroxide and aldehyde, the tetrahedral intermediate and the enolate. We shall see later that there may also be further reaction products.

How do we know that both reactions are occurring? Experimental evidence for this comes from studies using isotopically labelled reagents, details of which are given in the next section.

Isotopic labelling

It is possible to prepare water or hydroxide where the normal ^{16}O atoms or hydrogen atoms are replaced with their heavier isotopes ^{18}O and deuterium (D or ^{2}H). When ^{18}O-labelled hydroxide and water are mixed with ^{16}O-containing aldehydes and ketones, it is found that the heavier oxygen isotope is readily incorporated into the carbonyl compounds. The mechanism by which this occurs in shown in Fig. 14.8.

Similarly, if deuterated water (D_2O) and deuteroxide (OD^-) are used, the methyl hydrogens are rapidly exchanged for deuterium atoms. The mechanism for this reaction is shown in Fig. 14.9.

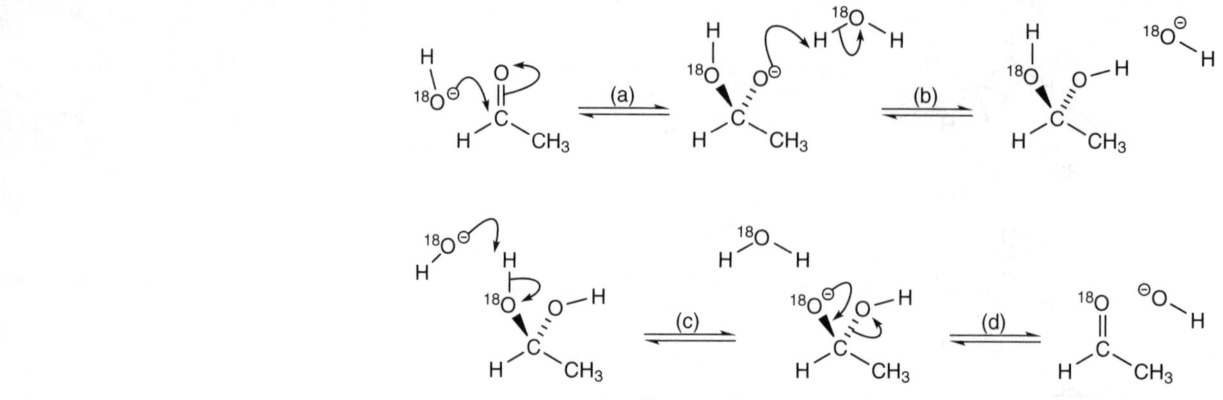

Fig. 14.8 Illustration of how ^{18}O from $^{18}OH^-$ can be incorporated into carbonyl compounds. Step (a) shows the initial attack of the labelled hydroxide on the ethanal. The best leaving group after this step is still the labelled hydroxide that just attacked – we need to make the unlabelled oxygen into the best leaving group. In water, there is extensive hydrogen bonding between the solvent and any $-O^-$ ions; this aids the tetrahedral intermediate in protonating the $-O^-$ and deprotonating the OH group. Thus step (b) shows the intermediate deprotonating the solvent and (c) shows the labelled hydroxide deprotonating the other OH group. The net effect of these two steps is that the unlabelled oxygen now becomes the better leaving group (as OH^-) and is then pushed off in step (d) by the O^- to yield the ^{18}O-labelled product. All of the steps are reversible, so given the large excess of ^{18}O-labelled water and hydroxide, eventually almost all of the initial ^{16}O in the ethanal will be replaced.

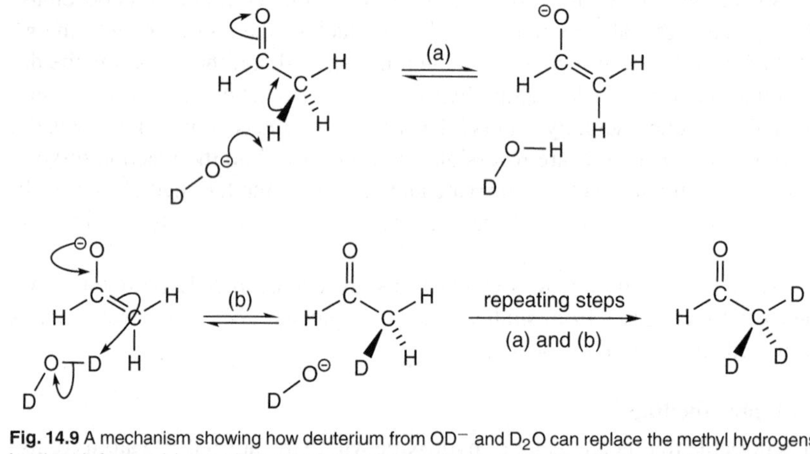

Fig. 14.9 A mechanism showing how deuterium from OD^- and D_2O can replace the methyl hydrogens in ethanal. In step (a) one of the out-of-plane hydrogens is removed by the deuteroxide. This step is reversible so the enolate could pick up this hydrogen again but, given the large excess of the deuterated water, it is more likely that the enolate will react with a fully deuterated water molecule instead, as shown in step (b). Rotation of the C–C bond means that any of the methyl hydrogens could eventually be replaced in this manner by repeating steps (a) and (b). Note that the aldehyde hydrogen could *not* be replaced in this manner since, as explained in the text, it is not significantly acidic.

Energy profiles for the alternative reactions

The isotopic labelling experiments show that *both* reactions are taking place in solution and that all the reactions are reversible. A schematic energy profile for both of these reactions is shown in Fig. 14.10.

The attack of the hydroxide into the $\pi^\star$ MO to form the tetrahedral intermediate is fast (as shown by the small activation energy for this step, $E_{a,tet}$) but we see from the profile that the collapse of the intermediate back to the alde-

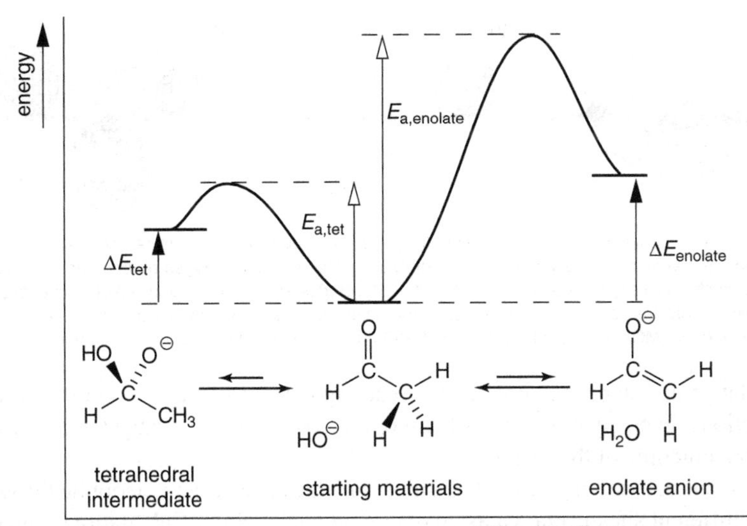

Fig. 14.10 A schematic energy level profile for the equilibria between ethanal, the tetrahedral intermediate (formed by the hydroxide attacking the C=O π^*) and the enolate (formed by the hydroxide removing one of the methyl hydrogens). Overall, the lowest energy species are the starting materials. As $E_{a,tet}$ is smaller than $E_{a,enolate}$, the tetrahedral intermediate forms faster than the enolate. Both intermediates are higher in energy than the reactants, so at equilibrium the concentration of these intermediates will be small.

hyde and hydroxide is faster still. The intermediate is higher in energy than the starting materials and so at equilibrium there will be rather little of it present.

The energy profile also shows that the enolate is slower to form than the tetrahedral intermediate as the activation energy for enolate formation, $E_{a,enolate}$, is higher than for formation of the tetrahedral intermediate, $E_{a,tet}$. Formation of the enolate is slow because the methyl hydrogens are only weakly acidic so there is hardly any hydrogen bonding to these hydrogens. As a result, a hydroxide is not always present in exactly the right position to take off the hydrogen.

Since both the enolate and the intermediate are higher in energy than the starting materials, the equilibrium mixture will contain more aldehyde and hydroxide than either of the intermediates.

The enolate anion undergoes many useful reactions and so it is often used in synthetic chemistry. Of course, at the same time as forming the enolate we almost certainly will form some of the tetrahedral intermediate, but usually this is not important as this intermediate does not lead to further reactions.

14.2 Reactions of enolates

The enolate has significant negative charge on both the oxygen and the carbon next to the carbonyl carbon (see Fig. 14.7 on p. 219). We might therefore expect that when the anion acts as a nucleophile it can do so either by attacking from the carbon or the oxygen; as the charge is not equally distributed, we would not expect the reactivity at these two positions to be the same. These expectations are borne out by experiment, and in addition it is found that exactly how the

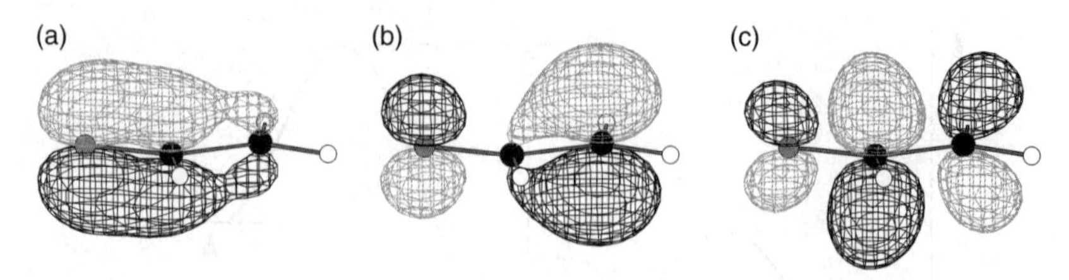

Fig. 14.11 The π MOs of the enolate anion. The lowest energy MO is all in-phase and is shown in (a). Unlike the allyl anion, the enolate anion is not symmetrical due to the oxygen on one end; this draws more electron density towards it in this MO. MO (b) includes an extra node and is higher in energy. It mainly has electron density on the two end atoms but with a greater contribution from what was the methyl carbon. MOs (a) and (b) are occupied; MO (c) is the π^* anti-bonding MO and is unoccupied.

enolate reacts depends on what it is reacting with and on the conditions of the reaction – even changing the solvent or the counter-ion can have a large effect on the outcome of the reaction.

A species such as the enolate anion which can act as a nucleophile through two different sites is known as an *ambident* nucleophile. Calculations suggest the charges are approximately -0.8 on the oxygen, $+0.3$ on the carbonyl carbon and -0.6 on what was the methyl carbon. The form of the π MOs, shown in Fig. 14.11, is consistent with the idea that most of the charge is on the two end atoms. This is similar to other cases in which a π system is formed from the overlap of three p orbitals, e.g. the carboxylate anion (Fig. 10.16 on p. 159) and amides (Fig. 10.18 on p. 161).

The MOs in Fig. 14.11 are for the free anion. In practice, however, a counter-ion must also be present, and one of the most commonly used is Li^+. The first question is which end will the counter-ion be most associated with?

Lithium will not form a strong covalent bond with either carbon or oxygen, simply because the 2s orbital on lithium is not well matched in energy with either the carbon or oxygen AOs (see the graph of orbital energies on p. 57). Of the two, the better match is with the carbon orbitals so we might expect the lithium to be more associated with the carbon end of an enolate as shown in Fig. 14.12 (a).

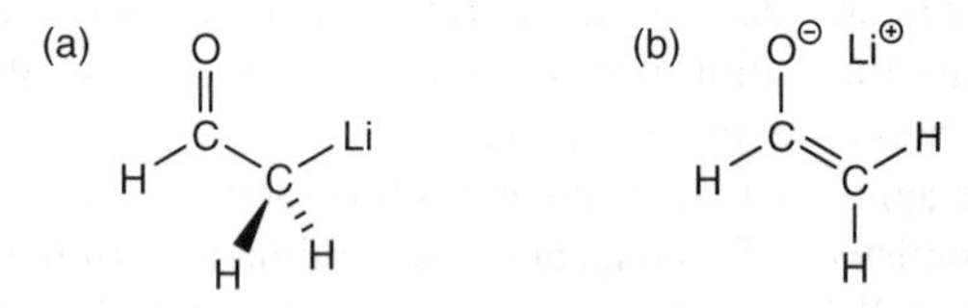

Fig. 14.12 In a lithium enolate we might expect the lithium to be covalently bound to the carbon, as in (a), on the grounds that the carbon AOs are better matched in energy to those on lithium than are the oxygen AOs. However, the electrostatic interaction, shown in (b), turns out to be much stronger, and so (b) is the favoured structure.

However, there is an alternative which is for the cation simply to have an electrostatic interaction with the enolate. Such an interaction will be strongest between the Li^+ and the oxygen on the enolate as this is the position with the highest electron density; the resulting structure can be represented as in Fig. 14.12 (b). Experimental work confirms that this structure is indeed the one adopted.

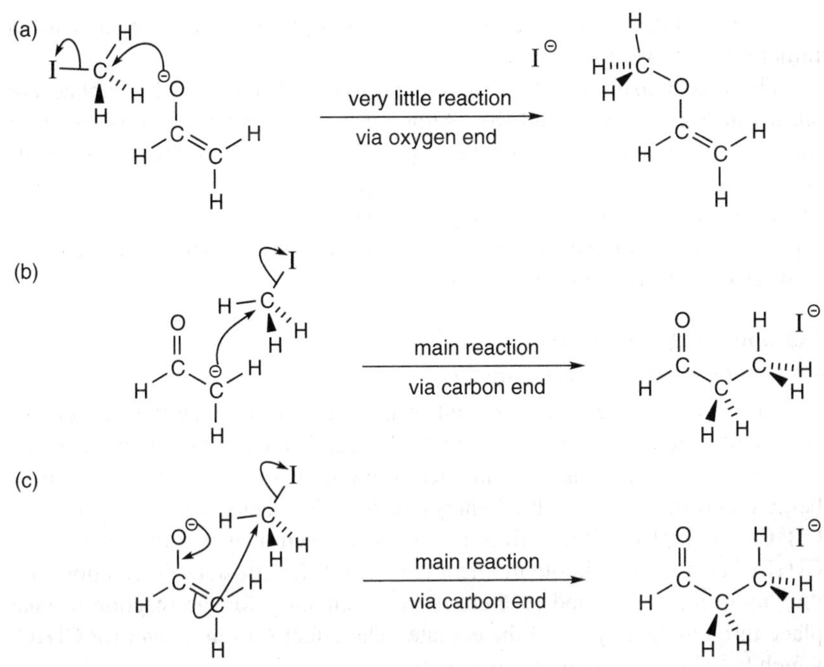

Fig. 14.13 Possible mechanisms for the reaction between the enolate from ethanal and CH_3I. In (a) the reaction is via the oxygen end of the enolate, whereas in (b) the carbon end of the enolate is involved; experimentally it is found that most of the reaction takes place through the carbon. CH_3I is uncharged so there is very little electrostatic interaction and as a result the reaction is controlled by orbital interactions. The HOMO of the enolate has the greatest contribution on the end carbon and therefore the reaction is expected to take place through this position. Mechanism (c) has the same outcome as (b), i.e. attack through the carbon, but in (c) the enolate has been drawn in a more realistic manner with the charge shown on the oxygen.

Suppose that we now allow iodomethane, CH_3I, to react with the enolate. In principle, either end of the enolate could attack the CH_3I in an S_N2 mechanism to form two alternative products as shown in Fig. 14.13 (a) and (b). As the reaction ultimately involves adding a methyl group to the enolate, CH_3I is called a *methylating agent* or, more generally, an *alkylating agent*. Experimentally what is observed is that the *only* product is the one in which the reaction is via the carbon end of the enolate; our task is to explain why this is.

Iodomethane is not charged so there is going to be very little electrostatic attraction between it and either end of the enolate anion. As outlined in Section 7.5 on p. 107, the poor energy and size matches between the carbon and the iodine AOs means that the C–I $\sigma^\star$ orbital will be particularly low in energy for an unoccupied MO. Thus a good orbital interaction is possible between the enolate and CH_3I; this is in contrast to the case of Li^+ interacting with the enolate where there is a good electrostatic attraction but poor orbital interaction.

As we saw on p. 101 the most significant orbital interaction is between the HOMO of one species and the LUMO of the other. For iodomethane reacting with an enolate, this interaction will be between the HOMO of the enolate and the LUMO of the iodomethane. Figure 14.11 shows that the HOMO of the enolate (orbital (b)) has a greater contribution on the end carbon than on the oxygen, so the best orbital interaction is from the carbon rather than the oxygen end of the enolate. Therefore the reaction takes place through the end carbon rather than through the oxygen.

The mechanisms for the reaction through both ends of the enolate are shown in Fig. 14.13 (a) and (b). Although most of the reaction takes place through the carbon we do not have to draw the enolate with the charge localized on the carbon. We mentioned earlier that probably the best representation of the enolate ion is to have the negative charge on the oxygen end; using this representation of the enolate we can still draw a mechanism for the reaction through carbon, as shown in Fig. 14.13 (c).

Manipulating the reaction

– by changing the alkylating agent

It is possible to increase the amount of the product in which the oxygen becomes alkylated, i.e. in which the methyl group is attached to the oxygen. To do this, we need to make the interaction more charge controlled rather than being controlled just by orbital energy factors. Replacing CH_3I by CH_3Br or CH_3Cl accomplishes this. Bromine is more electronegative than iodine, so CH_3Br has a greater dipole moment than CH_3I. Electrostatic interactions are thus more important, and so there is more tendency for the reaction to take place through the oxygen of the enolate. The effect is even greater for CH_3Cl which has a greater dipole moment still.

As an example, we will look at how the ratio of carbon- to oxygen-alkylated product varies in the reaction shown in Fig. 14.14 as the halogen, X, in the alkylating agent is changed.

When iodoethane is used as the alkylating agent, the ratio of oxygen- to carbon-alkylated products is 13:87. With bromoethane the ratio becomes 39:61 whilst with chloroethane the ratio increases further to 60:40. Clearly the amount of the oxygen-alkylated product increases as the halide becomes more electronegative.

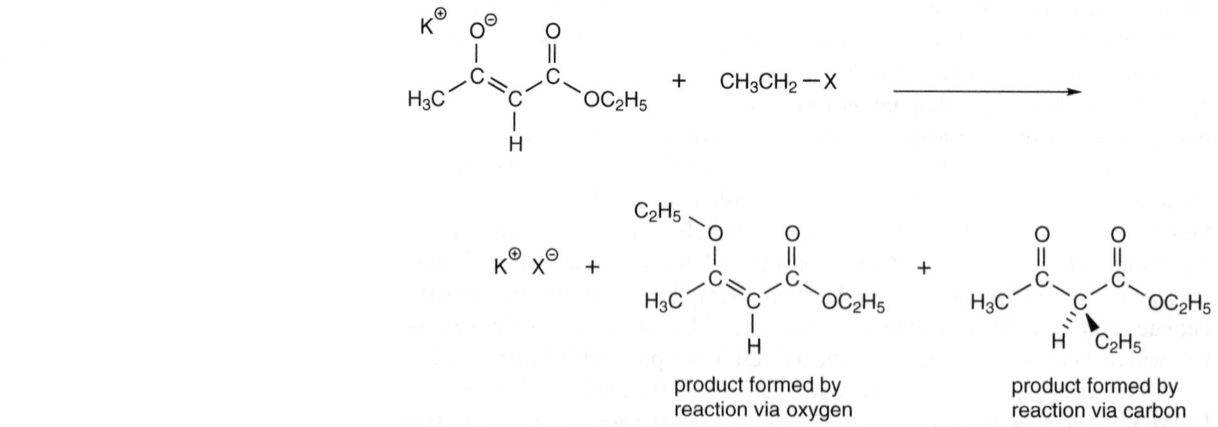

Fig. 14.14 The reaction between an enolate and a halogenoalkane can result in alkylation at either oxygen or carbon. The ratio between these two products depends on, amongst other things, the halogen X. It is found that the proportion of the oxygen-alkylated products increases as the electronegativity of the halogen increases; we interpret this as the increased dipole moment of the halogenoalkane leading to the reaction becoming more controlled by electrostatic interactions.

The amount of oxygen-alkylated product may be increased still further when ethyl tosylate, C_2H_5OTs, is used in place of the halogenoethane. The structure of ethyl tosylate and the curly arrow mechanism for the reaction is shown in Fig. 14.15. As can be seen in the diagram, it is the CH_2 carbon of the ethyl group which is attacked by the enolate in an S_N2 mechanism. However, instead of a halogen, this carbon is now attached to an even more electron-withdrawing oxygen, which increases the partial positive charge on the carbon still further. This results in the formation of even more of the oxygen-alkylated product – the oxygen-:carbon-alkylated product ratio is now 88:12.

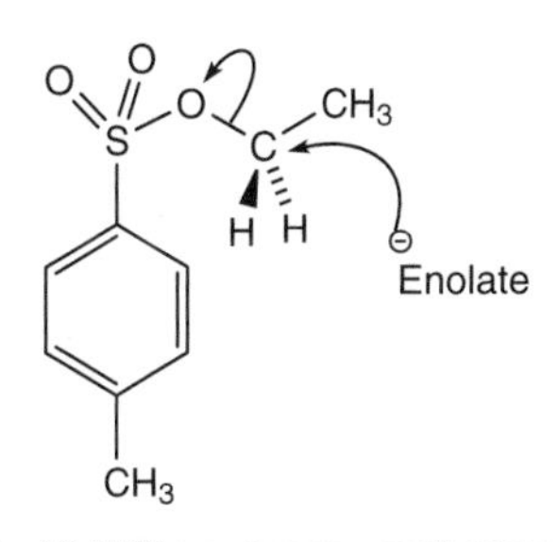

Fig. 14.15 The curly arrow mechanism for the reaction between ethyl tosylate and an enolate; in this S_N2 reaction the enolate can attack through carbon or oxygen.

Notice that although we wanted to have an oxygen attached to the ethyl group to increase the charge on the carbon, it would not have been sufficient simply to use an OH group for this purpose. As we saw in Section 13.2 on p. 213, OH^- is not a good leaving group but the tosylate anion, TsO^-, is. Thus, using ethyl tosylate provides an increased positive change on the carbon as well as a good leaving group.

– by changing the solvent

Changing the solvent can also influence the outcome of the reaction, i.e. the ratio of the carbon- to oxygen-alkylated products. The data quoted so far for the reaction shown in Fig. 14.14 was obtained in the solvent hexamethylphosphoramide (HMPA) whose structure is shown in Fig. 14.16. This solvent is classed as *polar aprotic* (see Section 12.1 on p. 190). It is very good at solvating cations but, as it has no acidic hydrogens, it does not solvate anions so well, as no hydrogen bonding to them is possible.

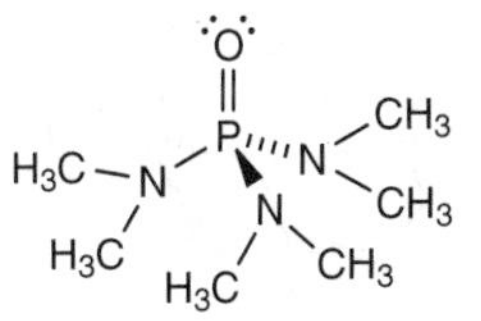

Fig. 14.16 The structure of the solvent hexamethylphosphoramide (HMPA).

Using HMPA as the solvent means that the K^+ ion is strongly solvated whereas the enolate anion is not so strongly solvated and so is free to react. In contrast, if the reaction is carried out in the polar *protic* solvent *t*-butyl alcohol (2-methyl-2-propanol), even the ethyl tosylate forms exclusively the *carbon*-alkylated product.

The reason for this change is because the polar protic solvent forms hydrogen bonds to the *oxygen* end of the anion. These hinder the approach of the alkylating agent to the oxygen and so reaction through the carbon is favoured.

14.3 Unsymmetrical enolates

There is only one enolate which can be formed from ethanal, but for unsymmetrical ketones two different enolates can be formed, depending on which hydrogen is removed. An example is shown in Fig. 14.17; in this ketone there is an extra methyl group on one of the carbons adjacent to the carbonyl, so two different enolates can be formed.

By careful control of the reaction conditions we can favour one of the enolates over the other. The first point to note is that we will need to use a strong base to take off one of the acidic hydrogens; however, we do not want the base to act as a nucleophile and attack the carbonyl group. One way to avoid this is to use a bulky base which is simply too large to get in close enough to attack the carbonyl; an example of such a base is triphenylmethyllithium, whose structure is shown in Fig. 14.18.

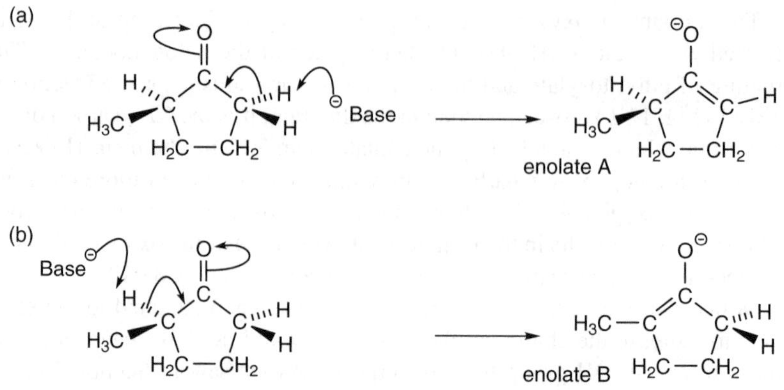

(a)

(b)

Fig. 14.17 Two different enolates can be formed from an unsymmetrical ketone, depending on which hydrogen is removed by the base.

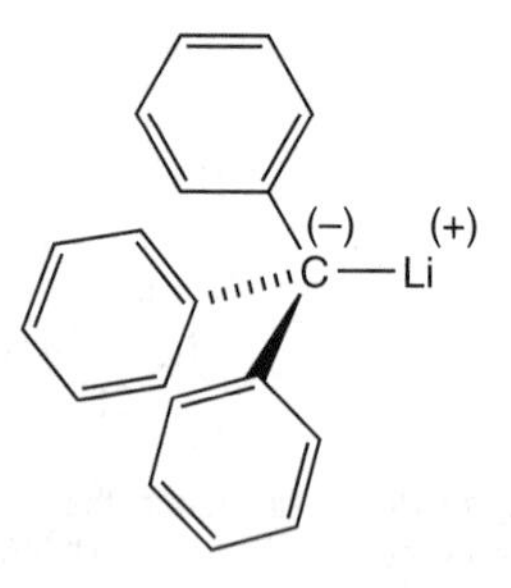

Fig. 14.18 The structure of triphenylmethyllithium, Ph_3CLi, which is a strong base but, on account of its bulk, a poor nucleophile.

A consequence of using such a bulky base is that it is easier for it to remove the hydrogens from the carbon which does *not* have the methyl group attached, simply because these protons are less sterically hindered. The result is that enolate A is easier to form than enolate B.

However, although enolate A is easier to form than enolate B, it turns out that enolate B is actually lower in energy than enolate A. The reason is to do with different amounts of σ conjugation present in the two enolates.

We have seen how any out-of-plane hydrogens can σ conjugate with a neighbouring π system. In enolate B the hydrogens on the methyl group are able to σ conjugate with the π system, whereas in enolate A the methyl hydrogens cannot be involved in such conjugation. As we have seen, conjugation generally lowers the energy of the species, so we therefore expect enolate B to be lower in energy than enolate A.

This σ conjugation means that the electrons in the methyl group are, to a certain extent, spread over the whole π system. This can be seen in Fig. 14.19 which compares the HOMOs of the two enolates; the extra delocalization in enolate B is clear.

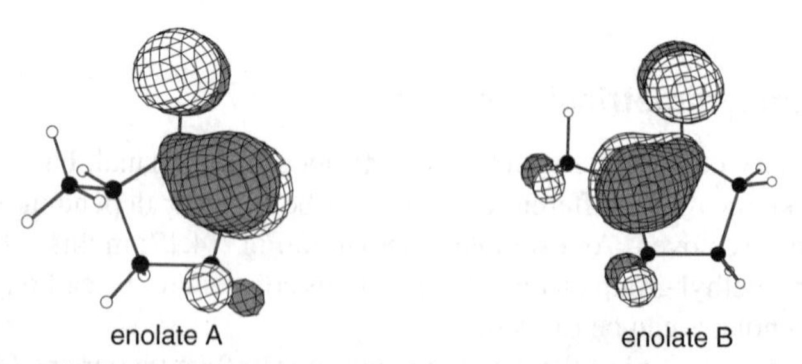

Fig. 14.19 Surface plots of the HOMOs of enolate A and enolate B shown in Fig. 14.17. In enolate B the hydrogens on the methyl group are involved in σ conjugation with the π system, something which is evident from the way in which the HOMO is spread over all of these atoms. In contrast, no such conjugation is possible for enolate A, as evidenced by the HOMO. This extra delocalization makes enolate B lower in energy than A.

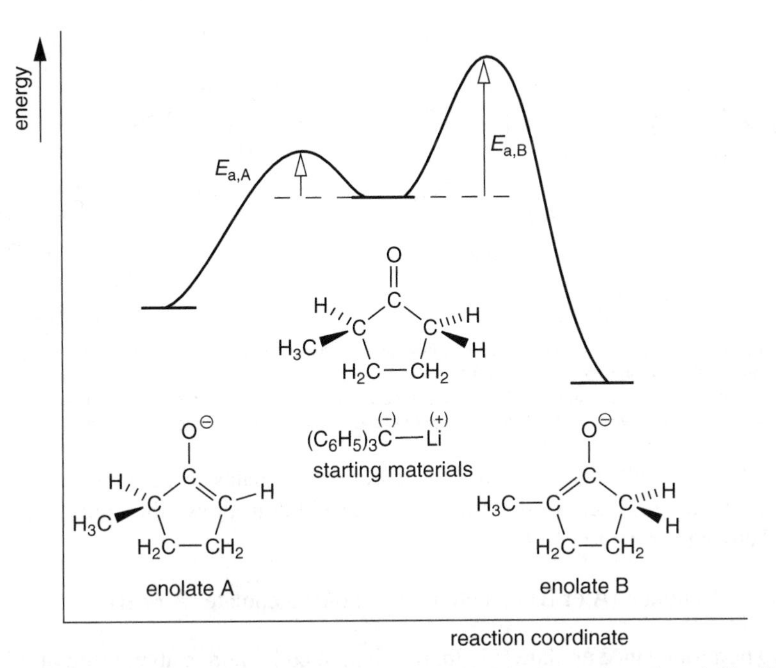

Fig. 14.20 Energy profile for the formation of enolates A and B by the reactions shown in Fig. 14.17. As explained in the text, enolate A forms more rapidly than enolate B on account of the steric crowding caused by the methyl group; on the profile this difference in the rates of reaction is shown by the smaller activation energy for the formation of enolate A. However, enolate B is actually lower in energy than enolate A, on account of the σ conjugation which is possible for B and not A.

Thermodynamic and kinetic products

As we have seen whilst it is *easiest*, and hence quickest, to form enolate A, enolate B is lower in energy; this is illustrated in the energy profile shown in Fig. 14.20. Enolate A is described as the *kinetic product* as it is formed fastest, whereas enolate B is described as the *thermodynamic product* as it is thermodynamically the most stable enolate and so present in the greatest amounts at equilibrium. The question is, how can we manipulate the reaction conditions to favour one enolate or the other?

When an excess of the base is added to the ketone at room temperature, the ratio of the enolates formed is 72:28 A to B. Once formed, these enolates cannot return to the ketone since in the presence of excess base there is nothing acidic for the enolates to pick up a hydrogen from. Now if an alkylating agent is added which reacts with the enolates the ratio of the two alkylated products formed will be the same as the ratio of the two enolates.

Returning to the formation of the enolates, if instead of having excess base we have a slight excess of ketone, then under the same conditions (i.e. same time and temperature) the ratio of enolates A to B becomes 6:94. Now, the lower energy (more thermodynamically stable) enolate B is the dominant species. This comes about because it is now possible for the enolates to return to the ketone by deprotonating some of the excess ketone present; this is shown in Fig. 14.21.

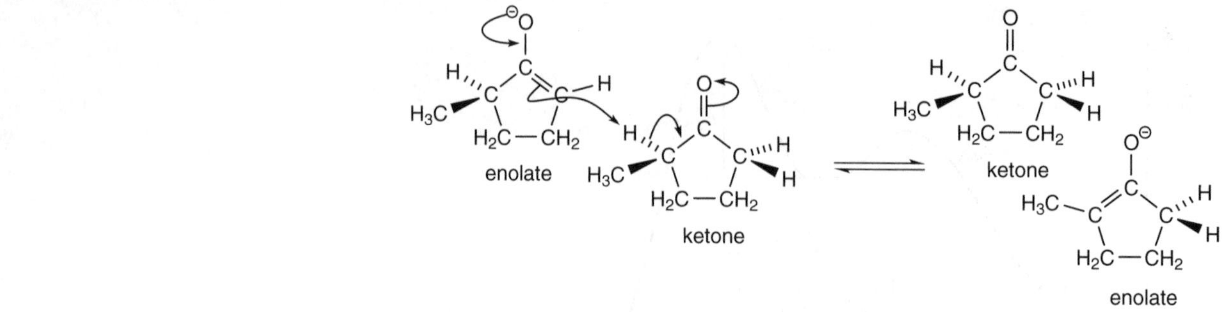

Fig. 14.21 Curly arrow mechanism showing how enolate A can return to the ketone by picking up H^+ from another ketone, forming enolate B in the process; it could just as well have formed another molecule of enolate A. Similarly, enolate B could react with a ketone to form enolate A or B. In other words, there is a pathway which allows both enolate and the ketone to come to equilibrium.

Of course, all this reaction does is convert enolates to ketones and vice versa. However, because the reaction is reversible it allows the enolates and the ketone to come to equilibrium:

$$\text{enolate (A or B)} + \text{ketone} \rightleftharpoons \text{ketone} + \text{enolate (A or B)}.$$

Therefore, since enolate B is thermodynamically more stable, more of this enolate will be present at equilibrium. Subsequent reaction with an alkylating agent will then lead to more of the final product being formed from enolate B than from enolate A, provided that the alkylation reaction does not perturb the equilibrium significantly.

14.4 Substitution versus elimination revisited

The reaction between ethanal and hydroxide is relatively straightforward to understand since there are only two reasonable possibilities for the initial interaction between the two reagents. For the majority of reactions carried out in the laboratory, many outcomes are possible and it is the task of the chemist to try to optimize the conditions to produce the desired product.

We have already seen in Chapter 11 how a base/nucleophile reacts with halogenoalkanes to form either substitution products or elimination products. Thus, as shown in Fig. 14.22, 1-bromopropane could react with hydroxide to form either propan-1-ol or propene. For less symmetrical halogenoalkanes,

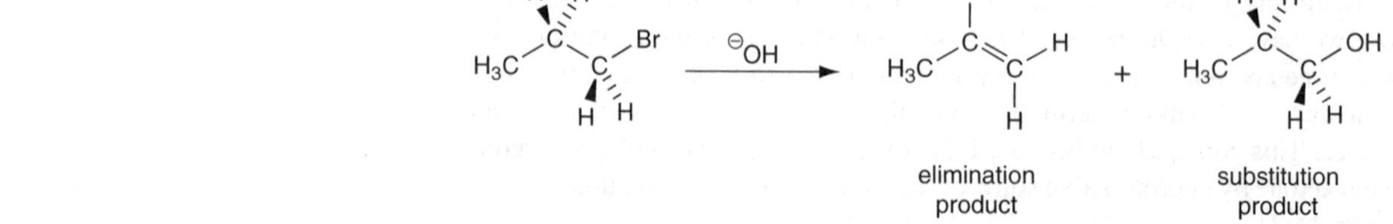

Fig. 14.22 The halogenoalkane 1-bromopropane can react with OH^- as a nucleophile to give the substitution product propan-1-ol, or with OH^- as a base to give the elimination product propene.

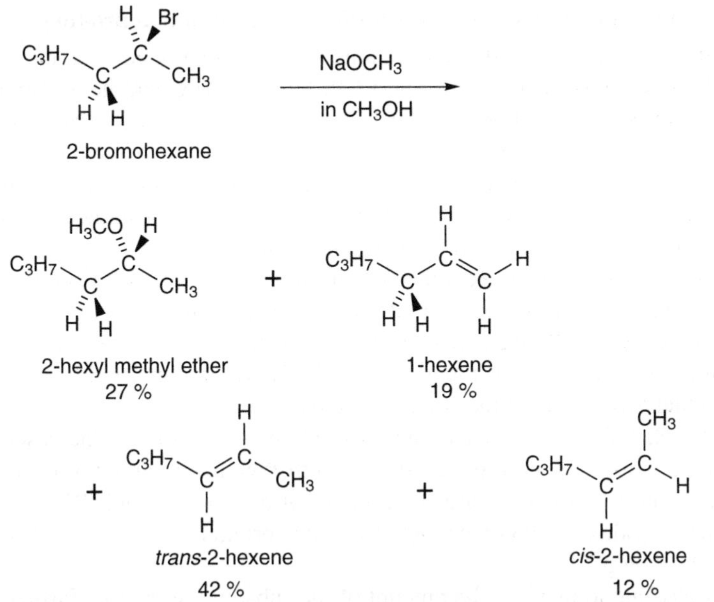

2-bromohexane

$NaOCH_3$ in CH_3OH

2-hexyl methyl ether 27 %

1-hexene 19 %

trans-2-hexene 42 %

cis-2-hexene 12 %

Fig. 14.23 The reaction between 2-bromohexane and methoxide, CH_3O^-, gives one substitution product (the ether) and three different alkenes which arise from elimination reactions. The proportions of each product are shown.

more than one elimination product is possible; for example, the reaction between 2-bromohexane and methoxide, CH_3O^-, gives four different products as shown in Fig. 14.23.

The ether is formed by a substitution reaction between the bromohexane and methoxide, following the S_N2 mechanism as shown in Fig. 14.24 (a). The elimination products are formed via the E2 mechanism, which is shown in Fig. 14.24 (b) for the case of formation of *trans*-2-hexene.

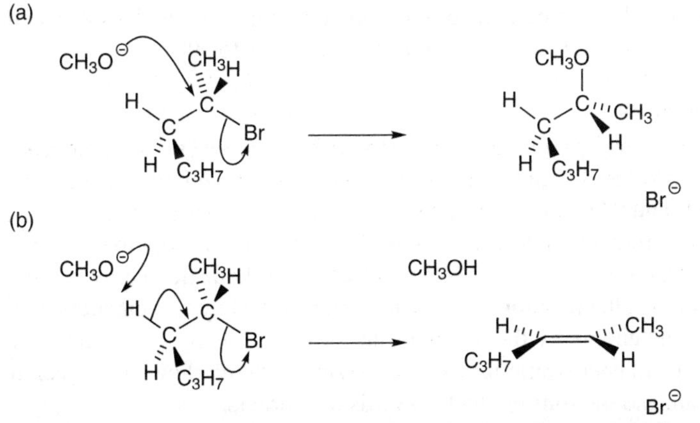

Fig. 14.24 Curly arrow mechanisms for the reaction of 2-bromohexane with methoxide via (a) an S_N2 substitution reaction and (b) an E2 elimination reaction. There are three possible products from elimination reactions (see Fig. 14.23), of which only one is shown here.

Altering the ratio between the substitution and elimination products

How can we make sense of the relative amounts of these products and increase the yield of one product over the others? In forming the ether, the methoxide ion is acting as a nucleophile; it is firstly attracted towards the partial positive charge of the carbon with the attached bromine atom, and then attacks into the C–Br $\sigma^\star$ orbital, thereby breaking the C–Br bond. If we can increase the partial positive charge on the carbon, there will be greater electrostatic attraction to the CH_3O^- group, thus favouring the substitution reaction.

This is exactly what we see if the bromine atom is replaced by a chlorine or fluorine atom; whereas the reaction of 2-bromohexane gives 27% of the ether, 2-chlorohexane gives 38% and 2-fluorohexane more still. In contrast, replacing the bromine by iodine means the partial charge on the carbon is reduced and it is found that the amount of the ether decreases to 16%.

However, whilst the amount of the substitution product formed increases on replacing the bromine atom by the more electron-withdrawing fluorine atom, the *rate* of the reaction decreases by a factor of over a thousand. This is because the fluoride is a poorer leaving group than bromide, as was described in Section 13.2 on p. 213.

It is possible to increase the amount of the substitution product further by replacing the bromine by a really good leaving group which is still capable of polarizing the carbon to a significant extent. An example of such a leaving group is the tosylate (Fig. 13.7 on p. 214); if this is used in place of bromine the fraction of the substitution product increases to over 70%. Previously, when looking at the reactions of enolates, we saw how the tosylate group could be used in a similar way to both increase the partial positive charge of the attached carbon and also provide an excellent leaving group.

The percentage of the substitution product can be increased still further by making the elimination reactions more unfavourable. In these reactions, CH_3O^- is acting as a base, so not surprisingly changing to CH_3OH, which is a much weaker base, reduces the proportion of the elimination products. If use of this weaker base is combined with replacing the bromine by the excellent tosylate leaving group, the fraction of the ether rises to 90%.

Energy profiles

Drawing an energy profile in the case where there are several different reaction pathways is quite difficult since we need to show the relative energies of all the species and their transition states. Let us consider the case where reactants X can form four possible products A, B, C, and D, as was the case in the reaction shown in Fig. 14.23. Figure 14.25 is a schematic energy profile for these reactions; the reaction coordinate is represented by movement on the *xy*-plane and the energy by the height of the surface. The reactants and the four products are all energy minima, with the reactants being shown in the centre of the diagram and the four products towards the corners.

On the surface, the dotted lines trace the lowest energy paths between the reactants and each of the products. The highest energy point along such a path for each of these paths is the transition state; $A^\ddagger$ for forming A, and so on.

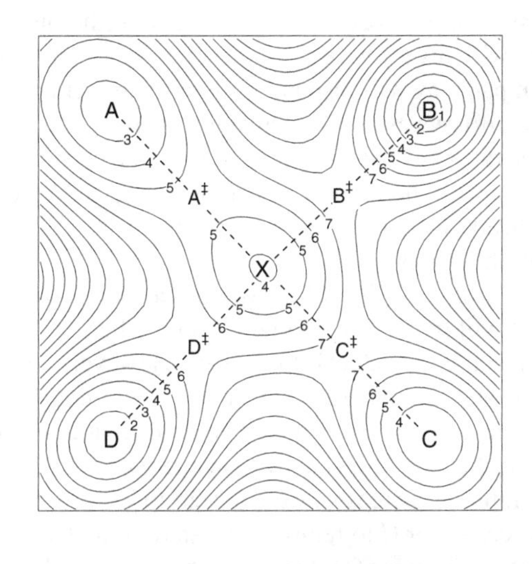

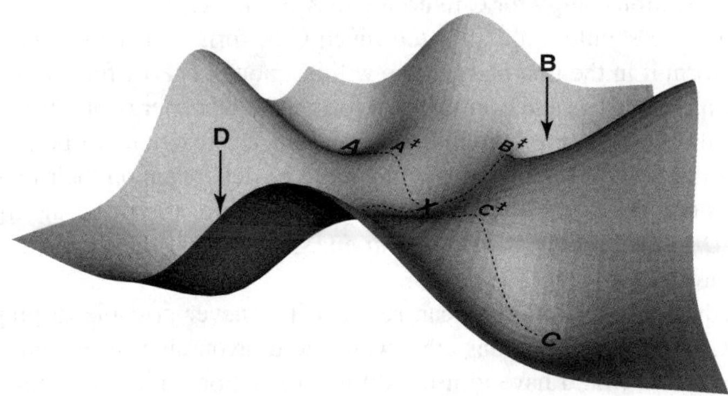

Fig. 14.25 The lower figure shows a perspective view of a schematic energy profile for the situation where reactants X can form four possible products A, B, C, and D. The height of the surface indicates the relative energy of the species and movement in the *xy*-plane represents the reaction coordinate. The dotted lines over the surface indicate the paths of minimum energy between reactants and particular products; the transition state (denoted by ‡) is at the maximum energy point along each path. The upper figure shows a contour plot for the same energy profile; the contours are labelled in order of increasing energy.

The reactants can form any of the products and given enough energy, any of the products could return to the reactants. In order to understand which will be the main products, we need to consider the relative depths of the energy minima, i.e. the relative energies of the reactants and products, and also the energy barriers between these minima which must be overcome if the product is to form. The energy profiles along the paths between the reactants and each of the products are shown in Fig. 14.26.

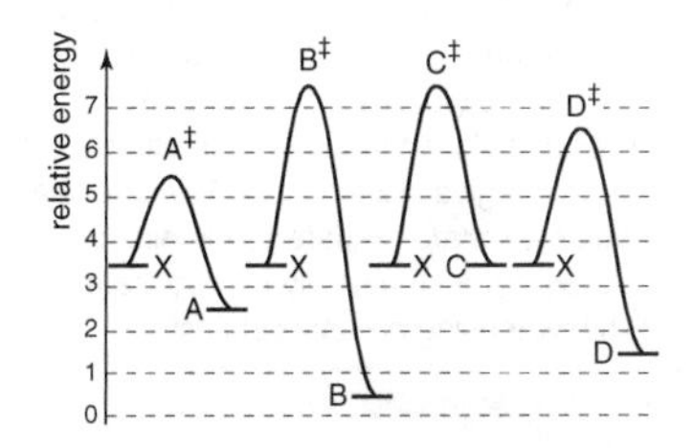

Fig. 14.26 Energy profiles for the paths between the reactants and each of the products shown in Fig. 14.25.

As we have seen in Section 9.4 on p. 137, as there is an energy barrier (the activation energy) which reactions have to overcome, changing the temperature has a large effect on the rate constant. We also saw (Fig. 9.8 on p. 138) that the *greater* the activation energy, the *larger* the change in rate constant for a given change in temperature.

If there are several pathways open to a reaction, the one with the lowest activation energy will always have the largest rate constant. However, the rate constant for each pathway will respond differently to a change in temperature on account of the differences in their activation energies. As the temperature is lowered, all of the rate constants will decrease, but the decrease will be greatest for those pathways with the highest activation energies. At low enough temperatures, the pathway with the lowest activation energy will therefore come to dominate over all the others.

Looking at the profiles in Fig. 14.26 we can see that at low temperatures the main product that will be formed will be A, simply because the activation energy required to reach this product is the lowest. However, A is not much lower in energy than the starting materials and so we would not expect the reaction to go to completion. The next most abundant product at low temperatures will be D since this has the next lowest activation energy.

As the temperature is increased, we will start to see more of products B and C. However, it is easier for C to return to the starting materials than it is for B since the activation energy for C returning to X is smaller than for B returning to X. Hence we should not expect too much C to form – if there is enough energy to form it in the first place, there will be enough energy for it to return to the starting materials and eventually to form the lower energy products.

When the temperature is high enough for all the reactions to be readily reversible, the ratio of the products will simply be dependent on their relative energies: since B has the lowest energy, at equilibrium it will be the major product. D is the species next lowest in energy, so it will be the next most abundant product.

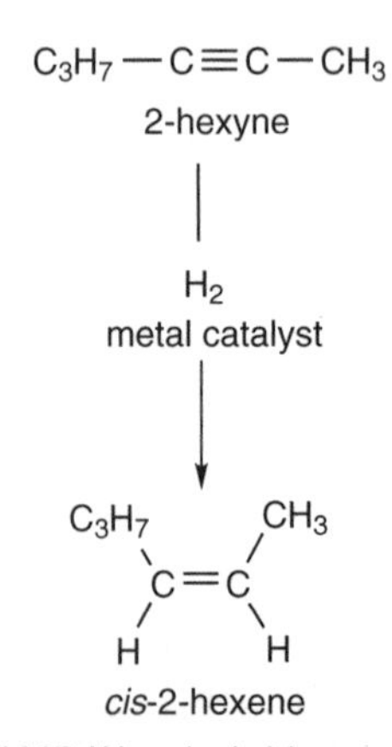

Fig. 14.27 Although *cis*-2-hexene cannot be prepared in good yield from the reaction between a 2-halogenohexane and base shown in Fig. 14.23, other routes do give a good yield. For example, a simple way to prepare predominantly this isomer is to partially hydrogenate 2-hexyne using a metal catalyst.

From this energy profile we can see that it is never possible to prepare much of C since other reactions are always more favourable; if we did want to prepare C, we would have to use a different reaction which would have a different energy profile. Altering the energy of the reactants (for example by changing a leaving group as we did for the reaction in Fig. 14.23 or even by changing the solvent) will change the whole energy surface. The energies of the reactants and products will change, as well as the energy barriers. It might be that on the new surface C is a more favourable product.

For example, in the reaction shown in Fig. 14.23, it is never possible to generate more *cis*-2-hexene than *trans*-2-hexene no matter how we alter the temperature or change the leaving group. This is because the activation energy for forming the *cis* isomer is always higher than the activation energy for forming the *trans* isomer; also the *trans* isomer is lower in energy than the *cis*.

This does not mean that it is impossible to generate the pure *cis* isomer, just that it is not possible using this particular reaction. In fact, *cis*-2-hexene may be prepared in good yield by partial hydrogenation of 2-hexyne using an appropriate metal catalyst as shown in Fig. 14.27. Under these conditions it is *much* easier to form the *cis* isomer than the *trans* form.

14.5 So why do chemical reactions happen?

The simple answer to this question is that reactions happen because they result in an increase in the entropy of the Universe – which is also the reason why *anything* happens. We should really ask:

Why does this particular chemical reaction increase the entropy of the Universe?

For an endothermic reaction, it *must* be because the entropy change of the reaction is positive. However, for the majority of reactions the answer will be because the reaction is exothermic and so gives out heat to the surroundings thus increasing their entropy.

Why is a reaction exothermic?

This is because the particular arrangement of atoms in the products is in some sense lower in energy than in the reactants. We must remember to consider *all* of the reactants and *all* of the products. It may well be that the deciding factor is something as subtle as the way in which the species interact with the solvent.

Why is one arrangement of atoms lower in energy than another?

When there is a particularly favourable interaction between two species (atoms or molecules) we say there is some sort of bond. These bonding interactions may be electrostatic or covalent in origin. We need to understand the factors which favour strong ionic interactions and those which favour strong covalent interactions. In most reactions, both are involved to some degree.

How can we make the product we want?

If the reaction is favourable, all that is needed is to supply enough energy to get over the barrier. If it is unfavourable (i.e. the products are higher in energy than the reactants) we will need to couple it to another reaction which is strongly favoured, or simply not allow the reaction to come to equilibrium. We also need to think about reactions which will compete with the one we want and how these can be avoided or minimized.

... and finally

Chemists are always trying to make new molecules, or devise better ways of making them – they do this partly out of curiosity and partly because new compounds are needed in every aspect of our life, from pharmaceuticals to novel materials such as ceramics and semiconductors. To be successful, the chemist needs to understand why and how reactions occur.

Index